# Three-Dimensional Analysis of Human Movement

**Paul Allard, PhD**
Sainte-Justine Hospital, Montreal, Quebec

**Ian A.F. Stokes, PhD**
University of Vermont, Burlington, Vermont

**Jean-Pierre Blanchi, PhD**
Joseph Fourier University, Grenoble, France

Editors

Human Kinetics

**Library of Congress Cataloging-in-Publication Data**

Three-dimensional analysis of human movement / [edited by] Paul
Allard, Ian A.F. Stokes, Jean-Pierre Blanchi.
    p.    cm.
    ''Selected papers from an invited symposium on three-dimensional
analysis held in Montreal in July, 1991''.
    Includes index.
    ISBN 0-87322-623-2
    1. Human locomotion--Analysis--Congresses.  2. Human mechanics--
Analysis--Congresses.  3. Three-dimensional imaging in biology--
Congresses.  4. Three-dimensional imaging in medicine--Congresses.
I. Allard, Paul, 1952-   . II. Stokes, Ian A.F.  III. Blanchi, Jean-
Pierre, 1938-
QP303.T59   1995
612.7'6--dc20                                     93-37710
                                                                    CIP

ISBN: 0-87322-623-2

**Acquisitions Editor**: Richard D. Frey, PhD; **Developmental Editor**: Marni Basic; **Assistant Editors**: Jacqueline R. Blakley, Ed Giles, Dawn Roselund, and Myla Smith; **Copyeditor**: Dianna Matlosz; **Proofreader**: Pam Johnson; **Indexer**: Theresa J. Schaefer; **Typesetting and Pagination**: Angela K. Snyder; **Text Designer**: Keith Blomberg; **Layout Artists**: Denise Lowry and Laura Jolley; **Cover Designer**: Jack Davis; **Illustrators**: Studio 2D and Denise Lowry; **Printer**: Braun-Brumfield

Printed in the United States of America    10  9  8  7  6  5  4  3

**Human Kinetics**
Web site: www.HumanKinetics.com

*United States:* Human Kinetics, P.O. Box 5076, Champaign, IL 61825-5076
800-747-4457
e-mail: humank@hkusa.com

*Canada:* Human Kinetics, 475 Devonshire Road, Unit 100, Windsor, ON N8Y 2L5
800-465-7301 (in Canada only)
e-mail: orders@hkcanada.com

*Europe:* Human Kinetics, 107 Bradford Road, Stanningley
Leeds LS28 6AT, United Kingdom
+44 (0) 113 255 5665
e-mail: hk@hkeurope.com

*Australia:* Human Kinetics, 57A Price Avenue, Lower Mitcham, South Australia 5062
08 8277 1555
e-mail: liahka@senet.com.au

*New Zealand:* Human Kinetics, P.O. Box 105-231, Auckland Central
09-523-3462
e-mail: hkp@ihug.co.nz

# Dedication

This book is dedicated to our good memories of Herman J. Woltring (1943-1992) and his pioneering work on biomechanics and on three-dimensional analysis of human motion.

# Contents

# Forewords

The uniqueness of *Three-Dimensional Analysis of Human Movement* is its focus on the three-dimensional nature of body segments and of the Euclidean space in which they move. As opposed to simpler, and often simplistic, ways of representing the intriguing phenomenon of biomovement, 3-D analysis requires significant data reduction and computation. This did not frighten our Prussian predecessors, Braune and Fischer, but it did frighten many other biomechanists who came later. This is the case even with Eberhart and collaborators who, just after World War II, carried out the well-known gigantic fundamental work on human locomotion by looking at the movement of body segments as projected onto three planes.

Part I illustrates well how technology has finally provided equipment that permits researchers to overcome the problems associated with 3-D analysis.

In the past robotics borrowed knowledge from biomechanics to imitate natural systems and to create automata, manipulators, and locomotor capacities similar to those of human beings. Now, as is made evident in Part II, biomechanics is enjoying in its turn the theoretical body developed by that science and is using it to improve its description of biomovement. The rest of the part explores how large quantities of accurate information about human and animal movement can be obtained both through direct observation and through the use of appropriate calculations.

The third and final part of the book builds on the previous chapters to address applications of 3-D analysis in rehabilitation, sports, and ergonomics. It also suggests that future research will emphasize the interpretation of the observed phenomena to identify the laws that govern biomovement in general as well as to solve specific individual problems.

I am sure I express the sentiments of readers by thanking, on their behalf as well as mine, the editors and the individual authors for this outstanding work. They have given us the opportunity to learn more about biomovement analysis, modeling, and interpretation.

**Aurelio Cappozzo**
**Universita "La Sapienza"**
**Rome, Italy**

Despite the fact that we live in a three-dimensional world, nearly all published biomechanics research has concentrated on planar movement. Why is this? Until fairly recently we lacked the necessary technology to study 3-D movement in a routine manner. Furthermore, mechanical analysis of 3-D phenomena is not simply half again as difficult as 2-D analysis—it is an order of magnitude more challenging!

Musculoskeletal biomechanics has the potential to make significant contributions to medicine, ergonomics, and sport performance. One of the other keys that will unlock the mysteries of 3-D human movement is pertinent and readily accessible knowledge. Until now, no such compendium has existed. Students of human movement are fortunate indeed that the editors of this text have brought together such a complete review of the field.

Whether your focus is applying the intricacies of the direct linear transformation approach to measuring 3-D coordinates, modeling 3-D motion using the Lagrangian formulation, or understanding how knowledge of 3-D dynamics can be crucial to treating children with cerebral palsy, you will find something of interest here.

**Kit Vaughan**
**University of Virginia**
**Charlottesville, Virginia**

# List of Contributors

**Rachid Aïssaoui**
Research Center
Sainte-Justine Hospital
University of Montreal
Montreal, Quebec
Canada

**Paul Allard**
Research Center
Sainte-Justine Hospital
University of Montreal
Montreal, Quebec
Canada

**Kai-Nan An**
Biomechanics Laboratory
Department of Orthopedics
Mayo Clinic/Mayo Foundation
Rochester, Minnesota
USA

**James G. Andrews**
Department of Mechanical
  Engineering
The University of Iowa
Iowa City, Iowa
USA

**Jean-Pierre Blanchi**
RE.S.ACT.-SPORT Laboratory
Joseph Fourier University
Grenoble
France

**Armin Bruderlin**
School of Computing Science
Simon Fraser University
Burnaby, British Columbia
Canada

**T.W. Calvert**
School of Computing Science
Faculty of Applied Sciences
Simon Fraser University
Burnaby, British Columbia
Canada

**Edmund Y-S. Chao**
Biomechanics Laboratory
Department of Orthopaedics
Mayo Clinic/Mayo Foundation
Rochester, Minnesota
USA

**Laurence Chèze**
Biomechanics Group
University of Lyon I
Villeurbanne
France

**Joannes Dimnet**
Group of Biomechanics
University of Lyon I
Villeurbanne
France

**Alain Durey**
Inter-university Laboratory of
  Research on Teaching Sciences
  and Technology
Ecole Normale Supérieure de Cachan
Cachan
France

**Giancarlo Ferrigno**
Centro di Bioingegneria
Fondazione Pro Juventute
Milan
Italy

**James R. Gage**
Orthopaedic Surgery
Gillette Children's Hospital
St. Paul, Minnesota
USA

**John O.B. Greaves**
Motion Analysis Corporation
Santa Rosa, California
USA

**Steven E. Koop**
Orthopaedic Surgery
Gillette Children's Hospital
St. Paul, Minnesota
USA

**R. Journeaux**
Inter-university Laboratory of
   Research on Teaching Sciences
Universite Paris Sud
Orsay
France

**Kenton R. Kaufman**
Motion Analysis Laboratory
Children's Hospital and Health Center
San Diego, California
USA

**Zvi Ladin**
Biomedical Engineering Department
   and Neuromuscular Research Center
Boston University
Boston, Massachusetts
USA

**Steven A. Lavender**
Department of Orthopedic Surgery
Rush-Presbyterian-St. Luke's Medical
   Center
Chicago, Illinois
USA

**Antonio Pedotti**
Centro di Bioingegneria
Fondazione Pro Juventute
Milan
Italy

**Michael Raymond Pierrynowski**
School of Occupational Therapy
   and Physiotherapy
McMaster University
Hamilton, Ontario
Canada

**Sudhakar L. Rajulu**
Ergonomics Section-Human Factors
   Engineering
Lockheed Engineering and Sciences
   Company
Houston, Texas
USA

**Ian A.F. Stokes**
Department of Orthopaedics and
   Rehabilitation
University of Vermont
Burlington, Vermont
USA

**Michael W. Whittle**
Cline Chair of Rehabilitation
   Technology
The University of Tennessee
   at Chattanooga
Chattanooga, Tennessee
USA

**Jack M. Winters**
Biomedical Engineering Program
Department of Mechanical
   Engineering
The Catholic University of America
Washington, District of Columbia
USA

**Herman J. Woltring**
Brussellaan 29
NL 5628 TB Eindhoven
The Netherlands

# Preface

From the earliest times, scientists have been fascinated by human motion. Aristotle and Leonardo DaVinci are two scientists whose written records testify to this. By the beginning of the 20th century, new techniques became available for studying motion. Biomechanical pioneers such as Marey, Muybridge, Braune, and Fischer emerged to exploit these new techniques. In the 1950s the advent of high-speed photography, together with the emerging possibilities of digital computation, opened up new horizons for study of normal human locomotion, amputee gait, as well as sports biomechanics and investigations of trauma. Most recently, the advent of real-time data acquisition has led to an explosion of possibilities in this field.

Human motion is especially fascinating. Apart from our own self-interest, the extraordinary adaptability of the human body and its control results in an exceptional wealth of activities that we wish to study. Therefore, this book focuses on human motion, although the same principles could be applied to studying animals, plants, or even single cells.

Why does this book emphasize three-dimensional motion? Certainly, the newly developed data acquisition technologies permit and encourage three-dimensional study. However, the ability to measure motion in three dimensions does not automatically give us the ability to analyze or even visualize these measurements adequately. Biplanar analysis is simply the sequential analysis of two-dimensional projections of motion in various planes (the sagittal plane, the frontal plane, etc.), but a truly 3-D analysis takes into account *all* of the complex morphology and asymmetries of the human body and its movements, which is much more challenging.

Our principal objective is to present the state of the art of 3-D analysis to those who have been working in this field but who may be unfamiliar with some aspects. Researchers concerned with either normal or pathological human movement and those practitioners who wish to understand the concepts underlying this type of analysis will also find the book valuable. Readers should be familiar with the contents of an undergraduate course in biomechanics and have a working knowledge of calculus, mechanics, and anatomy. Academic programs in bioengineering and rehabilitation engineering as well as graduate studies in physical education, kinesiology, neurophysiology, and biomechanics should all be able to profit immediately from the contents of this book.

The book is a technical manual and reference for the scientific field of biomechanics, demonstrating both the possibilities and the potential pitfalls of 3-D analysis of human movement. It contains in a single volume the many technical elements required to conduct a three-dimensional analysis. It is organized into three sections, which can be consulted independently. The first part reviews the principal methods of 3-D data capture, reconstruction, and processing techniques. It also covers some technical details of workstations and computer graphics techniques that can be helpful in representing the 3-D behavior of the musculoskeletal system. The second part presents the fundamental concepts of mechanical and neuromuscular modeling. The traditional Newtonian and Lagrangian models are presented first, followed by a chapter showing the parallels between robotics and human motion analysis. The last chapter in this section gives the fundamentals of neuromuscular modeling, which is probably the most scientifically challenging area covered by this volume. The final section of the book illustrates the potential of these new technologies in the clinical study of locomotion, ergonomics, and sporting activities.

Inevitably, some issues remain unresolved. For example, the difficult area of standardization of axial systems is not addressed here. Should the global (laboratory) coordinate system or a local (segmental) coordinate system be used to describe joint motion? Is joint motion better described by the use of Euler angles or by a helical axis of rotations? These unresolved questions and others that need further clarification are not discussed in this edition.

Although the reader may consult specific chapters independently, the sections and chapters are arranged in a natural sequence from selection of instrumentation and data acquisition to illustration and interpretation of recordings. Each chapter presents the most important material first, and though presented in a readily understandable form, the depth of material covered is sufficient for an advanced course in biomechanics. Because of the complexity and breadth of the work, extensive use has been made of illustrations and tables. Each chapter ends with a bibliography that provides key references for further reading.

Editing this book has been enormously instructive and enjoyable for us. The various authors have submitted material that is both challenging and intriguing. Establishing the connections between the chapters and dealing with possible areas of overlap forced us to confront fundamental issues in the field and stimulated us to marvel at the complexities of human motion and its control. We sincerely hope that the readers of this book will enjoy it at least as much.

# Part I

# Data Capture
# and Processing

# Chapter 1

# Three-Dimensional Instrumentation

*Zvi Ladin*

This chapter describes the existing instrumentation for three-dimensional measurements. It first describes devices that record joint rotations: goniometers and electromagnetic and acoustic sensors. Each device is described in terms of its principle of operation and the advantages and limitations to its use in a clinical environment. The photogrammetric reconstruction that serves as the common approach to all camera-based systems is described next. The DLT algorithm and the derivation of complete spatial kinematics of a rigid body from two-dimensional projections are also discussed. The accelerometric approach is described last, considering its advantages and limitations as applied to kinesiological studies conducted in an environment that includes gravity. The chapter concludes by describing the procedure for estimating joint loads, and the estimation improvements entailed in combining kinematic information from both position *and* acceleration measurements.

## SEGMENTAL KINESIOLOGICAL ANALYSIS

The human musculoskeletal system is composed of a series of jointed links, which can be approximated as rigid bodies. Six independent parameters (the *degrees of freedom*, DOF) are needed to describe the location and orientation of such a link in space. Consider for example the femur illustrated in Figure 1.1. Its spatial position in the laboratory coordinate system (LCS) can be determined by attaching to it a body coordinate system (BCS). The BCS can then be fully characterized by specifying the position of its origin (x, y, z), and sequentially rotating it around its axes by the amount of $(\theta_x, \theta_y, \theta_z)$. The six coordinates (x, y, z, $\theta_x$, $\theta_y$, $\theta_z$) constitute the degrees of freedom of the femur and therefore uniquely define its spatial location and orientation at any time instant.

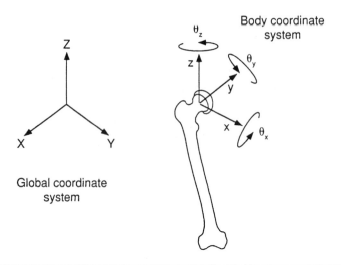

**Figure 1.1**   The determination of the six degrees of freedom of a rigid body in space using three translation parameters (x, y, z) to determine the location and three rotation parameters ($\theta_x$, $\theta_y$, $\theta_z$) to determine the orientation.

Kinesiological measurements are aimed at quantitatively describing the spatial motion of body segments and the movements of the joints connecting those segments. The results can be used for the objective determination of the kinematics (the changes of spatial coordinates with time) and for the calculation of the forces and moments that are associated with the motion (i.e., the kinetics). Both aspects of movement studies are also referred to as biomechanical analysis. Modern studies of human motion are based on three elements that can be traced back to the end of the 19th century, when three major publications appeared in press during a span of less than 10 years:

- the objective and accurate capture of motion (Marey, 1873; Muybridge, 1887),
- the measurement of the forces between the moving body and the environment (Marey, 1873), and
- the approximation of skeletal segments by rigid links interconnected through a series of low-friction joints (Braune & Fischer, 1895/1987).

The American photographer E. Muybridge (1887) pioneered the use of successive-exposures photography by triggering a series of cameras located along a racetrack to study horse locomotion. His studies (first published in a book by Willmann in 1882) showed that photography could serve as a tool for capturing accurate information needed to characterize the kinematics of the animal's limbs. Illustrating the power of his approach, Muybridge was able to prove that a trotting horse had a flight phase, when all of its limbs were in the air. Muybridge then applied his technique to the study of human motion.

E.J. Marey (1873), a French physiologist who served as a professor at the College of France and a member of the Academy of Medicine, also studied animal and human locomotion. He developed a different technique for studying the interaction between the human body and its environment as well as the motion of certain elements of the body. Using *pneumatic* sensors and pressure chambers connected to the soles of the shoes, he measured the pressure under the foot, the acceleration of the head, and the spatial position of the pelvis. By examining the timing and the amplitudes of the signals, he observed that the pelvic motion during running reached the highest point of its trajectory during the stance phase (i.e., while the body was supported by only one of the legs) and not during the airborne phase of running (i.e., when the legs are not in contact with the ground). He therefore concluded that the airborne phase of running did not correspond to the common perception of a projectile being launched during toe-off but should be viewed as controlled falling rather than intentional jumping.

Marey also improved the photographic methods introduced by Muybridge. By modifying the design of the camera to enable it to follow a moving target (he dubbed the device the *photographic rifle*) and by introducing a rotating photographic plate, he was able to film at a frequency of 12 Hz. He later improved the performance of his photographic equipment by introducing the first *passive* kinematic markers: White stripes attached between the main joints of the extremities reflected light onto a photographic plate. To improve the contrast of the image, the stripes were attached to a black suit worn by the subject.

Wilhelm Braune and Otto Fischer described an improved process for studying human motion in a series of publications starting in 1895. They attached long, thin light-tubes to the body segments (i.e., the head, upper and lower arms, thigh, shank, and foot) and used pulses of electric current to generate short bursts of light, synchronously photographed by four cameras. This approach represents the origin of the *active marker* systems used today in many kinesiological studies. By reconstructing the three-dimensional coordinates of the markers and using simplifying assumptions on the location of the joints, Braune and Fischer were able to study both the spatial orientation and the time derivatives of the spatial coordinates of the segments of interest. They used the time derivatives to solve the dynamic equations of the body segments, treating each segment as a free body and generating the first estimates of the lumped joint forces experienced in gait. It is instructive to note that it took them about 12 hours to complete a single set of experiments and up to 3 *months* to analyze the data.

# GONIOMETERS

Goniometers are devices designed to measure the relative rotation of a given joint. The simplest form of a goniometer is a single-axis potentiometer, illustrated in Figure 1.2, a and b during the measurement of knee flexion. As the shank flexes with respect to the thigh, the arm (C) moves along the resistor (AB), thereby changing the output resistance of the potentiometer. By providing a constant voltage-drop across the fixed ends of the resistor ($V_i$), the output voltage

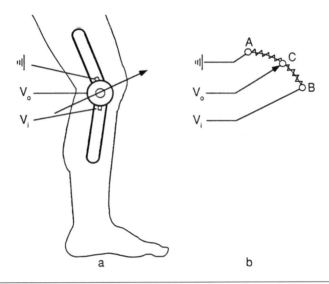

**Figure 1.2**   (a) A goniometer attached to the shank and thigh for measuring knee rotation. (b) The circuit schematic of the potentiometer; $V_i$ indicates the input voltage and $V_o$ indicates the output voltage.

($V_o$) will change, tracking the change in the flexion angle of the knee. Clearly, the goniometer's center of rotation must match the joint's center of rotation to yield a valid measurement.

A different design, based on a flexible cable instrumented by strain gauges, was recently introduced by the Penny & Giles Company (Blackwood, Gwent, UK). This design does not require the specification of the joint center and can provide one or two rotational degrees of freedom for each joint; however, detailed studies of its accuracy, reliability, and reproducibility are not yet available.

Human joints are usually more complex than a single-axis joint. The knee joint, for example, accommodates the small contact area between the femur and the tibia, performing a combination of rolling and sliding. The shape of the femoral condyles leads to a complex rotation that includes components of abduction/adduction, internal/external rotation, and flexion/extension. Using a single-axis goniometer to study the rotation of the knee joint presents two major difficulties:

1. cross talk—the contamination of the measured angle of rotation by an angle that is perpendicular to the measured angle
2. mechanical constraint—modification of the joint's angular rotation due to the mechanical coupling of the adjacent links by the goniometer

The first difficulty arises from the artifactual motion of the goniometer due to an angular rotation unrelated to the rotation of the monitored degree of freedom. An example of such an artifact could be the rotation of the goniometer depicted in Figure 1.2a (measuring flexion/extension of the knee) as a result of an internal or external rotation of the tibia. The second difficulty arises from the inability of the goniometer to follow the true rotation of the joint, thereby hindering the natural motion and introducing an artifactual measurement. An accurate measurement of the complete joint rotation requires that the goniometer measure the full six degrees of freedom that characterize such a joint.

Multidegree-of-freedom devices were described by Lamoreaux (1971), Townsend, Izak, and Jackson (1977), Sommer and Miller (1981), Lewis, Lew, and Schmidt (1988), and others. The first designs of goniometric systems that could accommodate more than a single degree of freedom were based on the use of parallelogram linkages to transmit perpendicular rotations to the monitoring potentiometer. Such designs can accommodate some degree of linear and translational misalignment between the joint's momentary center of rotation and that of the potentiometer, but still represent a measurement system that is sensitive to the attachment of the goniometer. More recent designs use a serial attachment of single-axis potentiometers, interconnected by small, rigid links. By connecting the two endlinks of the goniometer to the jointed links, and using six potentiometers, interconnected by five rigid links, it was possible to develop a true, six-DOF measurement system. Such a system requires a calibration procedure in order to relate the output of the individual potentiometers to either an inertial or a body-centered anatomical reference system.

Although there have been studies that described small errors obtained in a goniometric system (less than 1 mm in translation and less than 1° in rotation, Lewis et al., 1988; Suntay, Grood, Hefzy, Butler, & Noyes, 1983), there are some practical limitations that prevent such systems from gaining widespread acceptance in clinical studies.

- The relative joint rotations measured by such systems do not allow the incorporation of the measurements directly into the dynamic equations of the multilink system. Therefore, additional information is needed in order to obtain the joint loads.
- The tight attachment of the goniometer across the joint presents a mechanical constraint that limits the motion of the soft tissues, thereby possibly modifying the natural motion of the joint.
- The goniometric system is cumbersome (and sometimes heavy) and has difficulty accommodating different size limbs.
- There are difficulties in monitoring joints that are surrounded by large amounts of soft tissue (e.g., the hip) or that involve large structures with relatively small attachment areas (e.g., the ankle/subtalar complex).
- There are inherent, nonlinear effects, such as stick-slip and backlash problems, in the mechanical linkage system.
- Specific transducers for different joints must be developed (e.g., goniometers for the lower limbs could not be used for the upper limbs, the back, or the neck).

For these reasons, goniometric systems are used primarily for obtaining rough (and inexpensive) approximations of some joint rotations or for detailed studies of a single joint in a controlled laboratory setting. The need to measure joint rotations directly, while eliminating the mechanical-link coupling introduced by any goniometer, has led to the development of alternative transduction methods: for example, electromagnetic and acoustic transduction.

# ELECTROMAGNETIC AND ACOUSTIC SENSORS

Biomechanical application of electromagnetic measurement of the six degrees of freedom of rigid body motion was described by An, Jacobsen, and Chao (1988). Their device (available from Polhemus Navigation, Colchester, VT, USA) was based on two small units—one serving as the source and the other serving as the sensor, both tethered to a system electronics unit. Both the source and the sensor contain three sets of orthogonal coils. By exciting each loop (coil) in the source with an identical low-frequency magnetic signal, a pattern of three states is generated in the source (Krieg, 1984). This yields a set of three linearly independent vectors, which can be measured on the sensor, that contain enough information to discern the relative position and orientation of the sensor with respect to the source. The algorithm necessary to determine the kinematic variables is based on the linear small-angle rotation approximation of the trajectories traversed over a short period of time (such as between sampling points).

An et al. (1988) found the device to be accurate for monitoring the general rigid body motion around a single joint. However, using the system for monitoring the motion of a multiple number of rigid bodies may be limited by several factors:

- Electromagnetic interference from metallic objects in or near the sensor distorts the output signal.
- The relatively slow sampling rate (60 Hz), coupled with the linear kinematic approximation technique, limits the speed of joint rotation that can be accurately tracked.
- The small number of sensors (a maximum of four sensors and two sources) limits the number of joints that can be tracked.
- The range of angular motion of any rotational degree of freedom is limited to 180°.

A similar system developed by Ascension Technology Corporation (Burlington, VT, USA) is described as being able to monitor up to eight segments at rates of up to 140 Hz; however, detailed studies of its performance and application to biomechanical studies are not yet available.

Acoustic sensors have been reported by some authors in 3-D kinematic studies of human joints such as the wrist (Andrews & Youm, 1979), the ankle (Siegler, Chen, & Schneck, 1988), and the knee (Quinn & Mote, 1990). The system usually includes an array of acoustic sources (e.g., spark gaps) and an array of at least three noncolinear receivers (microphones), which define a body coordinate system (BCS). Acoustic waves are generated and transmitted by the sources, travel through the air, and are received by the microphones. Because the speed of sound is known, the system can calculate the location of the receiver relative to the sources. By fixing the sources (or the receiver) within a laboratory coordinate system, the complete three-dimensional location of the receiver (or the source) in the inertial reference system is determined. When the spatial location of all the markers in the array is known, the position and orientation of the BCS can be calculated, thereby fully characterizing the six DOF of the rigid body.

In a well controlled experimental setup, Quinn and Mote (1990) reported errors smaller than 0.5 mm in translation and 0.5° in rotation. However, the soft tissue motion encountered in dynamic studies, the relative size and orientation of the source and sensor array segments (needed to create a good signal-to-noise ratio), and the acoustic echoes, interference patterns, and sparking introduced by multiple acoustic source/sensor pairs make this approach difficult to apply to dynamic motion studies of human subjects' multiple body segments. Because of the inherent difficulties in developing a general, easy-to-use (and calibrate) multibody measurement system, based on transduction methods, camera systems are used widely to monitor the motion of multiple body segments.

# PHOTOGRAMMETRIC RECONSTRUCTION

The image created by a camera represents a two-dimensional projection of a *three*-dimensional object. Therefore, the challenge in using cameras for studying motion is to recreate the three-dimensional object that gave rise to the two-dimensional projection. This process is called photogrammetric reconstruction. A simple two-dimensional example serves as a good illustration of this process.

Consider the point object A and the two cameras depicted in Figure 1.3. In order to find the global coordinates $(X_A, Y_A)$ that represent the spatial location of the object, we use the one-dimensional images of the object on the two cameras. Rays of light originating at A and passing through the optic centers of the cameras create an image on the photosensitive surface of each camera. These images can be characterized as having the local (camera) coordinates $U_1$ in one camera and $U_2$ in the other camera. By finding the intersection of the two rays originating in the images and passing through the optic centers of the cameras, the spatial location of the object can be determined. Therefore, knowledge of the image coordinates and the locations of the optic centers are the only pieces of information needed for the determination of the two-dimensional global coordinates of the object $(X_A, Y_A)$. The graphic solution illustrated above for the two-dimensional case can be expanded to handle the three-dimensional reconstruction of the spatial coordinates of a point $(X, Y, Z)$ by either of two methods:

- Combining the information originating from projections of the point onto two planar cameras (i.e., using the local—camera—coordinates $[U_1, V_1]$ and $[U_2, V_2]$)
- Combining three independent linear coordinates from one-dimensional cameras (i.e., using the local—camera—coordinates $[U_1, U_2, U_3]$)

The tedious process of extracting the image coordinates from the film of the cameras using manual digitization has been largely replaced by the automated

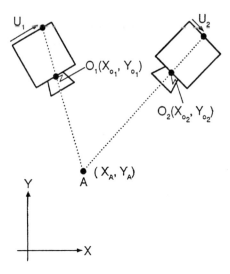

**Figure 1.3** A two-dimensional illustration of the photogrammetric approach. The location of the image on the camera detector is given by a single coordinate, U. The intersection of the two lines, each passing through the camera image and its optic center, results in the global coordinates of the marker, $X_A$, $Y_A$.

electronic extraction of that information. This procedure requires attaching markers to the moving subject and then identifying the markers in the viewing volume of the cameras. The markers can be divided into two groups:

- Active markers—Light emitting diodes (LEDs) that usually emit infrared light
- Passive markers—Light reflecting devices that reflect ambient or projected light

Active marker systems pulse the LEDs in a sequential manner; therefore at any instant in time there can be only one LED turned on. The cameras then uniquely identify the momentary spatial location of that marker. Passive systems use markers that reflect projected light and appear on the screen of a video monitor as bright spots. The spatial location of each marker is determined by identifying (automatically or by user input) the bright spots that belong to the same physical marker, and then applying the photogrammetric approach to the two (or more) reflected images of the marker.

Each of the imaging approaches has its own advantages and disadvantages. Active marker systems have the advantages of user-selected sampling rate and automated identification of the markers. However, they require that wires be attached to each marker, which makes measurement cumbersome and difficult to conduct outside a laboratory setting. Reflective marker systems usually have a fixed sampling frequency (the frequency of the electrical network), though some newer versions offer the option of a higher sampling frequency (up to 200

Hz). These systems have the advantage of not having the subject tethered by wires, since the markers do not require any power input. However, they do carry the penalty of not being able to associate the bright projections on the two (or more) cameras with the physical markers that created those projections. This problem is especially severe if the distance between the markers is small, and the sampling frequency is high. Under these conditions a large measure of user intervention may be required to resolve any errors in the physical identification of the markers.

When the projections of the markers on the cameras have been identified, the photogrammetric approach can be applied to calculate the 3-D spatial coordinates of each marker. As discussed earlier, this procedure requires knowledge of the optical parameters of the lenses of each camera, such as the focal length of the lens and the location of the optic center. These parameters can be difficult to measure and are not generally available for most cameras. Furthermore, the nonlinearities of the lenses and the electronic noise in the circuitry that measures the projections of the markers create hazy areas on the detector plates of the cameras. Therefore, an analytic solution that will accurately (i.e., with *no* error) reconstruct the 3-D position of the marker cannot be obtained.

The most common approach used to calculate the three-dimensional marker coordinates is based on direct linear transformation (DLT, Abdel-Aziz & Karara, 1971). This reconstruction algorithm can be reduced to a set of two linear equations with a minimum of 11 parameters that characterize the calibration, position, and orientation of a single camera. Such a set can therefore be viewed as a procedure that maps the two image coordinates to the three global marker coordinates. Because there are only two equations and three unknowns, another set of equations is required. This additional information is provided by a second camera. The data provided by the two cameras result in four equations having three unknown spatial coordinates of the marker. The redundant equation can be used as an additional source of information, applying an iterative process that will converge on a better estimate of the marker location.

The determination of the camera parameters necessary for the implementation of the DLT algorithm requires a calibration procedure. This step is performed using a set of precalibrated markers, usually attached to known locations on a rigid frame. A set of two cameras targeted at a typical calibration frame is shown in Figure 1.4. Each marker provides one set of two equations. The global coordinates of the marker are known, and the image coordinates are measured by the camera. Because the equations are linear, a minimum of six markers (generating 12 independent equations) is required to determine the camera parameters. In order to reduce the errors and provide a degree of redundancy to the process of parameter determination, more markers are often used. Once the calibration procedure has been completed for both cameras, the parameters are stored and the image coordinates can then be transformed into the global coordinates for any marker. *It is imperative that the position of the cameras and their spatial orientation be unchanged during the experiment.* Any relocation or reorientation of the cameras requires recalibration.

The spatial reconstruction of a single marker represents the first step in determining the six degrees of freedom of a rigid body in an inertial reference

**Figure 1.4**   A setup showing a two-camera system and the calibration cube in the center of the viewing volume.
*Note*. Photo courtesy of Northern Digital, Inc.

system. By identifying at least three noncolinear markers on a rigid body, it is possible to establish uniquely its location and orientation in space. This approach was developed at MIT and recently described by Antonsson and Mann (1989) and Rowell and Mann (1989). Small rigid fixtures (referred to as *arrays*) that contain multiple markers (i.e., more than three noncolinear markers) are attached to the monitored link. As the link moves through space, so does the rigid array, thereby accurately monitoring the spatial motion of the body under study. A computer program called TRACK (distributed by OsteoKinetics, Newton, MA, USA) calculates the spatial translation and orientation of the array and describes it in an inertial frame of reference. The complete kinematic characterization of two jointed links results in an accurate determination of the kinematics of the joint that connects the two links. Because the motion of each link is completely described in an inertial reference system, the correct joint rotation is given by the relative rotation of one link with respect to the other. Therefore, the complete 3-D joint rotation can be determined without hindering the actual motion of the joint and with an accuracy that depends only on the accuracy of the spatial reconstruction of the individual markers. This approach has been successfully used both with active (Ladin, Flowers, & Messner, 1989) and with reflective marker systems.

## ACCELEROMETERS

Accelerometers are small devices that measure acceleration. Their principle of operation is illustrated in Figure 1.5, a and b. A small mass is attached by a cantilever beam to the moving body. As the body accelerates, the beam becomes deflected, and the amount of deflection is related to the acceleration of the mass. A set of strain gauges attached to the beam is used to transduce the beam deflection into

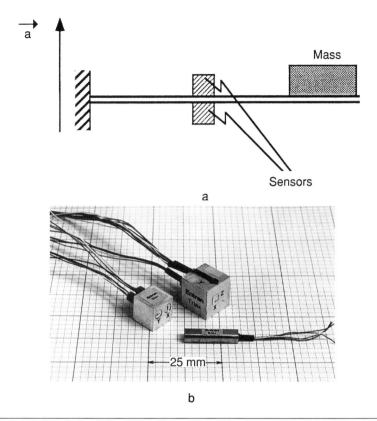

**Figure 1.5**    (a) A schematic description of the accelerometer's principle of operation. As the body accelerates (with an acceleration of $\vec{a}$) in the direction indicated by the arrow, the small mass attached to the accelerating beam causes deflections that are measured by the sensors (strain gauges, or piezoresistive or piezoelectric devices). (b) A set of small multiaxial accelerometers (courtesy of Entran Devices, Inc.).

an electrical signal. Such an approach was adopted by the manufacturers of piezoresistive accelerometers (e.g., Entran Devices, Fairfield, NJ, USA) in which the strain gauges are made of silicon glass, allowing miniaturization of the sensor to approximately 1 *cm* in size. Three sensors, whose sensitive axes are mutually perpendicular, combine to form a small triaxial accelerometer, which measures the three-dimensional acceleration. With the advent of silicon and the many improvements in the fabrication of integrated circuits, this design has been further miniaturized and implemented on a silicon chip (ICSensors, Milpitas, CA, USA). These silicon accelerometers cost substantially less than other accelerometers. In the future, electronic conditioning will be implemented on the sensor itself, which will reduce the noise and the wiring currently associated with accelerometer output.

The primary advantage of this class of accelerometers is its broad frequency band, covering the range of 0 to 1,000 Hz. The accelerometer also responds to the field of gravity, and therefore its output represents the vector sum of the kinematic acceleration and the acceleration of gravity. A different category of accelerometers is represented by piezoelectric devices (Kistler Instrument, Winterhur, Switzerland). These devices use piezoelectric crystals as sensors and are therefore insensitive to constant acceleration. Thus, they are unaffected by the field of gravity; however, they are also insensitive to slow kinematic processes and so are of limited use in kinesiological measurements.

Time integration of the acceleration signal yields the momentary velocity of the point to which the accelerometer is attached. This computational approach has the advantage of reducing the high-frequency noise that exists in most measurements and, therefore, will produce a "cleaner" estimate of the velocity, compared to the estimate generated by differentiating the trajectory of a marker attached to the same point. A second time integration will lead to the spatial displacement of that point, potentially providing an alternative measurement to that generated by a more expensive, position measurement system.

Some practical limitations have prevented the wider use of accelerometers:

- The determination of the segment's initial conditions
- The effect of the field of gravity
- The need to identify the segment rotational degrees of freedom
- Low-frequency drift of the accelerometer's output

Using the accelerometer's output to calculate a segment's velocity and position requires knowledge of the initial values of these variables. Because these values cannot be determined by the accelerometer, the researcher must either use a different measurement system or simply assume these values. The effect of gravity presents still another limitation: Since the measured acceleration represents the vector sum of the gravity and kinematic accelerations, additional information describing the segmental orientation in the field of gravity is needed to derive the kinematic acceleration.

Even if these problems are resolved, one is still left with the inherent limitation of not being able to determine the segmental six degrees of freedom merely from the acceleration of a single point on a segment. Assuming that the accelerometer is attached to the segment at a point away from its center of mass, the measured acceleration will include terms that arise from the translational and rotational motion of the segment. The rotational components are nonlinear and include terms related to the rotational velocity and acceleration of the segment. In order to calculate these terms, so that the rotational degrees of freedom can be later determined by time integration, one needs a large number of accelerometers, attached to different points on the segment. Solution of a set of nonlinear equations (Hayes, Gran, Nagurka, Feldman, & Oatis, 1983; Morris, 1973) can then extract the rotational velocity and acceleration.

Finally, the accelerometer signal tends to "drift" over time, causing a low-frequency noise whose effect increases with time. Such an artifact will contaminate both the translational and rotational degrees of freedom. Hayes et al. (1983) and Seemann and Lustick (1981) showed that low-frequency noise can lead to substantial errors in the long-term determination of the segmental kinematic degrees of freedom.

# DYNAMIC FORCE ESTIMATES

Analysis of joint dynamics (i.e., the joint forces and moments produced by the muscles in the process of a given musculoskeletal activity) is complementary to the task of deriving accurate kinematic information. The estimation of joint dynamics based on the motion of the adjoined links involves solution of the inverse dynamics problem. The computational process requires knowledge of the link linear acceleration and angular velocity and acceleration, in order to estimate the complete set of joint forces and moments responsible for the monitored motion.

Ladin and Wu (1991) described an approach that combines position and acceleration measurements. This process is based on the integrated kinematic sensor (IKS), a rigid array that includes markers for position measurements (which are monitored by the WATSMART optoelectronic system in Ladin and Wu's application), and a triaxial accelerometer for direct measurements of the segmental acceleration. The position and orientation of the rigid array in the laboratory coordinate system (determined by TRACK) were used to establish the momentary orientation of the accelerometer in the field of gravity. This information was used to subtract the effect of gravity from the accelerometer output, leaving the true kinematic acceleration (i.e., the second time-derivative of the position vector). This value, when substituted into the dynamic equations that describe the motion of the rigid body, yields estimates of the joint forces.

The system was tested on a two-DOF mechanical pendulum, where its estimates were compared to joint forces directly measured by an array of strain gauges. The estimated forces based on the acceleration measurements showed excellent correspondence to the measured values. The integrated estimator was able not only to trace the basic oscillations but to capture even the high-frequency vibrations that were observed by the strain gauge measurements. The error in the estimates was less than 2%, creating a reliable and accurate assessment of the actual joint forces. The force estimates generated exclusively by the position measurements (i.e., using the position measurements to estimate the segmental acceleration and the estimated acceleration to calculate the joint forces) were highly dependent on the filtering scheme. The basic low-frequency oscillation was captured accurately using a 5-Hz, low-pass filter, but this cutoff frequency eliminated the high-frequency vibrations. Increasing the cutoff frequency made the estimates much noisier, which illustrates the inherent tradeoff in such an estimation process.

The integrated kinematic sensor described in Ladin and Wu's study was designed to improve the quality of joint force estimates that are based on the solution of the inverse dynamics problem. The comparison of the joint force measurements by the strain gauges to the estimates based on the integrated sensor and those based on the position measurements alone clearly showed the benefit that arises from the integration of position and acceleration measurements. The close correspondence between the estimated forces and the measured ones suggests that by integrating the three elements that comprise such an estimation, namely accurate position measurements, rigid body kinematic analysis that extracts the six degrees of spatial motion, and the use of this information to dynamically calibrate the acceleration measurements of a linear accelerometer, one can obtain high-quality noninvasive estimates of joint forces.

## SUMMARY AND CONCLUSIONS

The selection of a measurement system for studying human motion requires carefully matching the nature of the motion, the environment in which the motion is performed, and the properties of the measurement system. The high cost of some of the more general systems usually means that a system is purchased for a laboratory with the expectation that it can be used in a variety of studies. In order to make optimal use of the measurement system, it is imperative to try to project the most demanding studies in which such a system could potentially be used. Some factors that should be considered include

- the frequency range needed to accurately describe a given physical activity (e.g., walking vs. running or jumping),
- the nature of the physical variables to be measured (e.g., position and rotation vs. acceleration),
- the environment in which the activity is performed (e.g., a clinical laboratory vs. an Olympic arena),
- the number of limbs or joints to be studied,
- the accuracy and resolution needed for the measurement, and
- the nature of the user interaction needed to operate the system.

Once the technical parameters have been specified, the selection of an appropriate system can proceed by examining those systems that satisfy them. The final determination of the best system will obviously depend on financial constraints.

## REFERENCES

Abdel-Aziz, Y.I., & Karara, H.M. (1971). Direct linear transformation from comparator coordinates into object space coordinates in close-range photogrammetry. *Proceedings of the Symposium on Close-Range Photogrammetry* (pp. 1-18). Falls Church, VA: American Society of Photogrammetry.

An, K.N., Jacobsen, L.J., & Chao, E.Y.S. (1988). Application of a magnetic tracking device to kinesiologic studies. *Journal of Biomechanics*, **21**, 613-620.

Andrews, J.G., & Youm, Y. (1979). A biomechanical investigation of wrist kinematics. *Journal of Biomechanics*, **12**, 83-89.

Antonsson, E.K., & Mann, R.W. (1989). Automatic 6-d.o.f. kinematic trajectory acquisition and analysis. *Journal of Dynamic Systems, Measurement, and Control*, **111**, 31-39.

Braune, C.W., & Fischer, O. (1987). The human gait. (P. Marquet & R. Furlong, Trans.). Berlin: Springer-Verlag. (Original work published in 1895.)

Hayes, W.C., Gran, J.D., Nagurka, M.L., Feldman, J.M., & Oatis, C. (1983). Leg motion analysis during gait by multiaxial accelerometry: Theoretical foundations and preliminary validations. *Journal of Biomechanical Engineering*, **105**, 283-289.

Krieg, J.C. (1984). A feedback mechanism for use in paraplegic stimulation techniques. In *Proceedings of the 2nd International Conference on Rehabilitation Engineering* (pp. 413-414).

Ladin, Z., Flowers, W.C., & Messner, W. (1989). A quantitative comparison of a position measurement system and accelerometry. *Journal of Biomechanics*, **22**(4), 295-308.

Ladin, Z., & Wu, G. (1991). Combining position and acceleration measurements for joint force estimation. *Journal of Biomechanics*, **24**(12), 1173-1187.

Lamoreaux, L. (1971). Kinematic measurements in walking. *Bulletin of Prosthetic Research*, **BPR 10-15**, 3-84.

Lewis, J.L., Lew, W.D., & Schmidt, J. (1988). Description and error evaluation of an in vitro knee joint testing system. *Journal of Biomechanical Engineering*, **110**, 238-248.

Marey, E.J. (1873). *Animal mechanism: A treatise on terrestrial and aerial locomotion*. New York: Appleton. Republished as Vol. XI of the International Scientific Series.

Morris, J.R.W. (1973). Accelerometry—A technique for the measurement of human body movements. *Journal of Biomechanics*, **6**(6), 729-736.

Muybridge, E. (1887). Animal locomotion. Reprinted in Brown, L.S. (Ed.) (1957). *Animal in motion*. New York: Dover.

Quinn, T.P., & Mote, C.D. (1990). A six-degrees-of-freedom acoustic transducer for rotation and translation measurements across the knee. *Journal of Biomechanical Engineering*, **112**(4), 371-378.

Rowell, D., & Mann, R.W. (1989). Human movement analysis. *SOMA—Engineering for the Human Body*, **3**(2), 13-20.

Seemann, M.R., & Lustick, L.S. (1981). Combination of accelerometer and photographically derived kinematic variables defining three-dimensional rigid body motion. *SPIE—Biomechanics Cinematography*, **291**, 133-140.

Siegler, S., Chen, J., & Schneck, C.D. (1988). The three-dimensional kinematics and flexibility characteristics of the human ankle and subtalar joints—Part I: Kinematics. *Journal of Biomechanical Engineering*, **110**, 364-373.

Sommer, H.J., & Miller, N.R. (1981). A technique for the calibration of instrumented spatial linkages used for biomechanical kinematic measurements. *Journal of Biomechanics*, **14**, 91-98.

Suntay, W.J., Grood, E.S., Hefzy, M.S., Butler, D.L., & Noyes, F.R. (1983). Error analysis of a system for measuring three-dimensional joint motion. *Journal of Biomechanical Engineering*, **105**, 127-135.

Townsend, M.A., Izak, M., & Jackson, R.W. (1977). Total motion knee goniometry. *Journal of Biomechanics*, **10**, 183-193.

# Chapter 2

# Bases of Three-Dimensional Reconstruction

*Paul Allard*
*Jean-Pierre Blanchi*
*Rachid Aïssaoui*

Many biomechanical three-dimensional (3-D) analyses of human movement start with data capture by an imaging device. Still or high-speed cameras, video cameras, or radiographic systems are the most common of these data capture systems. With a minimum of two different perspectives or views, the spatial coordinates of object points or body markers can be determined using photogrammetric principles. Object points are specified because they can be specifically and uniquely identified in each view and cannot be confused among themselves or with adjoining surfaces or landmarks.

Photogrammetry, defined by the American Society of Photogrammetry, is "the art, science, and technology of obtaining reliable information about physical objects and the environment through the processes of recording, measuring, and interpreting photographic images and patterns of recorded radiant electromagnetic energy and other phenomena" (Wolf, 1983). Within photogrammetry, metric and interpretative photogrammetry are two distinct areas. *Metric photogrammetry* consists of precise measurements from photos and other sources to determine the relative locations of points or markers. *Interpretative photogrammetry* consists of using systematic analysis to recognize and identify objects. Automated gait-tracking systems involve both the recognition and identification of object markers, as well as their three-dimensional reconstruction.

This chapter covers basic 3-D reconstruction techniques. Classical methods as well as two-step reconstruction techniques are reviewed. The most frequently used of the latter is the direct linear transformation (DLT) method. Sources of errors from different camera types, lenses, and calibration objects are also discussed.

# SPATIAL COORDINATE SYSTEMS

Three-dimensional analysis requires several coordinate systems. In a planar analysis, there is normally a fixed, or laboratory, coordinate system (FCS), like that shown in Figure 2.1, from which the body marker coordinates are calculated. The axes of the FCS should be labeled unambiguously to avoid confusion with the local coordinate systems, which define joint movement. The origin can be located at any practical point in the laboratory.

A local, or relative, coordinate system (LCS) may be located on a moving body segment. For example, if a kinematic description of the upper limb is sought, an LCS can be located on the shoulder even though the entire upper limb is in motion. Then the body marker coordinates are calculated relative to those of the shoulder. The marker velocities and accelerations are also relative to the shoulder and do not take into account the absolute displacement of the shoulder with respect to a fixed coordinate system located somewhere in the laboratory.

In a three-dimensional reconstruction, two or more different views of the subject are required. These views can be photographs, video or camera frames, and so on. Each view must have its own camera coordinate system (CCS) based in the imaging device.

Figure 2.2 illustrates four camera coordinate systems (CCSs) in which the optical axis defines the $Z$ direction and the $X$ and $Y$ axes are the horizontal and the vertical, respectively, when the camera image is leveled. The origins, $O_i$ (where $i$ is the camera number), are located at the focal point of each camera.

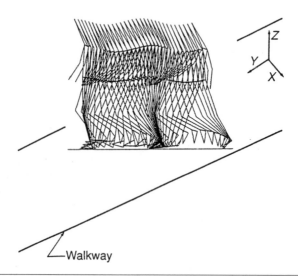

**Figure 2.1**   Stick diagram of a person walking from right to left. The origin of the FCS can be located at any convenient point in the laboratory. The axes should be identified according to the right-hand rule. The vertical and the direction of progression corresponds to the $z$ and $y$ axes, and the $x$ axis is in the direction perpendicular to the $yz$ plane.

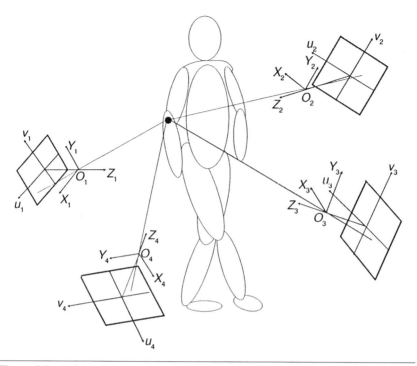

**Figure 2.2**   A four-camera system where the image of all the body markers are identified on the CCS located at the focal point of each camera.

The projection of these coordinate systems onto the image forms the image coordinate system (ICS) whose origin is the principal point, which is aligned with the lens focal point. Assuming the image is flat, one axis is eliminated and the other two axes are labeled $u_i$ and $v_i$ respectively, to avoid confusion between the image and the camera coordinate systems.

## RECONSTRUCTION TECHNIQUES

Once the coordinate systems have been identified, equations are used to determine the object point coordinates from their image coordinates. (In gait analysis, the object points are either passive or active markers that identify specific anatomical landmarks.) These equations relate the external parameters (i.e., the spatial orientation of the camera) and the internal parameters, such as lens characteristics or camera type, to the object point or marker's position. Equations and solution methods have been developed for both fixed and mobile camera setups. Most three-dimensional gait measurements are carried out with cameras fixed and facing the subject or patient. In aerial photography, or in machine vision systems such as self-guided vehicles, the cameras are mounted on a carrier and move along with it. The camera displacement is then taken into account in the calculation.

The standard photogrammetric procedure is based on the exact and precise knowledge of all the internal and external camera parameters. An error in any one of these parameters results immediately in an erroneous coordinate value. Nevertheless, these techniques are extremely good when using metric cameras designed for stereo-photogrammetry and when each camera's spatial position and attitude has been determined accurately.

Using nonmetric or off-the-shelf cameras whose internal and external parameters have been well estimated can yield acceptable results. Because the internal and external parameters have been obtained by an indirect means, this approach is called an *implicit* method as opposed to the direct, or *explicit*, method.

## Fixed Cameras

Three-dimensional reconstruction techniques for fixed cameras can also be used with more than two camera views. The standard technique is based on using two images; any additional images are used to create additional camera pair combinations. Thus, three images form three pairs, whereas four images make six pairs, and so on. This applies to still photography as well as high-speed cinematography or videography. With videography, the image pairs must be synchronized beforehand. However, a single moving camera can also be used if the object is relatively stationary, as in serial photography.

One of the guiding principles of photogrammetric analysis is the colinearity condition. This condition, illustrated in Figure 2.3, requires that an object point

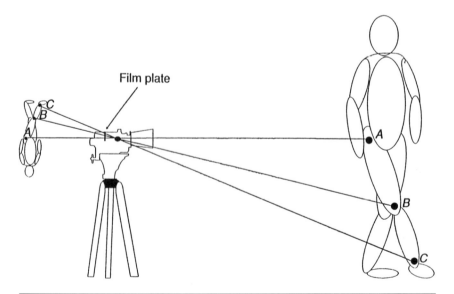

**Figure 2.3**    The basic colinearity condition required for 3-D reconstruction is shown by the lines joining Markers *A, B,* and *C* to their corresponding image crossings at the camera's focal point.

or marker point and its corresponding image form a straight line that passes through the focal point. (The simplest example of this is the pinhole camera.) If this condition is met, then geometric and trigonometric relationships can be applied to determine the spatial position of the object or marker. In Figure 2.3, the object points, $A$, $B$, and $C$ lie on the same straight lines as their corresponding image points, $a$, $b$, and $c$. The colinearity condition equations are developed from similar triangle relationships and are fully detailed in Wolf's (1983) textbook on photogrammetry.

Equally important is the coplanarity condition, which requires that an object point, $A$, its corresponding image pair, $a_1$ and $a_2$, as well as the cameras' focal points, $f_1$ and $f_2$, lie in a common plane (see Figure 2.4). Whereas the colinearity condition relates the object point to its image, the coplanarity condition links the images (at least two) to the object's point. With these two conditions fulfilled, the spatial coordinates can be analytically reconstructed from image representations.

## Classical Reconstruction Techniques

Most of the implicit techniques were derived from classical techniques. A good understanding of the basic approach leads to better comprehension of current practices. Among the many 3-D equations derived from the colinearity condition, the basic formulas of aerial photogrammetry are similar to those used in human motion analysis. The difference is that the optical axis is usually horizontal in a human motion laboratory rather than vertical as in aerial photography.

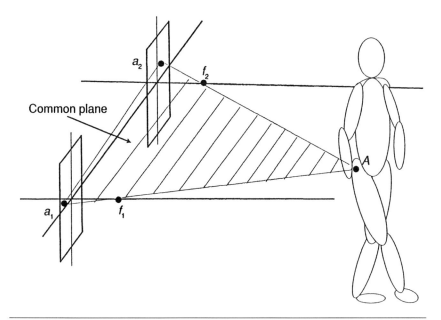

**Figure 2.4**   The coplanarity condition is illustrated in that each body marker lies in a common plane formed by its corresponding image positions and the focal point of each camera.

A stereo-photogrammetric setup is shown in Figure 2.5, with typical distances given for illustrative purposes. The important assumptions are that the external camera parameters are known and that the film plates lie in the same plane. This implies that their optical axes are parallel and perpendicular to the film plane. The $u_i$ and $v_i$ axes originate at the film principal point, $P_i$. Each camera's optical axis passes through the focal point and is aligned with the respective camera's principal point. The FCS has been arbitrarily set at the focal point of Camera 1 at $O_1$, and the cameras are separated by a base distance, $B$, of 1.200 m. The focal length, $C$, is 0.300 m. In this example, the coordinates of an object point, $A$ (0.600, 1.300, 0.300), expressed in meters with respect to $O_1$, are known a priori. They can be determined analytically by applying the following equation (Hallert, 1960).

The equations are

$$X_1 = (B \cdot u_1)/p , \tag{2.1}$$

$$Y_1 = (B \cdot c)/p , \tag{2.2}$$

$$Z_1 = (B \cdot v_1)/p , \tag{2.3}$$

where $p = u_1 - u_2$ and is used to correct for parallax in the $O_1$ coordinate system. Using these equations, the object coordinates are as follows:

$$X_1 = (1.200 \cdot - 0.140)/(- 0.140 - 0.137) = 0.606 \text{ m}$$

$$Y_1 = (1.200 \cdot 0.300)/(- 0.140 - 0.137) = 1.300 \text{ m}$$

$$Z_1 = (1.200 \cdot 0.690)/(- 0.140 - 0.137) = 0.299 \text{ m}.$$

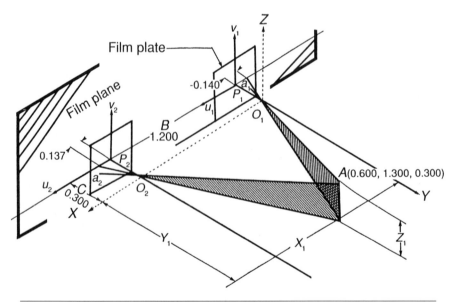

**Figure 2.5**    Basic stereo-photogrammetric setup with the optical axes parallel to each other. The camera base is 1.200 m and the focal length is 0.300 m. Colinearity and co-planarity conditions are met. All dimensions are expressed in meters.

Although the mathematics are quite simple, the technique requires metric cameras accurately positioned. Ayoub, Ayoub, and Ramsey (1970) describe in detail their photographic system and supporting assembly as well as the correction procedure to determine the principal distance, the principal point, and lens distortion.

Today's low-distortion lenses are readily available for stereophotogrammetry with high-speed and video cameras. But we recommend you check the lens for distortion errors. First ask whether the manufacturer can supply you with the lens characteristics. Alternatively, you can do your own calibration with an acceptable accuracy using a sheet of graphic paper taped to a granite measuring table. These tables are usually flat to ±2 μm. Fiducial marks on the graphic paper are used to test for lens distortion. The granite table is positioned vertically with the use of a good quality level and set as perpendicular as possible to the camera's optical axis. Although the table can be slightly oblique, this bias should be constant.

Equations 2.1 to 2.3 can be expressed in a more general form:

$$
\begin{bmatrix} u - u_p + \Delta u \\ v - v_p + \Delta v \\ -C \end{bmatrix} = \lambda[\mathbf{M}] \begin{bmatrix} X - X_o \\ Y - Y_o \\ Z - Z_o \end{bmatrix} \tag{2.4}
$$

where $u$ and $v$ are the image coordinates, $u_p$ and $v_p$ are the principal point image coordinates, $\Delta u$ and $\Delta v$ are the $u$ and $v$ errors, $C$ is the principal distance of the camera, $\lambda$ is the linear scale factor, $\mathbf{M}$ is a $3 \times 3$ transformation matrix from the image to the laboratory coordinate system, $X$, $Y$, and $Z$ are the object laboratory coordinates, and $X_o, Y_o$, and $Z_o$ are the focal point coordinates in the $X$, $Y$, and $Z$ system. It is from this general form that many implicit three-dimensional equations are derived.

## Implicit Reconstruction Techniques

Close-range photogrammetry, or the application of photogrammetry where object point distance is less than 308 m (1,000 ft), became possible with the emergence of suitable instrumentation, methodologies, and techniques for data reduction (Marzan, 1975). The departure from metric to nonmetric cameras using new analytical representations for determining the camera's internal and external parameters is an important development in photogrammetry to those involved in the study of human movement. The immediate benefits are in the low cost and ready availability of off-the-shelf cameras, which are not suitable for the classical reconstruction approach.

Today, implicit techniques have proliferated. Many have been developed by rearrangement, expansion, and simplification of Equation 2.4. The classical equation is totally transformed into a series of unknown parameters, which are functions of constants and object point coordinates. The constants are complex mathematical relations involving one or more camera parameters. To determine these parameters experimentally, a calibration device consisting of accurately measured object points is used to determine the unknowns in a two-step approach.

In the first step, the calibration device is photographed and the image and real-object point coordinates are used to determine the unknown constant values in the mathematical expression. This is a form of camera calibration. Then, the calibration device is replaced by the unknown object points or body markers to be measured. Their coordinates are determined from the image's object coordinates and the previously calculated analytical constants. The number of calibration points required to solve the unknown parameters vary according to the implicit technique applied.

Several excellent three-dimensional techniques have been published in the field of biomechanics of human movement. To review them is outside the scope of this work. Generally, the error varies between 0.01 mm and 5 mm. However, the reported values are not always based on the same definition of error.

Among these three-dimensional reconstruction methods, the most widely applied and discussed probably is the direct linear transformation (DLT) technique, developed by Marzan (1975). It has been used with many types of imaging devices. Stokes, Bigalow, and Moreland (1987) applied the DLT in an X-ray photogrammetric technique to measure scoliotic spines, while Allard, Duhaime, Labelle, Murphy, and Nagata (1987) used still cameras to determine the location and orientation of the ankle and subtalar joint axis of rotation. The DLT errors associated with the use of high speed photography (Shapiro, 1978) and videography (Leroux, Allard, & Murphy, 1990) have been assessed.

The DLT method is based on the classical technique and is expressed as

$$u + \Delta u = \frac{L_1 X + L_2 Y + L_3 Z + L_4}{L_9 X + L_{10} Y + L_{11} Z + 1},$$

$$v + \Delta v = \frac{L_5 X + L_6 Y + L_7 Z + L_8}{L_9 X + L_{10} Y + L_{11} Z + 1}, \tag{2.5}$$

where $u$ and $v$ are the image coordinates and $\Delta u$ and $\Delta v$ are the image coordinate correction for lens distortion. The object point coordinates are $X$, $Y$, and $Z$, whereas the constants $L_1$ to $L_{11}$ are the DLT parameters, which define the camera position and orientation as well as the camera internal parameters and linear lens distortion factors. See Marzan (1975) for a detailed description and two Biomch-l (E-mail) files, which are available to the membership of Biomch-L with the following commands to ListServ@hearn.bitnet or to ListServ@nic.surfnet.nl: SEND DLTDSP README BIOMCH-L and SEND DLTDSP FORTRAN BIOMCH-L.

In the example based on the classical reconstruction technique (Figure 2.5), the optical axes must be parallel to each other. If they are not parallel, Equation 2.4 cannot be applied, and new sets of equations must be developed to correct for angled cameras. With the DLT equations, the cameras can converge, but care must be taken so that the relationship between the convergence and the overlap angles defined in Marzan (1975) is respected to minimize any reconstruction error. This is not always possible in a laboratory environment, and a compromise between camera setting and an acceptable error must be reached.

Equation 2.5 has been implemented on a few commercial video-based kinematic systems to track body markers. For example they have been applied to the study of normal and pathological gait. Figure 2.6, a through c, and Figure 2.7, a through c, illustrate two different representations of a normal gait pattern where the corresponding stick diagrams are shown (a) for the plane of progression, (b) the frontal plane, and (c) from above.

## Moving Cameras

Moving cameras increase the complexity of calibration and the demands placed on the reconstruction technique. When video cameras are mounted on a moving vehicle, the computer image analysis and 3-D reconstruction techniques push the computing devices to their extreme limits. Simpler applications can be used to study human movement. For example, the camera can be allowed to rotate about a vertical axis while tracking a subject. The tilt-angle (vertical) and the pan-angle (horizontal) can be obtained from the projection of the reference markers' images in the projection plan of the cameras. De Groot, de Koning, and van Ingen Schenau (1989) modified the technique developed by Dapena,

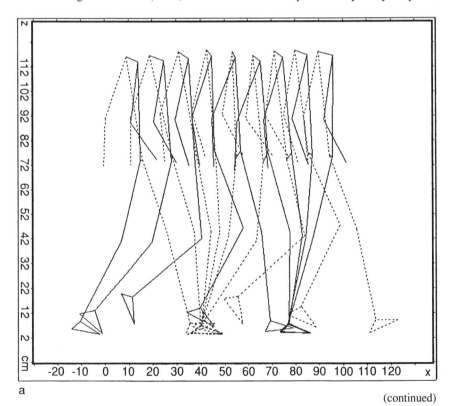

a

(continued)

**Figure 2.6**  Overlaid stick diagrams for a person walking from left to right as seen in (a) the sagittal plane, corresponding to the generalized direction of progression, (b) the frontal plane (see next page), and (c) the horizontal plane (see next page).
*Note.* Figures courtesy of Motion Analysis Corporation.

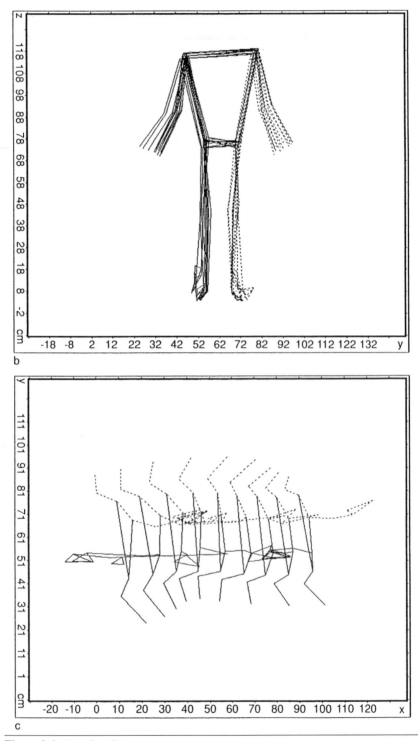

b

c

**Figure 2.6**    (continued)

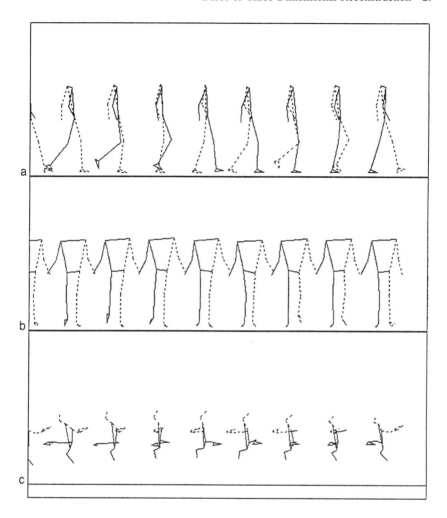

**Figure 2.7**   Spaced stick diagram for a person walking from left to right as seen in (a) the sagittal plane, corresponding to the general direction of progression, (b) the frontal plane, and (c) the horizontal plane.

*Note.* Figures courtesy of Motion Analysis Corporation.

Harman, and Miller (1982) for handling this effect. This technique requires precalibration to determine the transformation matrix.

De Haan and den Brinker (1988) outline the important factors that influence measurement accuracy for subject tracking. Continuous tracking caused some blurring of the reference markers. This was related to both exposure time and marker size. Additionally, the position of the cameras relative to each other can generate substantial error.

Tsai (1987) reported a new 3-D camera calibration technique for machine vision metrology using off-the-shelf TV cameras and lenses. This two-stage method aims at efficient computation of the camera's internal and external parameters, which, within limits, can be processed in real time using standard video cameras. Efforts are under way to improve existing 3-D reconstruction techniques or to develop new ones to account for moving cameras. Among the problems that have been identified are the relative camera positioning and computation time limitations. Reliable techniques are still needed, but there are ways to cope with the errors that arise.

# MINIMIZING RECONSTRUCTED COORDINATE ERROR

Each measurement has an associated error that makes it more or less inaccurate. Bertil Hallert (1960) states that "true" errors are fiction because exact values are seldom known. Thus, it is preferable that the term *reference*, rather than *exact* or *true*, measure be used in this context.

Many factors can influence the quality of the reconstructed coordinates. The major ones are the imaging device, marker identification, and the camera setup and calibration. Before attempting any discussion on these sources of error, a few words of caution about the reporting of the reconstructed coordinate error: Errors are given in the literature as means, standard deviations, root mean squares, percentages, estimates, and so on. These terms *must* be carefully defined and used consistently.

## Error Definition Terminology

One should distinguish between accuracy and precision. The former refers to systematic differences between the reference (true) and measured values. The latter refers to the repeatability with which a measured value can be obtained. These can be thought of as systematic and random error components, respectively.

We recommend that only the root mean square (RMS) error be given and that it be expressed as a fraction (percentage) of the mean camera base-to-object distance. The RMS error is expressed as

$$E_{\text{RMS}} = \frac{\sqrt{\Sigma(x_R - x_i)^2}}{N} \, , \tag{2.6}$$

where $N$ corresponds to the number of observations, $x_R$ to the reference value, and $x_i$ the error values. Furthermore, it must be applied only to object point coordinates that have *not* been used to determine the analytical constants.

To illustrate the difference between the mean error, the absolute mean error, and the RMS error, an example is given for 12 measured and reconstructed markers obtained by one of the authors. Table 2.1 shows that the mean value is very close to 0, as expected. There is a very small bias of about 0.11 mm in the instrumentation and a random error of about 0.25 mm in each coordinate. The mean error is inappropriate for expressing the instrument accuracy, because the negative errors are cancelled by the positive one. Here the absolute mean error

**Table 2.1   Mean, Standard Deviation, Absolute Error, and RMS Values**

| Marker | $\Delta X$ (mm) | $\Delta Y$ (mm) | $\Delta Z$ (mm) | Absolute difference |
|---|---|---|---|---|
| 01 | −0.21 | 0.18 | 0.43 | 0.51 |
| 02 | −0.18 | 0.30 | −0.20 | 0.40 |
| 03 | 0.05 | −0.26 | −0.02 | 0.27 |
| 04 | −0.09 | 0.18 | 0.52 | 0.28 |
| 05 | −0.07 | −0.08 | −0.34 | 0.36 |
| 06 | 0.28 | −0.23 | −0.45 | 0.58 |
| 07 | 0.19 | 0.11 | 0.15 | 0.27 |
| 08 | 0.14 | −0.32 | 0.24 | 0.42 |
| 09 | 0.07 | 0.03 | −0.67 | 0.28 |
| 10 | −0.04 | 0.27 | 0.27 | 0.38 |
| 11 | −0.35 | 0.33 | 0.09 | 0.49 |
| 12 | −0.00 | −0.02 | −0.14 | 0.14 |
| Mean error | 0.02 | 0.04 | 0.10 | 0.11 |
| (Standard deviation) | (0.18) | (0.23) | (0.35) | |
| Absolute error | 0.14 | 0.19 | 0.29 | 0.37 |
| (Standard deviation) | (0.11) | (0.11) | (0.19) | |
| RMS error | 0.17 | 0.26 | 0.31 | 0.44 |

*Note.* Values were calculated from the measured and reconstructed referenced markers that were not used to determine the DLT constants.

(0.37 mm) and the RMS error (0.44 mm) are three and four times greater, respectively, than the mean error. The RMS error is a conservative estimate of the instrument accuracy.

Often in photogrammetry, the error is expressed as a ratio of the camera base-to-object distance. In this case, the RMS/camera base-to-object ratio would be 1:2,951. However, if a different lens (a 12-mm or a zoom lens) had been used instead of the 8-mm lens, the error would have been different even with a similar camera position. Increasing the focal length of the lens is comparable to bringing the camera closer to the object. Using the same camera and lens, the error varies nearly linearly with the camera base-to-object distance.

Because the camera base-to-object distance can vary from one camera setup to another and considering that the focal distance used can be easily modified, it is better to express the error as a proportion of the field of view. This has the effect of lowering the ratio, because the width of the field of view is usually less than the camera base-to-object distance. For the above case, the ratio is now 1:1,141, shown in Table 2.2. The ratio is still a constant, because the field of view is in proportion to the camera base-to-object distance, shown in Figure 2.8. Triangle *ABC* is similar to *A'B'C*; thus *AB/A'B'* is the same as *AC/A'C*. How should the field-of-view length be defined? Is it the vertical, horizontal, or

**Table 2.2   Mean, Absolute Error, and RMS Values Expressed as
a Percentage of Different Distances**

| Distance | Mean | Absolute error | RMS |
|---|---|---|---|
| Camera-base-to-object of 1,298.3 mm | 1:11,803 | 1:3,509 | 1:2,951 |
| Field of view | | | |
| Horizontal (410 mm) | 1:3,727 | 1:1,108 | 1:931 |
| Diagonal (410 by 205 mm, 458.4 mm) | 1:4,167 | 1:1,239 | 1:1,042 |
| Spatial diagonal (410 by 205 by 205 mm, 502.2 mm diagonally) | 1:4,565 | 1:1,357 | 1:1,141 |

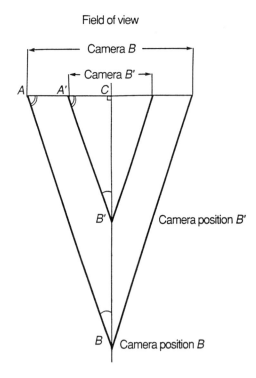

**Figure 2.8**   Similar trigonometric relationship between different camera
positions.

diagonal distance? Although any one of these is correct, the diagonal yields a better appreciation of the field of view, because it includes the height and width. However, because they are all related to the camera base-to-object distance, which can easily be measured compared to a camera's field of view, the mean camera base-to-object distance should be used. The diagonal of the calibration volume represents only a portion of the camera's field of view, and its use would yield a low ratio.

Any reference marker used previously in the calculation of the analytical constants will generally yield the smallest error, because the analytical parameters have already been optimized for such points. (That the "error" calculated using such reference markers is minimal only reveals the reliability of the optimization process.) Consequently, only the remaining reference marker coordinates should be used in the error calculation. Marzan's (1975) DLT computer program prints out the standard error or DLT error on the 11 DLT parameters. When we attempted to correlate the DLT error to the absolute error (i.e., the mean of the absolute differences between the reference and the measured coordinates), we found that the DLT error was always significantly lower (André, Dansereau, & Allard, 1990). In a simulation trial, when the reference marker error was increased from 0 mm to 5 mm, the DLT error increased from 0 mm to 0.4 mm, whereas, the absolute error reached a maximum of 1.5 mm with peak error close to 3.5 mm.

In summary, the RMS error should be calculated from marker coordinates that have not been used in the calibration process, and the ratio of the RMS or the absolute error should be expressed in relation to the mean camera base-to-object distance.

## Camera Types and Lens Distortion

The principal components of a stereo-photogrammetric optical system are the camera used, either metric or nonmetric, the film unflatness, and the lens. Each of these contribute to the measurement error, but basic three-dimensional recon-struction models do not correct for these sources of error.

Implicit techniques, such as the DLT, replace the painstaking steps of camera calibration. Karara and Abdel-Aziz (1974) compared the accuracy of the DLT method using a metric camera and four types of nonmetric cameras. The nonmetric cameras were a Honeywell Pentax Spotmatic, Crown Graphic, Hasselblad 500 C, and a Kodak Instamatic 154 camera; the metric camera was a Hasselblad MK 70. The standard deviation RMS error was calculated for reference points located at camera base-to-object distances of 4.00 m and 5.50 m. The RMS values were always smaller for the closest camera positions. The largest standard deviation of the RMS error was 3.1 mm (obtained with the Kodak Instamatic camera), representing a ratio of about 1:1,780 with respect to the camera base-to-object distance. The RMS error with the other nonmetric cameras were 55% to 75% smaller than the Instamatic camera error. These were not very much different from the metric camera error. The ratios calculated with respect to the camera base-to-object distance varied from 1:1,774 (Instamatic) to 1:5,200 (Pentax).

Our own work, using Nikon FE cameras fitted with Nikkor 55-mm macro lenses, yielded an average absolute error of 0.33 mm (1:1,540). Our accuracy was limited by the digital graphic tablet error of 0.4 mm and the manual digitization process.

Fraser (1982) was able to reach a precision of 1:10,000 or better with nonmetric cameras and a DLT-type method. A series of multistation phototriangulation adjustments was made using the concept of self-calibration with additional parameters set for each photograph. Among these are the photo-invariant parameters, which take into consideration film deformation particular to each photograph but assume a degree of film flatness that does not vary from photo to photo. According to Fraser (1982), lack of film flatness appears to be the most significant limiting factor on the attainable accuracy of nonmetric camera reconstruction techniques, especially for large-format cameras and short focal length lenses.

Reconstruction techniques involve the calculation of several constants related to the camera's internal and external parameters, lens correction factors, and other characteristics. Some of these constants are an integral part of the reconstruction technique, whereas others are additional correction factors added to the photogrammetric model. For example, a basic photogrammetric model assumes no lens distortions. If these are included, then specific analytical representations of the type of distortion must be added. Yet if this is done when little or no lens distortion exists, then random error or bias is introduced.

The full DLT three-dimensional algorithm contains 22 parameters. The first 11 are associated with the internal and external parameters and with linear lens correction factors. A polynomial of up to the seventh order can be added to take into consideration the symmetrical lens distortion, increasing the total number of coefficients to 16. The image refinement components $\Delta u$ and $\Delta v$ from Equation 2.5 can be expressed as

$$\Delta u = a_1 + a_2 u + a_3 v + u'(k_1 r^2 + k_2 r^3 + k_3 r^4 + k_4 r^5 + k_5 r^6),$$

$$\Delta v = a_4 + a_5 u + a_6 v + v'(k_1 r^2 + k_2 r^3 + k_3 r^4 + k_4 r^5 + k_5 r^6), \qquad (2.7)$$

where the $a_i$ are constants that reflect the linear components of lens distortion and film deformation. These are incorporated into the $L_1$ to $L_{11}$ coefficients, whereas $r$ is the length of the radial vector from the point of symmetry to the point under consideration $(u',v')$. The first five additional unknowns, the $k_i$, are the symmetrical lens distortion coefficients.

The asymmetrical lens distortion reflects the distortion caused by the decentering of lens elements and accounts for the selection of a point other than that of symmetry as reference. Conrady's model (1919) is used to express asymmetrical lens distortion. Equation 2.7 becomes

$$\Delta u = a_1 + a_2 u + a_3 v + u'(k_1 r^2 + k_2 r^3 + k_3 r^4 + k_4 r^5 + k_5 r^6) +$$
$$P_1(r^2 + 2u'^2) + 2P_2 u'v';$$
$$\Delta v = a_4 + a_5 u + a_6 v + v'(k_1 r^2 + k_2 r^3 + k_3 r^4 + k_4 r^5 + k_5 r^6) +$$
$$P_2(r^2 + 2v'^2) + 2P_1 u'v', \qquad (2.8)$$

where $P_1$ and $P_2$ are the asymmetrical lens distortion coefficients.

Four other coefficients can be added to Equation 2.8 to account for the nonlinear component of film deformation leading to the complete image refinement model with 22 coefficients.

$$\Delta u = a_1 + a_2 u + a_3 v + a_4 u^2\ a_5 v^2 + u'(k_1 r^2 + k_2 r^3 + k_3 r^4 + k_4 r^5 + k_5 r^6) +$$
$$P_1(r^2 + 2u'^2) + 2P_2 u'v'\ ;$$
$$\Delta v = a_6 + a_7 u + a_8 v + a_9 u^2 + a_{10} v^2 + v'(k_1 r^2 + k_2 r^3 + k_3 r^4 + k_4 r^5 + k_5 r^6) +$$
$$P_2(r^2 + 2v'^2) + 2P_1 u'v'\ . \tag{2.9}$$

Thus, the $\Delta u$ and $\Delta v$ of Equation 2.5 must be substituted for by their corresponding values in Equation 2.9.

Using video cameras, the film is replaced by a detector and electronic system, which should produce a linear relationship between a light spot's location on the image plane and the resulting data. A detailed mapping of video image-plane errors was carried out by Antonsson and Mann (1989), who used 12,000 points in an effort to calibrate the positional accuracy over the camera's entire field of view at a 3-m range. This extremely large number of reference points was generated by a 2-m-long channel section onto which 30 reference markers were positioned along the bar and measured to an accuracy of less than 1 mm. The rod was mounted on a precision motor-driven and computer-controlled rotary table with its axis of rotation aligned with the optical axis. The table was rotated in 0.9° increments to obtain 400 steps around the circle. Plots of isoradial and isocircumferential error contours for a video camera are far from being purely radial, as shown in Figure 2.9, a and b.

## Object Calibration and Reconstruction Error

There are many sources of error, but probably the most important and most often neglected is related to the quality and the care given to the measurement of the calibration object. It is impossible to calibrate a device or an instrument to an accuracy greater than that of the standard to which it is compared. Thus Doeblin (1975) recommends use of a calibration standard 10 times as accurate as the accuracy required of the device or instrument being calibrated.

In photogrammetry, this has further implications because we use an object that has been physically measured once to calibrate the cameras' internal and external parameters. Consequently, the accuracy of the instrument used in calibrat-ing the referenced object is crucial because it also affects the accuracy of the 3-D reconstruction. Figure 2.10 conceptually summarizes the sources of errors that can be carried in the different steps toward obtaining the reconstructed coordi-nates. Often, the coordinates of the reference markers on the calibration object are determined using a standard measuring tape that has an accuracy of 0.5 mm; the reader must judge the reported accuracy of the reconstructed 3-D coordinates. Some manufacturers of video-based systems supply laboratory-calibrated devices, which should contribute to error reduction.

Besides the attention given to the measurement of the calibration object, there are a number of other factors that influence the overall accuracy. These include the number and choice of reference markers used in the reconstruction technique,

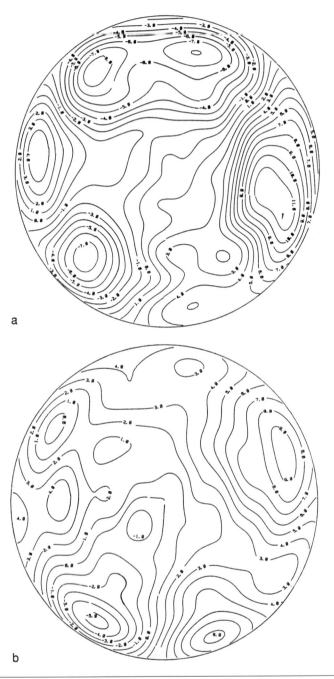

a

b

**Figure 2.9**   Plot of (a) isoradial and (b) isocircumferential correction contours.
*Note.* From ''Automatic 6-D.O.F. Kinematic Trajectory Acquisition and Analysis'' by E.K. Antonsson and R.W. Mann, 1989, *Journal of Dynamic Systems, Measurement, and Control,* **111**, pp. 31-39. Copyright 1989 by the American Society of Mechanical Engineers. Reprinted by permission of the American Society of Mechanical Engineers.

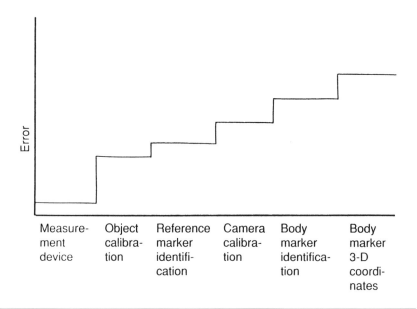

**Figure 2.10**   Propagation of errors as a function of the different steps taken to obtain reconstructed 3-D coordinates.

the shape of the calibration object, the relative position of the camera, and the reconstruction algorithm.

The minimum number of points selected for solving the unknown parameters in the reconstruction algorithm varies according to the technique used. For the DLT, the image coordinates of at least 6 reference markers are required. Shapiro (1978) recommends the use of 12 to 20 reference markers. If more than 6 are used, a least-squares fit of the parameters is performed. The error is not dramatically influenced by the number of points, as long as the calibration object is measured with good accuracy. Leroux, Allard, and Murphy (1990) measured marker positions to ±0.02 mm within a calibration frame of about 0.027 m³ and with a camera base-to-object distance of 0.600 m. With 8 and then 21 reference markers, the magnitude of the absolute error was 0.41 mm and 0.34 mm, respectively.

Caution must be exercised when increasing the number of points to reduce the reconstruction error. A least-squares approach should yield a best fit, and use of slightly more reference markers than the strict minimum is encouraged. However, although least-squares fit is used to minimize the random error, it should not be used to correct for an experimental bias that results from a poorly calibrated reference object. The error must be estimated not only on relative lengths, such as the distance between two markers, but also on the absolute coordinates of the markers themselves or the length of the vector originating at the laboratory coordinate system and the marker.

The distribution of the reference markers within the calibration is also important. Leroux et al. (1990) compared the errors after calibrating with additional markers biased toward the front, bottom, or right side of a reference cube. Initially, there were eight markers, one on each corner. Seven additional markers were added. Table 2.3 summarizes the results, which show that a biased distribution of reference markers roughly doubles the error, especially in the direction perpendicular to the film plane (Z direction).

Once the photogrammetric system is calibrated, then object marker positions can be determined. To avoid extrapolation errors, these should fall within the calibration frame. Table 2.4 presents the absolute error in the coordinates of object markers lying to the back, to the left, or higher than the calibration

**Table 2.3   Absolute Error (Standard Deviation) for a Nonuniform Marker Distribution**

| Marker position | Ex (mm) | Ey (mm) | Ez (mm) | Magnitude $\sqrt{Ex^2 + Ey^2 + Ez^2}$ |
|---|---|---|---|---|
| Bottom | 0.23 | 0.18 | 0.46 | 0.54 |
|  | (0.19) | (0.18) | (0.36) | (0.45) |
| Right | 0.22 | 0.23 | 0.48 | 0.58 |
|  | (0.17) | (0.22) | (0.36) | (0.45) |
| Front | 0.23 | 0.23 | 0.47 | 0.57 |
|  | (0.19) | (0.22) | (0.40) | (0.49) |

**Table 2.4   Absolute Error (Standard Deviation) in Coordinates of Object Markers**

| Marker position | Type | Ex (mm) | Ey (mm) | Ez (mm) | Magnitude $\sqrt{Ex^2 + Ey^2 + Ez^2}$ |
|---|---|---|---|---|---|
| Back | Intra | 0.16 | 0.17 | 0.31 | 0.39 |
|  |  | (0.13) | (0.15) | (0.20) | (0.28) |
|  | Extra | 0.21 | 0.33 | 0.46 | 0.60 |
|  |  | (0.24) | (0.26) | (0.31) | (0.47) |
| Left | Intra | 0.17 | 0.21 | 0.48 | 0.55 |
|  |  | (0.10) | (0.17) | (0.24) | (0.31) |
|  | Extra | 0.47 | 0.39 | 1.37 | 1.50 |
|  |  | (0.23) | (0.26) | (0.96) | (1.02) |
| Up | Intra | 0.14 | 0.26 | 0.44 | 0.53 |
|  |  | (0.08) | (0.17) | (0.19) | (0.27) |
|  | Extra | 0.54 | 0.39 | 0.43 | 0.79 |
|  |  | (0.27) | (0.26) | (0.29) | (0.47) |

*Note.* Intra—markers lying within the calibration space. Extra—markers lying outside the calibration space.

frame: The error has doubled or more. Wood and Marshall (1986) found that extrapolation errors were 50% to 100% greater.

Throughout this chapter, two assumptions have been maintained: a) that markers in multiple images are easily identifiable, and b) that they are representative of well-defined body landmarks or articulations. However, in practice, data files must be edited to eliminate noise from interfering light sources, to redefine markers after occultation or shadowing problems, to interpolate missing data, to correct for marker displacement resulting from skin movements, and so on. Woltring and Huiskes (1990) foresee the time when pattern recognition and artificial intelligence techniques may be more widely applied to data capture systems for the analysis of human movement.

## ACKNOWLEDGMENTS

The financial support by the Natural Sciences and Engineering Research Council of Canada is greatly appreciated, in particular for the equipment and operating grants to carry out the 3-D analysis of human motion. The technical support of Aliche Mongrain is also acknowledged.

## REFERENCES

Allard, P., Duhaime, M., Labelle, H., Murphy, N., & Nagata, S. (1987). Spatial reconstruction technique and kinematic modeling of the ankle. *Engineering in Medicine and Biology*, **6**, 31-36.

André, B., Dansereau, J., & Allard, P. (1990, November). Calibration object measurements and three-dimensional reconstruction accuracy of the DLT algorithm. Paper presented to the 14th annual meeting of the American Society of Biomechanics, Miami, FL.

Antonsson, E.K., & Mann, R.W. (1989). Automatic 6-D.O.F kinematic trajectory acquisition and analysis. *Journal of Dynamic Systems, Measurement, and Control*, **111**, 31-39.

Ayoub, M.A., Ayoub, M.M., & Ramsey, J.D. (1970). A stereometric system for measuring human motion. *Human Factors*, **12**, 523-535.

Conrady, A. (1919). Decentering lens systems. *Monthly Notices of the Royal Astronomical Society*, **79**.

Dapena, J., Harman, E.A., & Miller, J.A. (1982). Three-dimensional cinematography with control object of unknown shape. *Journal of Biomechanics*, **15**, 11-19.

de Groot, G., de Koning, J., & van Ingen Schenau, G.J. (1989). Method to determine 3-D coordinates with panning cameras. Paper presented to the XIIth International Society of Biomechanics Meeting, Los Angeles, 297.

de Haan, T., & den Brinker, B. (1988). Direct linear transformation method for 3-D registration using subject tracking cameras. In G. de Groot, P.A. Hollander, P.A. Huijing, & G.J. van Ingen Schenau (Eds.), *Biomechanics XI-B* (pp. 1051-1056). Free University Press.

Doeblin, E.O. (1975). *Measurement systems: Application and design.* New York: McGraw-Hill.

Fraser, C.S. (1982). On the use of nonmetric cameras in analytical close-range photogrammetry. *Canadian Surveyor*, **36**, 259-279.

Hallert, B. (1960). Photogrammetry. New York: McGraw-Hill.

Karara, H.M., & Abdel-Aziz, Y.I. (1974). Accuracy aspects of non-metric imageries. *Photogrammetric Engineering*, **40**, 1107-1117.

Leroux, M., Allard, P., & Murphy, N. (1990, November). Accuracy and precision of the direct linear technique (DLT) in very-close-range photogrammetry with video cameras. Paper presented to the 14th annual meeting of the American Society of Biomechanics, Miami, FL.

Marzan, G.T. (1975). *Rational design for close-range photogrammetry.* Doctoral dissertation, University of Illinois at Urbana-Champaign, Xerox University Microfilms.

Shapiro, R. (1978). Direct linear transformation method for three-dimensional cinematography. *Research Quarterly*, **49**, 197-205.

Stokes, I.A.F., Bigalow, L.C., & Moreland, M.S. (1987). Three-dimensional spinal curvature in idiopathic scoliosis. *Journal of Orthopaedic Research*, **5**, 102-113.

Tsai, R.Y. (1987). A versatile camera calibration technique for high-accuracy 3-D machine vision metrology using off-the-shelf TV cameras and lens. *Institute of Electrical and Electronic Engineering Journal of Robotics and Automation*, **RA-3**, 323-344.

Wolf, P.R. (1983). *Elements of photogrammetry.* New York: McGraw-Hill.

Woltring, H.J., & Huiskes, R. (1990). Stereophotogrammetry. In N. Berme & A. Cappozzo (Eds.), *Biomechanics of human movement: Applications in rehabilitation, sports and ergonomics* (pp. 108-127). Worthington, OH: Bertec Corporation.

Wood, G.A., & Marshall, R.N. (1986). The accuracy of the DLT extrapolations in three-dimensional film analysis. *Journal of Biomechanics*, **19**, 781-785.

# Chapter 3

# Instrumentation in Video-Based Three-Dimensional Systems

*John O.B. Greaves*

In 1892, John William Strutt (Lord Rayleigh, 1842-1919, English) asserted that the ratio of the density of oxygen to that of hydrogen is 15.882 rather than 16.0. Thus, a hundred years ago, Lord Rayleigh correctly foresaw that science in the 20th century would require accuracies into the fourth and fifth decimal places. Today, instrumentation for 3-D point measurements is preparing us for science in the 21st century.

## THE EVOLUTION OF DATA CAPTURE AND PROCESSING

Some clear tendencies have arisen since the first measuring sticks were applied to film in the 19th century. The following four sections give a brief synopsis of the scope of this evolution.

### From Film to Electronic Capture

The availability of high-resolution film provided an attractive medium from which to extract motion data. Filming the event and developing the film is easy (although time consuming) compared to the digitizing process. The "film analyzer" requires an accurate optical projection path, a projection screen, and a cursor mechanism that reads out the X and Y coordinates. These systems were fairly common in the 1970s; but the biggest drawback was the amount of time required to manually digitize the X and Y coordinates, as well as the lack of reliability of the human operator. For example, a golf swing recorded for a full-body 3-D analysis from a four-camera system with 17 markers, 4 cameras, and

2 seconds of data from each of four 200-frames-per-second cameras, would have 17 × 4 × 2 × 200, or 27,200 data points. Manual digitization at 1 "cursor-click" per second would have taken over 7 hours of nonstop clicks! Eventually, the cursor was connected to a computer. Figure 3.1 shows the stick figure for such a swing that was automatically digitized in real time in 1988.

## More Sophisticated Computations

The evolution of computer hardware and programming methodologies has been a necessary part of the progress in 3-D instrumentation. An example is the development of a concise mathematical model of measurement using the direct linear transformation techniques. Another example is the automatic solution to what is called the "correspondence problem," which solves for numerically reasonable intersections of 2-D rays projected into 3-D space. (Earlier solutions used manually identified 2-D path segments to reconstruct 3-D trajectories.) Nonlinear elements, such as lenses and nonlinear camera sensors, have benefitted from correction methodologies. With the recent availability of greater computing power, more sophisticated computational methods have been adopted to improve the robustness and accuracy of reconstructed 3-D data points.

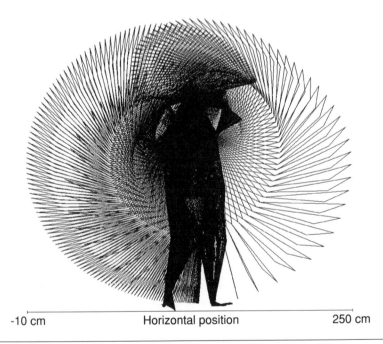

| -10 cm | Horizontal position | 250 cm |

**Figure 3.1**   3-D golf-swing stick figure automatically digitized from 200-frame-per-second video.

# Better User Control

Some of the first programs for computing 3-D coordinates from multiple 2-D views required that the X and Y data be punched into cards and then read into a mainframe computer. Errors were detected by trial and error and had to be corrected in a laborious manner. Later and more recent approaches allow much greater visualization of and direct interaction with the data for editing and manually reconstructing even somewhat sketchy data sets (e.g., those with missing or merging targets). The advent of graphical user interfaces (GUIs), the mouse pointer, and high-resolution computer screens has facilitated an even more interactive approach to reconstructing and visualizing complex 3-D data sets.

# "Off-the-Shelf" Applications Software

When early 3-D systems were brought to the market, they had little more than the barest of essential software for computing X, Y, and Z coordinates. Any applications software for modeling the human body's actions was left to the purchaser. A rich set of software to aid in the study of human movement has been gradually finding its way from the research lab to the commercial market. For example, "off-the-shelf" software exists that models the masses and lengths of segments of the human body. There is also software for collecting not only the digitized video, but the synchronized force plate measurements as well. Together, these programs allow the straightforward calculation of forces as they are transmitted up the lower kinetic chain and of the moments or torques about each of the modeled limb joints. All this can be accomplished with fewer button clicks than it formerly took to manually digitize the X, Y coordinates from a single camera's view of a single frame of data.

# HARDWARE AND SYSTEM DEVELOPMENTS

Although there are nonvideo-based 3-D systems in use, this chapter deals mostly with video-based systems. The current trend, suggested by the number of new products on the market, is to use video as the building block for motion tracking and analysis systems. Also, video-based measurement systems will benefit from the current international interest in and development of HDTV (high-definition television). Digitized signal-processing systems that are fast enough to handle real-time video information are now under development in many laboratories.

# Video Cameras

Much as the radar technology of World War II stimulated the development of commercial television in the 1940s and 1950s, the development of video cameras for commercial use proved a boon to electronic measurement technology. Significant developments in video camera technologies include the following:

- the videcon camera—This was one of the first, reasonably inexpensive, tube-type video cameras for scientific use.
- solid-state sensors—With the advent of MOS (metal oxide semiconductor), CCD (charge-coupled device), and other solid-state sensors, many of the problems of the tube-type cameras disappeared. For example, image smear, which occurs in a vidicon tube as the phosphor on the image plane is scanned by an electron beam, was eliminated. Also, warm-up problems (stabilizing the scanning electronics) and geometric distortion disappeared as sources of error. Solid-state cameras are said to have a geometric distortion of 0.0%, because precise photographic mask techniques are used to etch the array of solid-state photo sensors into the silicon. There are many advantages and disadvantages associated with different types of solid-state sensors, but these are beyond the scope of this chapter.
- mechanical and electronic shuttering—A shuttering mechanism in video cameras can greatly reduce the image smear associated with moving objects. If a camera has no shuttering, then the image sensor is exposed to the image for the duration of the video field time, or 1/60 s in a 60-Hz camera. Mechanical shutters typically employ a rotating disk with two slits, which are synchronized with the camera frame rate. The width of the pie-shaped slits determines the shutter exposure time (e.g., smaller slits give a faster shutter time). Shuttering reduces the amount of light that reaches the sensor, so shutter-equipped cameras must either have greater sensitivity or more light available.

Electronic shuttering is replacing the mechanical shutter in cameras. Electronic signals applied to the solid-state sensor activate or deactivate it with the same effect as mechanical shuttering. Electronic shuttering eliminates mechanical vibration, which can be an important consideration for 3-D camera configurations that require no camera movement after calibration. Also, electronic shutters are not as sensitive to damage from vibration and shipping.

The evolution of video cameras, recorders, playback units, and monitors was fueled by the commercial broadcast market and, later, by the home video market. High manufacturing volumes and international competition have produced extremely good quality units at reasonable costs. These units have been designed to present good color and resolution, given the limitations of the NTSC and PAL design standards. Because the human eye is the usual end user, these units are not necessarily designed for measurement purposes, where the end result is accurate X, Y, and Z coordinates. Fortunately for the instrumentation users, there is a significant amount of overlap between these two requirements.

## High-Speed Video

High-speed video generally implies video-framing rates that are faster than conventional broadcast rates. The 30-fields-per-second RS-170/RS-343 monochrome standard with 525 lines (480 visible lines) and the 25-fields-per-second PAL standard are the most commonly used video rates in video measurement instrumentation.

## 50-Hz and 60-Hz Video

Speed can be doubled by using the video signal as two separate video fields (denoted as the A and B fields) that comprise one interlaced video frame. For the 60-Hz system, there are 240 visible horizontal lines per field. These two interlaced video fields are reported sequentially at 60 fields per second. Caution must be observed when using measurement equipment, because a different set of sensors on the camera reports each field and there is displacement between them. This can introduce errors that can compromise the entire system's abilities to reconstruct 2-D or 3-D points (Greaves, 1986).

## Interlacing vs. Noninterlacing at 50/60 Hz

Interlacing is designed to provide a compromise between the human perception of "flicker" and a higher resolution image in the vertical direction. These compromises do not apply in video measurement systems. The vertical effect error between the A and B fields can be corrected to a certain extent (but not removed) by appropriate "deinterlacing" software. Also, the interlacing circuitry in certain video cameras can be disabled, thereby producing a noninterlaced or sequential scan that reports the A and B fields from the same set of sensors on the solid-state chip.

## Split-Image High-Speed Video

Another method of obtaining higher frame rates is to split the video image into two or more panels that divide the image into horizontal stripes. Each stripe contains fewer than the full number of horizontal lines associated with a normal video image, hence vertical resolution is severely degraded. For example, with a normal 60-Hz video image which has 240 viewable lines, the image can be split into three panels, each with up to 80 (i.e., 240 ÷ 3) viewable lines as shown in Figure 3.2. The electronics in these special cameras is configured to restart scanning the top 80 lines of the image sensor at 180 Hz, thus giving a 180-Hz scanning rate for one third of the vertical field of view. An additional few lines are usually lost in the switching process, so there may be fewer than the 80 viewable lines given in this example.

The main advantage of this split-image system is cost, because commercially available VCRs and monitors can be used. The camera still supplies a normal vertical synchronization at the usual 60-Hz rate. Thus one can view the results on a standard video monitor and the multiple panels are easily seen as horizontal stripes. Regular video recorders and playback units also work with this signal because it appears to be a normal 60-Hz signal. A noninterlaced camera works best for this application; otherwise the system would report three frames from the A field and three frames from the (offset) B field in the 180-Hz example cited previously.

## Above 50/60 Hertz

Frame rates above the conventional 50- or 60-Hz video field rates require nonstandard video cameras with special circuitry. High-speed cameras for instrumentation

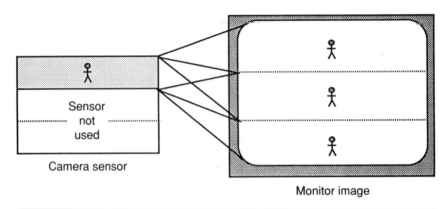

**Figure 3.2**    180-Hz split-image high-speed video camera and monitor.

purposes are almost all of the noninterlaced variety because the interlaced versus noninterlaced issue arises only at conventional rates with regard to compatibility with existing broadcast-type equipment. Thus, the distinction between a video *frame* and two superimposed video *fields* does not apply, and the term *frame rate* can be used unambiguously, as in cinematography. A frame denotes a complete image captured at an instant of time.

Two requirements of video cameras, recorders, and monitors for high-speed use are

- the ability to accept synchronization pulses (horizontal and vertical) at the higher line and framing rates; and
- the ability to produce or accept the gray-scale information at the faster pixel rates.

These requirements produce individual design issues for each hardware item.

## High-Speed Cameras

High-speed video cameras, like their lower speed counterparts, can have either analog output (as is the case with most of the current models) or digital output. The light that falls onto a single pixel of a typical 2-D camera sensor is integrated over its shutter exposure time and reported as an instantaneous analog signal. If the camera digitizes this signal, then it is a digital camera (i.e., the camera produces a digital stream of bits or bytes). If the camera amplifies an analog signal and sends it as an analog signal (the case with the RS-170/RS-343/PAL cameras), then it is an analog camera. The data stream from a conventional 525/30-Hz camera with 500 horizontal pixels could be simply calculated as

$$525 \text{ lines/frame} \times 500 \text{ pixels/line} \times 30 \text{ frames/s} =$$
$$7.88 \text{ million pixels/s.} \tag{3.1}$$

From a noninterlaced camera, the rate is the same:

$$262.5 \text{ lines/field} \times 500 \text{ pixels/line} \times 60 \text{ fields/s} =$$
$$7.88 \text{ million pixels/s.} \tag{3.2}$$

Because analog standards have both vertical and horizontal blanking intervals for "flyback," the instantaneous rate at which the analog camera must deliver the grey-scale information is higher than for the digital camera. The 500 pixels in each line must be sent in about 83% of each line time, so the instantaneous rate is closer to 9.4 million pixels/s.

The data rate for the digital camera is slower, if the synchronization information is carried on a separate wire or channel and does not need to be imbedded into the data channel. For an equivalent camera with 525 lines, only 480 are visible and need to be digitized:

$$480 \text{ lines/frame} \times 500 \text{ pixels/line} \times 30 \text{ frames/s} =$$
$$7.2 \text{ million pixels/s.} \tag{3.3}$$

At 8 bits per pixel (which is common for monochrome video), the information rate is

$$7.2 \text{ million pixels/s} \times 1 \text{ byte/pixel} = 7.2 \text{ million bytes/s,}$$
$$\text{or } 58 \text{ million bits/s.} \tag{3.4}$$

Conventional Ethernet channels of today offer only 10 million bits/s of raw bandwidth, whereas the newer fiber-optic computer network standard offers 100 million bits/s. We can see why digital video is not as common as the analog form, because a relatively inexpensive coaxial cable can easily carry hundreds of megahertz of analog information. This situation will change in the coming decade as the convenience and fidelity of digital channels gradually replace the analog media. Already, digital transmission schemes and fiber optics have gained acceptance in the telephone industry for long-haul networks. Voice bandwidth for telephone use is 64,000 bits/s, which is about three orders of magnitude smaller than that of monochrome video.

The limiting factor in designing high-speed cameras is the rate at which the grey-scale pixel information can be transferred from the sensor chip, whether it is in an analog or digital form. If the chip can move the information at 7.2 million pixels/s $\times$ 4, or 28.8 million pixels/s, then a $30 \times 4 = 120$-Hz full frame (480-line) camera can be built.

## Parallel Output Cameras

The limit on speed of the camera sensor (about 30 million pixels/s) can be increased by providing a camera sensor that has more than one output port. For example, if the image field of a 30-million-pixel/s (Mpixel/s) camera is sliced into eight vertical (or horizontal) stripes, each with its own output port, then the effective bandwidth of the camera can be multiplied by 8. Thus, the 120-Hz camera could produce 960 Hz (120 Hz $\times$ 8), or 960 full-frame pictures/s. Such parallel output cameras have been built and provide high-speed video pictures. Special provisions must be made for recording and viewing the output of these parallel output cameras, since common recorders and video monitors have only a single channel input.

As standard 30/60-Hz video cameras gain more horizontal pixels, the sensor and its associated electronics require a higher bandwidth to accommodate the higher number of pixels. For example, if the 500 horizontal pixel camera mentioned above had 1,000 pixels, then the required bandwidth figures above would double to 7.88 × 2, or 15.8 million pixels/s.

# High-Speed Recorder-Playback Units

Though there is a need for specialized recorder-playback units and high-bandwidth monitors for high-speed video applications, they are not part of the commercial broadcast market. Therefore, these high-performance units are produced in smaller volumes and cost more than their low-speed counterparts. Still there are two kinds of specialized record-playback machines that have been developed for high-speed recordings.

### Proportional Speed Recorders

Video cassette recorders (VCRs) capable of high-speed video records are less available and more expensive than the associated cameras. One common scheme used in high-speed recorders is to spin the entire recording mechanism (tape, record heads, etc.) at higher speeds. Thus, for recording at 180 frames/s, the camera output must be 180 frames/s and the recorder "paints" a proportional RS-170/RS-343 compatible signal using standard tape format at three times the standard speed. All of the data transfer times and synchronization pulses are compressed by a factor of 3. The tape recorder and the tape medium must have the capability to record at these higher rates, which would be 3 × 7.2 million pixels/s, or 21.6 million pixels/s, from the 500-horizontal-pixel, 60-Hz camera. The recording machine is designed to produce a 30/60-Hz compatible tape with a standard format, such as VHS or S-VHS. If it is played in a standard VCR playback unit at 60 fields/s, a slow-motion image appears on the standard monitor.

### Parallel Input/Output Recorders

For the highest record-playback speeds from parallel output cameras, a parallel input/output recorder is needed. Each output port of the high-speed camera (which depicts a slice of the overall picture) is associated with a synchronized channel with sufficient bandwidth to record the video for that port. A special tape is created with the parallel output channels. To play back the tape, the playback mechanism reads each channel, reassembles the partial frame, slices into a full frame, and displays it on the monitor.

# High-Speed Monitors

The function of the monitor in a measurement video system is to provide the user the information needed to assure a quality image for digitizing or recording.

Is the focus on the lens set correctly? Is the f-stop set correctly for the illumination? Is the field of view correct for capturing all of the events to be recorded or digitized? Because high framing rates are not needed for this, a normal 30/60-Hz monitor can be used along with suitable electronics to slow down the high-speed image stream from the camera—for example, a high-speed "frame grabber" to capture an image from the high-speed camera and play it back at the normal 30/60-Hz rate. In a 180-Hz camera, one image is captured and displayed on the 60-Hz monitor while the subsequent two images are discarded.

High-speed monitors that project every camera frame onto the CRT do exist for certain applications, though they are usually smaller and more expensive than their commercial 30/60-Hz counterparts.

## Compressed Video and the Digital VCR

At the time of this writing, there is a lot of development and standardization in progress for compressing and decompressing real-time video and audio signals. Two key international standards groups are charged with establishing industry-standard image-compression specifications. The JPEG (Joint Photographic Experts Group) and MPEG (Motion Picture Experts Group) are sections of the ISO (International Standards Organization) and work with the American National Standards (ANSI) group. Board-level products, application-specific integrated circuits (ASICs), and specialized boards are under development for "desktop multimedia" (video and audio editing), video mail, and video conferencing. One goal is to have desktop workstations capable of being "digital VCRs." But this compressed video technology may not become a tool for 30/60-Hz 2-D and 3-D motion tracking systems, because video-compression schemes compromise the quality and clarity of the compressed and decompressed images in order to achieve the bandwidth reduction (Rabbani & Jones, 1991). These characteristics are the opposite of those needed for more accurate and higher speed X, Y, and Z coordinate measurements.

## Video Processors: Needs and Uses

*Video processor* generally signifies a machine that processes or modifies a real-time video signal by either analog or digital means or some combination of the two. A video processor for measuring points in 3-D space is quite different in its architecture and implementation from a video processor whose job is to render an image "prettier" or to convey more information to the human eye, though more effort has gone into development of video processors with primarily visual output. Requirements for the measuring video processor are quite different from those for a visual image.

The task of most image processing systems includes (1) image or "blob" recognition: that is, what is a target (or marker) and what is nontarget (background), and (2) target location in a particular frame, reported with the highest possible accuracy allowed by the system. Most systems for human motion analysis permit *cooperative* targets, such as passive retroreflective dots or spheres. Other computer vision systems do not provide this luxury. Cooperative targeting greatly simplifies the task of the image recognizer, because the targets can be made much brighter than the background illumination. There are potential pitfalls to this approach: There may be restrictions on lighting (e.g., it is more difficult to maintain the background to target contrast when out-of-doors), and some events, such as athletic events, do not allow cooperative targeting.

## Color Cameras and Color Target Tracking

Highly saturated colored markers may facilitate target identification—subject to a few caveats. Both the NTSC and PAL commercial color video standards are modifications of their earlier monochrome standards, and both have added bandwidth for color video. The design was a compromise between what "looks good" to the human eye and what could be allocated within the available bandwidth of broadcast channels. So, color video that has been encoded into either of these formats suffers greatly from degradation in resolution of the color images. The image quality is similarly degraded by recording color onto video tape, although the method of recording and color rendition on playback varies among different tape formats.

With a live color camera, the prospects for effectively using color video information is greatly improved. Special color cameras with analog RGB (red, green, and blue) outputs are available, but these cameras require real-time operation, because recording equipment for analog RGB is not generally available. Single-chip RGB analog cameras exist, but they generally have reduced horizontal resolution. Three-chip color cameras exist, mainly for studio use. Although these give better resolution than single-chip cameras, they generally do not have noninterlaced options, and they are more expensive.

In addition to the RGB-type of video color cameras, the HSL (hue, saturation, and luminescence) model will also encode color from the camera. This type of camera may be preferable, because the hue of a target may be tracked more reliably than a particular blend of red, green, and blue values.

## SYSTEM AND SOFTWARE DEVELOPMENTS

In the 1980s, most human movement research laboratories had to do a great deal of computer programming at a level that required knowledge of compilers, graphics routines, linkers, libraries, and the use of an operating system. Fortunately, this, too, has changed. In the area of human gait analysis, normal patterns have been published for age groups from toddlers to adults. Software for measuring and integrating forces, EMG activity, and 3-D kinematics and kinetics is commercially available. Human modeling software, complete with segment masses and inertial characteristics, has been introduced, and there are other specialized applications packages for measuring spine motions and detailed foot motions. Gait analysis software is perhaps the most mature. The trend toward the use of application-specific software in all areas of human knowledge has been pronounced in the last decade. In motion analysis, the most critical component of such software is that which tracks the motion of body segments.

# Real-Time 3-D Tracking

The two problems that must be solved to achieve real-time tracking of targets in 3-D are (1) real-time 2-D tracking, while maintaining target identity, and (2) 2-D-to-3-D transformations in real time.

## Real-Time 2-D Tracking

Real-time tracking involves finding the centroids of markers and then following their paths in a series of frames. *Centroids* are computed target centers, and *paths* are the loci of centroids that have preserved their identity in successive frames. Thus, centroids are independent entities and computed anew in each video frame, whereas a path is characteristic of a particular set of centroids through time. Real-time centroids are easier to compute than real-time paths and can be computed for simple target schemes with integer-level hardware and software. (Furnée, 1989, gives a good account of the evolution of real-time centroid computations.) The problem of computing paths of identified centroids in real time can be solved with simple target schemes, but these fail in cases where several targets may disappear or even merge in a single camera's 2-D field of view.

Several methods have been proposed for maintaining target identity for each camera in 2-D:

- Solving the "correspondence problem" by searching for reasonable intersecting rays from two or more cameras
- Using different sized markers to facilitate identification
- Using different colored markers
- Using active LED targets with unique strobe "signatures" (Wilson, 1986)

**Marker Correspondence in Real Time.** In video-based systems, target identities in each camera's view generally are not known, unless a human operator names and tracks each target in each camera view. This generally precludes real-time operation. One way to know which targets in Camera A correspond to (or have the same identity as) targets seen in Camera B is to see if their "centroid rays" intersect close to each other in the measurement space (see Figure 3.3).

Imagine a ray starting at an image centroid in a camera and passing through the focal point of the camera's lens. If two or more cameras see the same point, then their corresponding rays should intersect or pass close to each other. Note that this method of identifying targets does not require a priori knowledge of the identity or name of the 2-D target centroids and is thus suitable for reasonably fast target tracking. However, it may not be a solution if there are many, close target points.

Eventually, the user will want to give the 3-D trajectories names (shoulder, elbow, etc.), but whether this is accomplished before the tracking begins or after

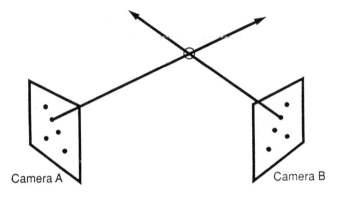

**Figure 3.3**   Rays from camera centroids intersecting in 3-D space.

the tracking is complete does not affect the quality of the data. There may be some benefit to visualizing the tracking as it proceeds, using a device such as a stick-man diagram. If the target identities are known at the start of the tracking, then it is possible to draw the stick diagram as the tracking proceeds because the linkages between targets in stick diagrams are defined (e.g., the hip is connected to the shoulder, the shoulder to the elbow, the elbow to the wrist, etc.)

**Different-Sized Targets.**   The different sizes of targets can be used to identify them. This could be useful in certain situations in both 2-D and 3-D target tracking, but it has limitations for the generalized case. Often the size of the marker is dictated by other constraints (e.g., three markers must fit on a person's foot). Some assistance can be given to a tracking algorithm by having the software keep track of the size of the marker dynamically ("dynamic sizing"). The software can then discard targets when they merge, detected by a sudden increase in the target size. Likewise, if a target becomes partially occluded from a camera's point of view, its apparent size will grow smaller and that target image can be discarded.

**Colored Targets.**   The use of color for target identification is theoretically possible, but this is complicated by the color encoding used in broadcast standards and video recording, mentioned previously.

**Active-Marker Strobing.**   The active target- (or strobed LED-) based systems identify targets in the hardware: Only one LED is on at any given instant. This luxury is not present in video-based systems because all targets that are illuminated are "on" in each video frame (if they are visible). However, these active-target systems have disadvantages: (1) There is no video image for viewing or recording; (2) the active markers require wires for power and synchronization from a master source; and (3) they are more sensitive to disruptive reflections, possibly giving the appearance of a target at two or more places in a camera's field of view.

## 2-D-to-3-D Transformations in Real Time

Once the targets have been correctly identified in each camera (2-D) view, formulating the 3-D coordinates involves mapping the corresponding targets for all cameras into a 3-D X, Y, Z value. *Corresponding targets* means the set of camera 2-D image points that are known to belong to a particular target. With a technique such as direct linear transformation (Abdel-Aziz & Karara, 1971; Walton, 1981), the transformation is straightforward mathematically and can be reduced to several floating point operations per frame per target. This is possible with existing high-speed computing chips.

# Tracking Without Reflective Markers

Reflective markers came into use for two reasons: (1) They allow fairly accurate marking of a point of interest on the human or other body, and (2) they provide a very high contrast against the background, which makes the task of developing video processors and tracking software simpler. The second reason may disappear with advances in color and shape recognition-capable hardware and software. The point-of-interest reason for using markers will remain longer.

It is more difficult for even the highly trained human eye and human perception system to pinpoint a shoulder joint or an elbow joint without an attached marker. Similarly, computer systems cannot function without markers. However, in the future, the machine may be able to learn the shape of the object being tracked and how to transform this image for different perspective views. If so, the computer also should be able to estimate six-degrees-of-freedom information about the body motion in the same way that the human vision system does. How its accuracy might compare to that of current schemes that use targets is difficult to predict.

# Shape Tracking: Dynamic Morphology

Morphology is the study of the size and shape of objects. *Dynamic* morphology involves the traditional measures (such as area, perimeter, lengths of major and minor axes, orientation of major axes, and other such 2-D shape classifiers) *tracked through time*. This requires that the perimeter of an object be differentiated from the background. The simplest and most robust method of object recognition is by contrast: The object must be either lighter or darker than its background, which allows use of binary image-processing techniques. In this case, the centroids of objects can be calculated from their boundaries, as is done with retroreflective targets. The centroids are tracked in time and connected into paths, but *shape* information also is retained and can be reported at each instant. For detecting nonbinary (gray-scale) images, methods such as the gray-level co-occurrence (GLC) matrix operators can be used. Reasonably good object-recognition results have been reported using these techniques (Trivedi & Harlow, 1985), but they require substantial computing resources for large, high-definition images.

## Volume Rendering and Volume Tracking

If shape changes of objects can be tracked through time, then multiple camera views of a solid object can give volumetric information about them. Different techniques have been used to solve this problem, including the use of structured lights (e.g., sheets of light) with a video camera sensor, laser scanners with a single- or multiple-spot pickup sensor, and conventional gray-scale video. In the simplest case, a single object is illuminated to be either brighter or darker than its background. Then each camera sees an ideal outline or silhouette of the perimeter of the object from which the volume calculation is made. In practice, this kind of illumination is very difficult to achieve.

## CONCLUSION AND CRYSTAL BALL

Techniques for measuring 3-D human motion have made tremendous strides since the 1970s. Hardware and software have improved, and we have a better understanding of 3-D systems, including their calibration requirements and the tracking of targets in 3-D space. Further advances in both instrumentation and software are imminent, driven by the needs of an expanding base of applications that must perform more exacting measurements and by the very-high-performance electronic vision and software technologies becoming available.

Visualization tools already have made big leaps forward: Stick-figure displays gave way, first, to wire frames; now such displays are merging with the tools and techniques of the computer animation industry, graphically representing human motion through images of "solid bodies" with reflections, shading, color, and textured surfaces. Synthetic actors with real, human motion scripts will "act out" full-length color feature films when, in the near future, HDTV (high-definition television) moves from the laboratory into the world of computerized motion applications. Connected to 1,000-×-1,000-pixel color video cameras, high speed instrumentation video processors will be optimized for extracting accurate X, Y, and Z coordinates, and "visualization" video processors for rendering the images from such data will be commonplace. The effectiveness of "smart weapons" that can "see" and home in on targets has already been demonstrated. As the accuracy of locating points, shapes, and volumes in 3-D space moves from the third decimal place to the fourth and fifth decimal places, specialized applications software will address an even greater diversity of motion applications in a variety of sports, medical, entertainment, life science, military, and industrial applications.

## REFERENCES

Abdel-Aziz, Y.I., & Karara, H.M. (1971). Direct linear transformation from comparator coordinates into object space coordinates in close-range photogrammetry. *Proceedings of the Symposium on Close-Range Photogrammetry* (pp. 1-18). Falls Church, VA: American Society of Photogrammetry.

Furnée, E.H. (1989). *TV/Computer motion analysis systems*. Unpublished doctoral dissertation, Delft University of Technology, Delft, Holland.

Greaves, J.O.B. (1986). State of the art in automated motion tracking and analysis systems. *Proceedings of Society of Photogrammetry and Instrumentation Engineers,* **693**, 277-281.

Rabbani, M., & Jones, P. (1991). *Digital image compression techniques*. Bellingham, WA: Society of Photogrammetry and Instrumentation Engineers (SPIE) Optical Engineering Press.

Trivedi, M.M., & Harlow, C.A. (1985). Identification of unique objects in high-resolution aerial images. *Optical Engineering,* **24**(3), 502-506.

Walton, J.S.W. (1981). *Close-range cine-photogrammetry: A generalized technique for quantifying gross human motion*. Unpublished doctoral dissertation, Pennsylvania State University, University Park, PA.

Wilson, R.S.W. (1986). Software for automatic tracking of moving targets in three dimensions. *Proceedings of the Society of Photogrammetry and Instrumentation Engineers (SPIE),* **693**, 269-276.

# Chapter 4

# Optoelectronic-Based Systems

*Antonio Pedotti*
*Giancarlo Ferrigno*

This chapter deals with the use of optical sensors for obtaining coordinate data on human motion. These systems have many potential advantages because they use optical (noncontacting) principles and can be automated to work rapidly and accurately without human intervention. The very rapid developments in this technology have revolutionized the scope of human motion analysis in recent years (Pedotti, 1990).

The first automatic systems for the measurement of movement were based on electrogoniometers applied to body segments. These presented the well-known problems of constraining the movements and only allowing measurement of angles between instrumented segments.

In the mid-1970s the first video camera-based systems were introduced. These systems detected active or passive markers on the basis of their brightness (Cheng, Koozekanani, & Fatchi, 1975; Furnée et al., 1974; Jarrett, Andrew, & Paul, 1976; Winter, Greenlaw, & Hobson, 1972). Thus, the markers had to be brighter than any other object on the scene to be recognized by a unique preset threshold on the video signal. Threshold detection requires that great care be taken to avoid detecting other bright objects in the field of view. Thus, to increase the total energy they reflect or emit, quite large markers are often used, which reduces the minimum allowable distance between them. The accuracy of such systems is limited also by the resolution of the sensor, because the information contained in the image is reduced to only black and white after thresholding.

The latest generation of systems, born in the mid-1980s, uses pattern recognition techniques for marker detection, thus solving the problems of threshold-based systems. First, the use of image-processing techniques allows the detection of the markers *only* if they match a predefined shape (Ferrigno, Borghese & Pedotti, 1990; Ferrigno & Pedotti, 1985); thus the reliability of detection is very high compared to that of threshold-based systems. Second, because all gray levels of the image are used, these systems allow us to retrieve more information about the marker position, which dramatically increases the actual resolution of the system.

# PRINCIPAL COMPONENTS OF A 3-D MEASURING SYSTEM

Three-dimensional measurement of human movement is reduced to the detection of the trajectories of several points that identify the positions of the body segments in space. Figure 4.1 illustrates the main blocks of a 3-D measuring system.

Two main levels of operations can be distinguished based on execution speed and complexity. The first, or low, level is devoted to the gathering of the 2-D coordinates of several landmarks in the reference frame of each sensor. In automatic systems, this task is performed in real time, and the complexity of the processing depends on the principle used for coordinate detection. This, in turn, affects the overall performance of the whole system.

The first level can be further split into two parts: the interface with the environment, which includes a set of markers placed on pertinent landmarks; and the image sensor together with the image processor, or marker detector.

The second, or high, level provides for more complicated operations that can be performed on-line or off-line. They include 3-D calibration, distortion correction, reconstruction of 3-D coordinates, digital filtering, modeling, and data representation. Processing of other biomechanical data including forces and EMGs is also performed at this level. The second level is usually software implemented on a general purpose computer.

Active markers, (those emitting some kind of energy) and passive markers have been described in chapter 1. The sensors we will address here are optoelectronic, which provide noncontacting measurement. They are video cameras (standard or custom designed), lateral effect photodiodes, and linear arrays.

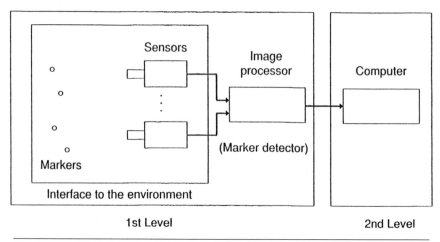

**Figure 4.1**   General structure of a 3-D motion analyzer.

The image processor, or marker detector, plays a fundamental role in the reliability and accuracy of the whole system. In automatic systems, it is usually implemented in hardware and can be based on principles ranging from simple threshold detection to pattern recognition procedures.

# Passive or Active Markers

Passive markers are usually made of retroreflective paper. Active markers need an energy supply to increase the contrast with respect to the background and may be activated stroboscopically to facilitate marker identification. Passive markers are less constraining for the subject, but they require a more intelligent data processing system. Active markers usually consist of low-powered (around 1 W) light-emitting diodes (LEDs). These present problems related to the limited cone within which they radiate an intensity of light that ensures an acceptable signal-to-noise ratio.

## Light-Emitting Diode Systems

Active markers require an external energy supply. Although flood lamps have been employed as active markers (Cheng et al., 1975), LEDs are used nowadays.

The optical wavelength of choice of such devices is usually in the near-infrared (IR), just above 800 nm, in order to avoid disturbances to the subject's sight while remaining in the range of the sensors' sensitivity.

The main advantage of active markers is their simpler labeling, described subsequently, and a high sampling rate when few markers are used. However, there are also problems with their use:

- They constrain the subject's freedom because of the wires connecting them to the power supply and to the synchronization and control unit. For applications that involve fast or wide movements, such as motion studies in sports, this disadvantage could be of paramount importance; furthermore, the energy drawn from the power supply (and thus its size) increases with marker activation and sampling rate.
- LEDs have a restricted light-emission angle. Even when using the larger available angles or using modified commercial devices, it is difficult to exceed 50° of half-power range. This creates a problem whenever there is rotation of the marker during movement. Also, it creates problems for 3-D analysis that involves at least two sensors. Accuracy along the depth of the working volume can be attained only if the angle between the sensors is wide enough. Unfortunately in these conditions, it is difficult for the two sensors to see the LED simultaneously and, even if this happens, the radiating power that reaches the sensors will be low, as is shown schematically in Figure 4.2a.
- The sensors employed with these markers usually require time-multiplexed lighting. This limits the number of markers that can be used on the subject and causes nonsimultaneous sampling of the markers' coordinates, which can raise problems with fast movements.

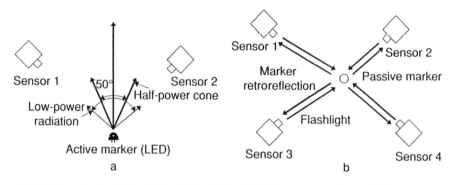

**Figure 4.2**   Effect of relative position between sensors and (a) active and (b) passive markers. The LED light intensity received by two sensors is small when the sensors are spaced for 3-D analyses, and this decreases the signal-to-noise ratio. Passive markers are always visible (if not hidden by other surfaces) with the same intensity, thanks to retroreflection.

## Retroreflective Markers

The alternative to the use of active markers is to employ passive retroreflective ones. These markers reflect light (within a narrow angle) in the direction from which it comes. This is the same principle demonstrated by the rear reflectors of automobiles and traffic signs (also passive devices), which appear to be lit up when illuminated by the headlights of approaching vehicles.

Advantages of these markers are that, unlike LEDs, they do not encumber the subject and are not limited by rotation. A passive spherical marker is visible from any direction within 360° as illustrated in Figure 4.2b. These markers usually consist of plastic supports coated with reflective paper and are fixed to the subject's body with adhesive tape. They do not require wiring and so can be used for studying broad, fast movements. Systems that, for the same working volume, use the smaller markers are more suitable than those that use bigger ones under these conditions.

The fact that passive markers are seen simultaneously by the sensors is an advantage from a signal-processing point of view (systems that use active markers have problems with nonsimultaneous sampling); but on the other hand, labeling software is required. This labeling procedure must assign, frame by frame, a landmark label to each pair of coordinates. This is usually accomplished by a tracking procedure; for example, once we have determined that the wrist of the subject corresponds to Marker 3, we must follow the same marker in successive frames by always labeling it "wrist" or "3."

Other advantages of passive markers include ease of attachment to the body segments, which allows the subject under analysis maximum freedom of movement, and, at least theoretically, use of a potentially unlimited number of markers.

The problems that arise with use of these markers are related to the difficult labeling in critical conditions. Overlapping of markers, for example, requires intelligent processing (Ferrigno et al., 1990) or interactive manual editing. Sometimes, if special tracking procedures are used, an overlapping or long missing marker requires the intervention of the operator, thus increasing the data-processing time. Moreover, the maximum sampling rate is limited by the cameras being used. Systems can reach rates of 100 Hz (with high resolution and accuracy) or 200 Hz (with a significant reduction of these). Of course, such frequencies are sufficient for most human movements, even in sports (D'Amico, Ferrigno, & Rodano, 1989, 1990).

## Addressable and Nonaddressable Sensors

Optoelectronic sensors may be divided into two classes, addressable and nonaddressable.

An addressable sensor is able to associate each picture element (pixel) with a pair of coordinates. Video cameras are addressable sensors because the image is coded in the so-called raster format. (An example of raster format is illustrated in Figure 4.3.) The image is scanned line by line on each frame and column by column on each line, so the signal is represented in a monodimensional way. The space relationships between different portions of the image are turned into time shifts. An appropriate decoding of these shifts allows recovery of exact spatial information. When several markers are simultaneously present in a picture, addressable sensors can provide coordinates of *all* of them without limits.

Nonaddressable sensors provide integral information on the picture they are picking up. For example, lateral effect photodiodes give, as output, the coordinates

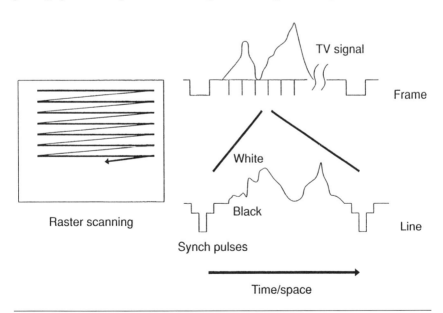

**Figure 4.3**    Raster scanning and frame-and-line electrical coding of the video signal. Black and white are turned into voltage levels, and space is turned into time lag.

of the centroid of the light distribution on the sensor. Therefore, if they are used for marker detection, only one marker—and no other sources of light—must be present for each sampling (Woltring, 1975). Between these two categories lie linear array sensors (Leo & Macellari, 1979).

The main difference between these classes of sensors is that stray light is detected as another pair of coordinates, or false markers, by addressable sensors, whereas it simply displaces the coordinates obtained from nonaddressable sensors. Software exists that can remove false markers from the data collected with video cameras, but it is impossible to do the same for data collected from lateral effect photodiodes. Theoretically, by carefully setting up the environment in which the tests are carried out, the only light source in front of the nonaddressable sensor should be the LED; but active markers can reflect on surrounding surfaces, including the subject's body, and such reflections corrupt the measurements.

## Sensor Synchronization

To make 3-D measurements, at least two 2-dimensional sensors must be used, thus the problem of synchronization of different sensors arises. This is generally resolved (for addressable sensors) by using a master synchronization generator to drive all the sensors. This device is included in the hardware processor and must also be connected to the coordinate generator. In the case of lateral effect photodiodes, the analog output of each sensor must be converted synchronously into digital form. Now another problem arises: The active markers are sequentially (noncontemporaneously) scanned in time, so their coordinates must be synchronized using suitable interpolating algorithms.

## Marker Detector

This part of the system must be capable of saving enough information for the reconstruction algorithms without overloading itself with a lot of false positives, which makes the work harder and increases the probability of either accepting a false marker (due to background noise) or discarding a true one. Marker detection in systems for motion analysis uses two approaches: threshold detection and image signal processing. These approaches apply only to addressable sensors and to linear arrays; lateral effect photodiodes simply provide the coordinates of the centroid of the impinging light.

### Threshold Detection

Threshold detection is the more popular and simplest way to detect markers that are brighter than the background. Figure 4.4 shows how the algorithm works. The input video signal, in raster format, is compared, pixel by pixel, to a reference threshold. Each time the threshold is overcome, a signal is sent to the coordinate generator and to the computer to signal that a bright pixel has been detected. Bright pixels are then grouped together into the markers to which they belong.

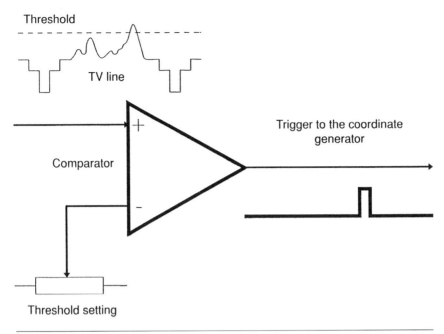

**Figure 4.4**   Threshold detector.

The major advantage of this approach is that it is very inexpensive and requires no special technology or know-how. On the contrary, the disadvantage of low discriminating power arises. Because detection is based only on the brightness of the target object, if other objects in the scene are as bright as a marker (e.g., reflecting surfaces, white areas, lamps, etc.), they too will be accepted and coded as markers. Moreover, because there are only two levels (either black or white) that code the image after detection, increments of resolution with respect to the sensor (about $500 \times 300$ pixels) are not allowed. Small improvements can be attained, though, by using oversized markers to generate a larger number of bright pixels whose coordinates may then be grouped and averaged, or by using models (e.g., circles) of the marker shape.

## Signal Processing

The recent developments in parallel computing, in fast, very-large-scale integration (VLSI) chips, and in computer vision make possible the design of hardware image processors for real-time pattern recognition in order to detect markers with high reliability. This can be accomplished by means of a cross-correlation technique that matches the expected marker shape with image features ( Ferrigno, Borghese, & Pedotti, 1990; Ferrigno & Pedotti, 1985). When the image best fits the expected shape, the cross-correlation function is maximized; when it differs significantly, lower, negative values are obtained. Essentially, the algorithm works as depicted in Figure 4.5. A kernel mask, $M$, is shifted pixel by pixel from left

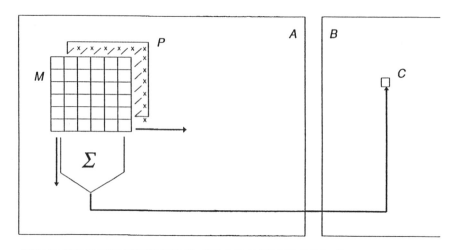

**Figure 4.5**   Cross-correlation algorithm.

to right and line by line from top to bottom on the incoming image. For each portion, $P$, of the image, the pixel-by-pixel product with the elements of $M$ is performed. The results are summed and give the value of each pixel, $C$, of the output image. By shifting the mask, all the pixels of the output image, $B$, are obtained.

The shape used for the markers must be symmetrical, otherwise it would change as the body landmarks rotate. (Spherical and hemispheric shapes are the most insensitive to rotation.) The mask is chosen to maximally reject background noise and to allow the size of markers to change as they move along the optical axes of the cameras. This variation is allowed in the range of 1 to 5 by using a mask of $6 \times 6$ pixels (Ferrigno, 1990). If wider variations are expected (or desired) either the size or the shape of the mask can be varied. If the shape is varied, the rejection of background noise is reduced to allow more size variations. However, variations in size of more than 1 to 5 are typically involved only when the markers leave the working space where measures are assumed to be accurate and where they are correctly focused.

The output of the cross-correlator is the transformation of the input image in which the signal-to-noise ratio (SNR) has been dramatically increased. For example, large white areas become darker than the black signal.

When the cross-correlation function overcomes a given threshold, the coordinate generator is triggered. Because the cross-correlation is coded on several levels (several bits), this information is also sent to the computer, which estimates the marker position more precisely by making assumptions about its brightness distribution. The increased information provided by the cross-correlation technique allows us to obtain high spatial resolutions even when using small markers (on the order of $60,000 \times 60,000$, with the cross-correlation values significant up to 8 bits). Shape recognition also allows the use of very small target objects, because it does not require that markers be the brightest objects on the image.

Figure 4.6 a and b shows an example of detection of very small (less than 1 mm) markers by a real-time image processor based on the cross-correlation algorithm. Figure 4.6a shows the surveyed image (the markers are placed on the lips), and in Figure 4.6b, the detected markers are represented by bright spots.

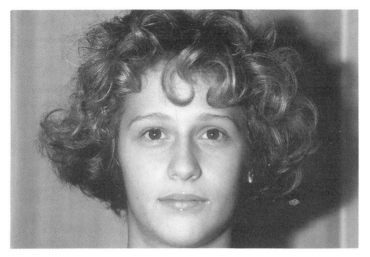

a

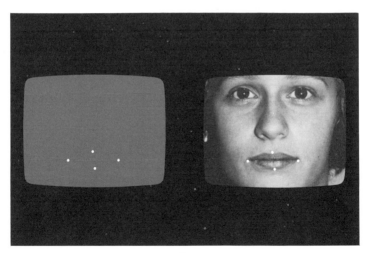

b

**Figure 4.6**   Marker recognition performed via bi dimensional cross-correlation. A subject wearing small markers (less than 1 mm) on the lips is shown in (a). The image (b) shows two monitors. On the monitor on the right, the objects detected as markers in the scene shown on the left monitor are highlighted as bright spots.

# Computer Level

This part of the system includes an interface with the dedicated hardware and a series of software procedures for on-line and off-line data processing.

The main requirement any computer interface must meet is the capability of accommodating the data of a real-time acquisition. The amount and the velocity of incoming data depend on the expected number of markers and on the number of sensors (addressable or not) and is proportional to the parameter, $T$:

$$T = NM \cdot NTV \cdot SR, \tag{4.1}$$

where NM = number of markers, NTV = number of sensors, and SR = sampling rate. Generally a dedicated direct memory access (DMA) protocol represents a good solution, but other general purpose interfaces, like GPIB (IEEE-488) and ETHERNET, have been employed.

Real-time data storage represents another big problem for motion analyzers because it constrains the acquisition parameters. It is proportional to the parameter, $D$:

$$D = NM \cdot NTV \cdot t \cdot SR, \tag{4.2}$$

where $t$ = acquisition duration. Combining Equations 4.1 and 4.2, the following relationship is obtained:

$$D = T \cdot t. \tag{4.3}$$

This means that the data storage is proportional to the throughput and to the acquisition time. For example, consider the case of 25 markers seen by four sensors at a sampling rate of 100 Hz. $T$ assumes a value of 10,000; that is, 10,000 coordinates (coded in a given number of bytes, depending on the system) are acquired in each second. Thus, the data storage required for a 30-s acquisition is 300,000 multiplied by the number of bytes required to code each coordinate. A system that codes each coordinate in 4 bytes requires a total of 1,200,000 bytes.

## Data Acquisition and On-Line Data Processing

During experiments, the computer must perform both data acquisition and some data processing to evaluate the reliability of the data and to compress them. During data acquisition, the number of markers, the number of sensors, sampling rate, and acquisition time must be specified at least in part. At this stage, the system can work without knowing the exact number of markers, and data acquisition can be stopped and started by the operator, by an internal timer, by an external trigger, or by some combination of these. In these cases, the parameters $t$ and NM will be an output of the system instead of an input, whereas NTV and SR must always be specified.

The basic on-line processing depends on the nature of the sensor. Nonaddressable sensors (like lateral effect photodiodes) produce analog output and do not require data processing. Because these sensors do not provide an image that can be seen on a standard monitor, the computer must produce a real-time or, at least, on-line representation of what is going on.

Linear CCD arrays and video cameras must have at least one pixel over threshold for each marker. Information redundancy increases the basic resolution of the sensor. The procedures used to enhance the resolution also allow data compression, because only one pair of coordinates is derived from all the over-threshold pixels of each marker. Figure 4.7 summarizes all these procedures and specifies the kinds of sensors for which they are mandatory.

## Basic Processing

Basic processing algorithms include all those procedures that are not required to be performed on-line but are still necessary to obtain usable measurements. Figure 4.7 shows these procedures in the processing line common to both addressable and nonaddressable sensor systems.

**Distortion Correction.** This compensates for optical and sensor nonlinearities. (The distortions introduced by lateral effect photodiodes and by cylindrical lenses used in conjunction with linear CCD arrays are greater; Leo & Macellari, 1979; Woltring, 1975, 1976.) Distortion correction is usually done by a polynomial expression with coefficients computed from measurements of a control object (see chapter 2). Marzan and Karara (1975) and Miller, Shapiro, and McLaughlin (1980) reported a global model for distortion, whereas Borghese and Ferrigno (1990) proposed a local one. The latter allows correction of local errors with fewer parameters than required with the global approach, which is less flexible. On the other hand, the piecewise approximation requires more precise acquisition of the coordinates of control grid markers.

**Marker Assignment.** This is a required procedure when using addressable sensors and passive markers. These systems transmit the coordinates of the markers in the order they are detected, specifying lines and rows in the raster

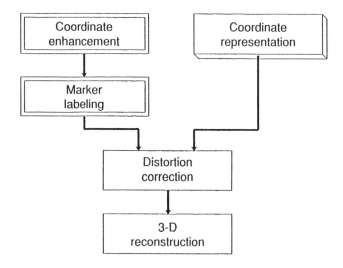

**Figure 4.7** Basic software processing to obtain the 3-D coordinates of the surveyed points. Double-framed boxes refer to procedures typical of addressable sensors; the 3-D box is used for procedures typical of nonaddressable ones.

format. If the movement is complex, this order may be reversed. A reshuffle is necessary to correct the order of representation and for the 3-D reconstruction. Several approaches to this problem have been proposed (Ferrigno et al., 1990; Ferrigno & Gussoni, 1988; Jarrett et al., 1976; Taylor et al., 1982). An initial assignment is required in the first, or the first few, frames and is generally accomplished manually, but knowledge-based procedures for automatic labeling are already under development. Once the markers of these frames have been correctly classified, their coordinates are used to track the rest of the trial and predict their trajectories. The latest techniques (Ferrigno et al., 1990) also introduce some a priori knowledge about the structure of the body under analysis, thus increasing the prediction reliability and automatically solving the problem of marker overlap. These goals are achieved by describing a model of the body in terms of links connecting the markers and the probability of the markers themselves being hidden by others.

An example of labeling is illustrated in Figures 4.8, a and b and 4.9. Figure 4.8, a and b shows the raw points acquired by two video cameras, and Figure 4.9 shows the 3-D stick diagram (projected on the sagittal plane) of the labeled movement (a jump from a platform). The markers were recognized by their shape using the ELITE system with a sampling rate of 100 Hz. It is evident that in such a complex movement the trajectories of the markers present various intersections and superimpositions that produce an apparently confused image in Figure 4.8, a and b, which is more clearly resolved in Figure 4.9. To obtain this resolution, the system had to perform several operations: (a) initial labeling

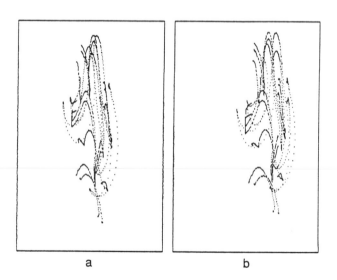

a                                b

**Figure 4.8**   (a and b) Raw points collected by two video cameras picking up a subject jumping from a platform: (a) TV1 and (b) TV2. The angle between the two cameras was 75°. The subject was wearing 10 markers: 2 on the head, and 1 each on the shoulder, elbow, wrist, hip, greater trochanter, knee, external malleolus, and fifth metatarsal head.

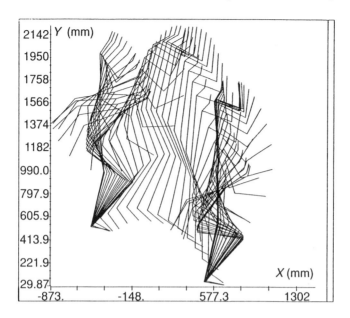

**Figure 4.9** Stick diagram of the jump shown in Figure 4.8, a and b, projected on the sagittal plane after 3-D reconstruction. For clarity's sake, a stick figure has been drawn every 50 ms only. The scales of the plot are in mm.

of the markers; (b) multipoint tracking; (c) reconstruction of hidden markers; (d) distortion correction; and (e) 3-D reconstruction.

**Three-dimensional Coordinate Reconstruction.** This is the last basic processing required before further analysis can be done, and several approaches have been proposed. Two of these procedures are addressed in this book: Direct linear transformation (DLT) is discussed in chapter 2, and the iterative least square solution of colinearity equations (ILSSC) will be discussed here. Both of these approaches rely on the same model of transformation between the object space (the world around the sensor) and the image space (the plane of projection on the sensor behind the lenses). Figure 4.10 shows these spaces with reference to the sensor-optics system.

The mathematical equation that transforms a point of object space into a point of image space is, for the $x$-coordinate,

$$x - x_0 = -c \cdot \frac{M_{11}(X - X_o) + M_{12}(Y - Y_o) + M_{13}(Z - Z_o)}{M_{31}(X - X_o) + M_{32}(Y - Y_o) + M_{33}(Z - Z_o)}, \qquad (4.4)$$

and for the $y$-coordinate,

$$y - y_0 = -c \cdot \frac{M_{21}(X - X_o) + M_{22}(Y - Y_o) + M_{23}(Z - Z_o)}{M_{31}(X - X_o) + M_{32}(Y - Y_o) + M_{33}(Z - Z_o)}, \qquad (4.5)$$

where $x$ and $y$ represent the coordinates of the projection of Point $P(X,Y,Z)$ on

Object space

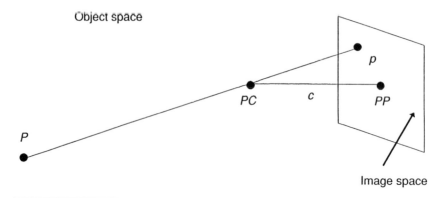

Image space

**Figure 4.10**   Perspective projection of a point, $P$, in the object space to image space, $p$, through the perspective center, $PC$. $PP$ is the principal point (the projection of $PC$ onto the image plane). The segment $PC$-$PP$ lies on the optical axis, and its length is the focal length, $c$.

the image space; $c$ is the focal length of the lens; $x_0$ and $y_0$ are the coordinates of the principal point (the intersection of the optical axis and the image plane); $X_0$, $Y_0$, and $Z_0$ are the coordinates of the perspective center; and $M_{ij}$ are the nine elements of the rotation matrix from the object space reference frame to the image space reference frame (Wolf, 1983). Although this makes a total of nine parameters, only three unknowns must be accounted for, because all the $M_{ij}$ are nonlinear functions of the roll, pitch, and yaw angles of the sensors $k$, $\omega$, and $\varphi$. The great problem in using such a model for space resection or calibration (determination of the nine parameters: $x_0$, $y_0$, $c$, $X_0$, $Y_0$, $Z_0$, $k$, $\varphi$, and $\omega$) and space intersection (determination of an unknown point $P$) is represented by the nonlinear nature of Equation 4.5.

DLT overcomes this obstacle by mathematically rearranging the terms of Equation 4.5 and introducing other parameters. The approach of ILSSC is based instead on the linearization of Equation 4.5 using the Taylor series in the proximity of a starting point. A new point in the unknown parameters space is computed using Equation 4.5 with respect to the recorded coordinates and minimizing the errors of the control points' $x$, $y$. The new point becomes the next starting point, and the iterations continue until the parameters update is small enough. Convergence occurs within five or six iterations. Subsequent space intersection is performed by searching for the shortest perpendicular distance connecting the two lines that join the perspective centers of two video cameras to the projections of the point on the sensor, as shown in Figure 4.11 (Borghese & Ferrigno, 1990).

## Deferred Analyses

Deferred analyses can be done at any time on data acquired and preprocessed by a motion analyzer. Joint angles, linear and angular velocities, and accelerations are computed for use in mechanical models (e.g., for the computation of internal

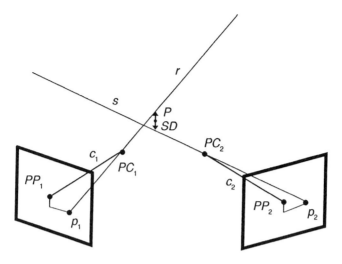

**Figure 4.11**   Space intersection performed by searching for the shortest distance ($SD$) between the lines $r$ and $s$ passing through perspective centers ($PC_1$ and $PC_2$) and point projections ($p_1$ and $p_2$) on the sensors.

forces and moments, which also requires knowledge of external forces). Data gathered with electromyographs may also be correlated with kinematics and dynamics to obtain a more complete description of a motor action. Thus, many motion recording systems require simultaneous data logging of analog signals.

# CLASSIFICATION OF 3-D KINEMATIC MEASUREMENT SYSTEMS

Three-dimensional kinematic measurement systems can be evaluated on the basis of their accuracy and precision and by their reliability.

## System Accuracy and Precision

*Precision* is the frame-to-frame repeatability of the measurement. It can be expressed as the standard deviation of the errors divided by the range of measurement, which is defined as the diagonal of the working volume (assuming it to be prismatic with a rectangular base). Thus, precision is measured over a time ensemble. It can be computed by using static markers and only accounts for the system noise; but precision cannot be used to characterize a measurement system, because it depends on the basic resolution of the sensor as well as on the positioning of markers used for the test. Paradoxically, a sensor having only $2 \times 2$ pixels would produce perfect precision for a very wide distribution of marker positions, whenever a marker lies completely within one of these four pixels. Only those markers lying on the borders of pixels would give a more realistic precision. Other quantities therefore must be taken into account.

*Accuracy* refers to those errors computed as the difference between the measured and the true values. This more consistent information reflects the errors encountered during the everyday use of a 3-D measurement system. To measure accuracy, compare the coordinates of several points obtained with the motion analyzer with their real positions. In practice, if we consider that the most advanced systems are able to measure deviations of less than 0.5 mm in a cubic volume with 1.5-m sides (e.g., the ELITE system; Ferrigno et al., 1990; Ferrigno & Pedotti, 1985), the equally accurate, independent measure of the positions of the markers used for the test is quite difficult.

An alternative is to use a statistical approach, such as moving a fixed-length stick along the whole calibrated field and assessing the variation of the measured distance between markers placed on its two ends (Borghese & Ferrigno, 1990). This kind of test gives information about the distance preservation, but what we are interested in is the accuracy of the measurement along each axis. Two steps are required to determine this: first, decomposition of the length variation of the two markers and second, the decomposition of the error in each marker along the three axes (see Figure 4.12). The error in the distance $(D' - D)$ is decomposed into the vector errors at $P_1$ and at $P_2$. These are subsequently projected onto the

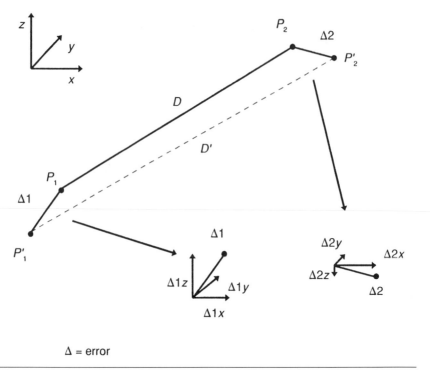

$\Delta = $ error

**Figure 4.12**    Error in the distance, D′–D, decomposed into the vector errors at $P_1$ and at $P_2$.

reference frame axes in order to obtain the three components of the error in $x$, $y$, and $z$.

The first step is straightforward (Borghese & Ferrigno, 1990), and the standard deviation of the error of each marker is found to be $1/\sqrt{2}$ times the standard deviation of the distance. In the second step, the sensor attitude and its position in space must be taken into account. Borghese and Ferrigno (1990) made a simulation showing that the accuracy along the three axes depends on the attitudes of the video cameras.

From the simulation it was also apparent that if good accuracy is desired along all three axes, the angle between sensors (assumed to be placed with only the yaw angle not $0°$) must be close to $90°$ (this can raise problems with active markers that do not have such a wide angle of emission). In this case, the error is distributed in the proportions $1:1$, $\sqrt{2}:1$, and $\sqrt{2}:1$ along the vertical, horizontal, and depth axes, respectively. Figure 4.13 shows the distribution of the standard deviation of the error of coordinate $z$ (depth) assuming a standard deviation of 2-D coordinates equal to 1 and that the video cameras' yaw angle ($\varphi$) around the $y$ (vertical) axis is the only nonzero angle. Note that between $40°$ and $90°$ no appreciable reduction of the depth axis accuracy is observed. This should be kept in mind when designing an experimental setup.

A final possibility is to evaluate the accuracy and precision of a system on a local basis. This can be accomplished by using a precision micrometer on which a marker has been placed. Figure 4.14, a and b shows an example of such evaluation conducted by the ELITE system. The test was carried out in the middle of the calibrated volume (1.7 m across the diagonal). A screw, which moved a spherical marker 0.01 mm for each turn, was used to produce a 3-D displacement of 2 mm (evenly distributed along the three axes $x$, $y$, and $z$). For each of the 200 positions of the marker, 100 samples were acquired with the ELITE system and the 3-D coordinates were computed and averaged. Figure 4.14a reports, on the vertical axis, the average value

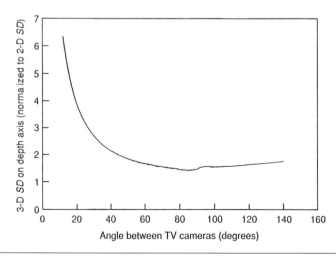

**Figure 4.13**   Standard deviation of the error in the depth axis, $z$, normalized by the standard deviation of the 2-D error of the sensors.

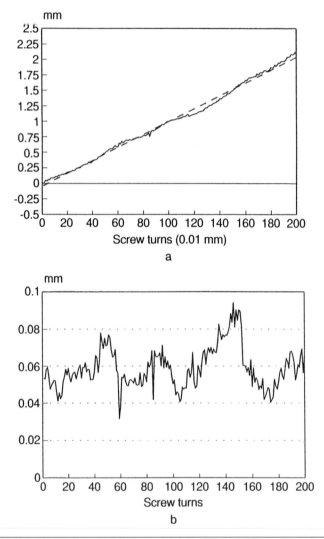

**Figure 4.14**   Local accuracy test carried out on the ELITE system: (a) The average value (static accuracy) for each position of the marker, the horizontal axis shows the number of screw turns (i.e., the number of displacements of 0.01 mm); (b) the standard deviation of each point in Figure 4.14a.
(The authors wish to thank Dr. P. Groppello and Mr. E. D'Amico of BTS for their help in making this test).

(static accuracy) for each position of the marker, while showing, on the horizontal axis, the number of screw turns (i.e. displacements of 0.01 mm). The maximum error found with respect to the true value was 0.17 mm, while 95% of the samples were accurate to within 0.07 mm, (1/24,000 of the diagonal). Figure 4.14b reports the standard deviation of each point in Figure 4.14a. This error is related to the precision and falls within 0.09 mm (1/19,000 of the diagonal).

## Systems Possibilities and Limitations

The performances of motion analyzers vary considerably according to the kind of sensor and signal processing used.

Systems using active markers, although they were introduced in the very early years of automatic movement analysis (Mitchelson, 1975; Woltring, 1974), are still used because they are relatively fast (i.e., they have a high sampling rate) and do not require tracking of the markers' trajectories. Despite the drawbacks in terms of reliability, accuracy, and flexibility of these systems, they allow highly automated data processing without requiring operator intervention. On the other hand, they are cumbersome, because the subject must wear cables, power supplies, and control units, often eliminating the analysis of fast movements (like those in sports) and constraining all other movements. The accuracy is not high, despite the high resolution, both for 2-D analog sensors and for linear array systems, for various reasons. In dynamic analyses, when derivatives of position measurement are used to calculate velocity and acceleration, the accuracy of the $k$th derivative of the data depends on the $k$th power of the error superimposed on the signal accuracy, but only on the power of $k/2$ of the sampling rate (Lanshammar, 1982). Therefore, accuracy can be more significant than sampling rate. Two-dimensional analog lateral effect photodiodes suffer accuracy reduction due to LED reflections and IR background lighting, whereas linear arrays are affected by the distortions introduced by the cylindrical lenses. The reduced viewing angle inherent in both systems can constrain the experimental setup. Neither kind of system provides a visible image of the measured scene.

Systems using passive markers, although requiring a trajectories-tracking procedure, present the advantages of freedom of movement (no subject-encumbering wiring or power supplies) and high flexibility in terms of experimental setup. Also, an image of the movement is visible on a monitor together with the recognizable markers, and this assures the possibility of retailoring the experimental setup on-line or in the field.

There are several good solutions to the problem of marker tracking. The limitations that still affect this approach are the difficulty of tracking markers when two or more of them disappear for a long time (such as when rotations of the subject in the plane of the video cameras occur) and the need for an initial labeling, which makes real-time data processing arduous. The tendency now is toward greater automation involving tracking of 3-D rather than 2-D trajectories. Other developments try to take advantage of the fact that the intersection of lines starting from two projections of a point $P$ (in 3-D space) to two different sensor images (2-D) should intersect almost exactly. Unfortunately, when the number of markers increases, many false intersections appear, which makes some kind of control necessary. Other approaches involve using neural networks to assign markers on the basis of previously tracked files. Also, sampling rates will probably increase thanks to the new solid-state technologies, which will make available faster sensors.

# CONCLUSIONS

Recent technical advances have made optoelectronic-based systems almost standard for human 3-D motion analysis. The increasingly higher speeds and lower costs of computers, together with hardware and software developments in marker detection and tracking, mean that the combination of passive marker and addressable video camera will likely be the future standard in this field. Thus, researchers should use the evaluation principles laid out in this chapter when selecting measurement equipment and markers and designing a configuration for their use.

# REFERENCES

Aristotle. (1945). *De motu animalium* (Progression of animals). (E.S. Foster, Trans.). Cambridge, MA: Harvard University Press. (Original work circa 304 B.C.).

Bernstein, N. (1967). *The Co-ordination and regulation of movements.* London: Pergamon Press.

Borelli, G.A. (1680). *De motu animalium* (*Progression of animals*). Lugdunium Batavorum.

Borghese, N.A., & Ferrigno, G. (1990). An algorithm for 3D automatic movement detection by means of standard TV cameras. *IEEE Transactions on Biomedical Engineering BME, 37,* 1221-1225.

Braune, C.W., & Fischer, O. (1987). *The human gait.* (P. Marquet & R. Furlong, Trans.). Berlin: Springer-Verlag. (Original work published in 1895).

Cheng, I.S., Koozekanani, S.H., & Fatchi, M.T. (1975). Computer television interface system for gait analysis. *IEEE Transactions on Biomedical Engineering, BME, 22,* 259-264.

D'Amico, M., Ferrigno, G., & Rodano, R. (1989). Frequency content of different track and field basic movements. In W.E. Morrison (Ed.), *Proceedings of the 7th International Symposium of the Society of Biomechanics in Sport* (pp. 176-193). Victoria, Australia: Footscray Institute of Technology.

D'Amico, M., Ferrigno, G., & Rodano, R. (1990). Statistical analysis of frequency content of selected anatomical landmarks in treadmill running. In M. Nosek, D. Sojka, W.E. Morrison, & P. Susanka (Eds.), *Proceedings of the 8th International Symposium of the Society of Biomechanics in Sport* (pp. 327-332). Prague, Czechoslovakia: CONEX.

Ferrigno, G. (1990). Elite system: The state of the art. In U. Boenick & E.M. Nader (Eds.), *Gait analysis* (pp. 51-59). Duderstadt, Germany: Mecke Druck und Verlag.

Ferrigno, G., Borghese, N.A., & Pedotti, A. (1990). Pattern recognition in 3D automatic human motion analysis. *International Society of Photogrammetry and Remote Sensing Journal of Photogrammetry and Remote Sensing, 45,* 227-240.

Ferrigno, G., & Gussoni, M. (1988). Procedure to automatically classify markers in biomechanical analysis of whole-body movement in different sports activities. *Medical and Biological Engineering and Computing, 26,* 321-324.

Ferrigno, G., & Pedotti, A. (1985). ELITE: A digital dedicated hardware system for movement analysis via real-time TV signal processing. *IEEE Transactions on Biomedical Engineering BME*, **32**, 943-950.

Furnée, E.H., Halbertsma, J.M., Klunder, G., Miller, S., Nieukerke, K.J., van der Burg, J., & van der Meche, F. (1974). Automatic analysis of stepping movements in cats by means of a television system and a digital computer. Proceedings of the Physiological Society, *Journal of Physiology*, **240**, 3-4.

Jarrett, M.O., Andrew, B.J., & Paul, J.P. (1976). A television computer system for the analysis of kinematics of human locomotion. *IERE Conf. Proc.*, **34**, 357-370.

Lanshammar, H. (1982). On practical evaluation of differentiation techniques for human gait analysis. *Journal of Biomechanics*, **15**, 99-105.

Leo, T., & Macellari, V. (1979). On-line microcomputer system for gait analysis data acqustion based on commercially available optoelectronic devices. In A. Morecki, K. Fidelius, K. Kedzior, and A. Wit (Eds.), *Biomechanics VII-B* (pp. 163-169). Baltimore: University Park Press.

Marey, E.J. (1902). *The history of chronophotography*. Washington, DC: Smithsonian Institution.

Marzan, G.T., & Karara, H.M. (1975). A computer program for direct linear transformation solution of the collinearity condition and some applications of it. In *Proceedings of the Symposium on Close-Range Photogrammetric Systems* (pp. 420-476). American Society of Photogrammetry, Falls Church, VA.

Miller, N.R., Shapiro, R., & McLaughlin, T.M. (1980). A technique for obtaining spatial kinematic parameters of segments of biomechanical systems from cinematographic data. *Journal of Biomechanics*, **13**, 535-547.

Mitchelson, D.L. (1975). Recording of movement without photography. In H.T.A. Whiting (Ed.), *Techniques for the analysis of human movement* (The Human Movement series) (pp. 33-65). London: Lepus Books.

Muybridge, E. (1901). *The human figure in motion*. London: Chapman and Hall.

Pedotti, A. (1977). Simple equipment used in clinical practice for evaluation of locomotion. *IEEE Transactions on Biomedical Engineering*, **24**, 456-461.

Pedotti, A. (1990). Future perspectives in Europe for quantitative analysis of movement. In V. Poenick & E.M. Nader (Eds.), *Gait analysis* (pp. 322-332). Duderstadt, Germany: Mecke Druck und Verlag.

Taylor, K.D., Mottier, F.M., Simmons, D.W., Cohen, W., Pavlak, R., Jr., Cornell, D.P., & Hankins, G.B. (1982). An automated motion measurement system for clinical gait analysis. *Journal of Biomechanics*, **15**, 505-516.

Winter, D.A., Greenlaw, R.K., & Hobson, D.A. (1972). Television computer analysis of kinematics of human gait. *Computers and Biomedical Research*, **5**, 498-504.

Wolf, P.R. (1983). *Elements of photogrammetry*. New York: McGraw-Hill.

Woltring, H.J. (1974). New possibilities for human motion studies by a real-time light spot position measurement. *Biotelemetry*, **1**, 132-146.

Woltring, H.J. (1975). Single and dual axes lateral photodetectors of rectangular shape. *IEEE Transactions on Electronic Devices*, **22**, 580-581.

Woltring, H.J. (1976). Calibration and measurement in 3-dimensional monitoring of human motion by optoelectronic means II. *Biotelemetry*, **3**, 65-97.

# Chapter 5

# Smoothing and Differentiation Techniques Applied to 3-D Data

*Herman J. Woltring*

This chapter addresses one of the problems that has always hampered routine application of movement analysis. The gap between the ultimate goal of the analysis and what can be measured directly is often wide and discourages many students of human movement. Though error propagation effects in this transformation can be considerable, special measures can be taken to minimize them.

Some errors are correlated to the process under investigation, and some are quite independent of it. Examples of the former are limitations of the functional model describing the movement (e.g., the rigid-body model ignores all soft-tissue deformation during movement) or uncompensated image distortion errors in measurement equipment. An example of essentially independent errors is wide-band measurement noise in optoelectronic equipment for 3-D kinematics. Systematic errors in measurement equipment have effects on measured variables similar to those of modeling errors in subsequent components of the data processing chain; thus, distinguishing between correlated and uncorrelated errors is more appropriate than, for example, distinguishing between measurement and modeling errors.

It might appear that there is no need to reduce wide-band measurement noise because of the limitations of the rigid-body model. Though this is true to some extent, different error types may have different effects on the variables being estimated. In particular, wide-band, "white" measurement noise has a deleterious effect on estimated derivatives, whereas model artifacts may have more influence in the smoothed data at low frequencies. As discussed later, improved lower frequency estimates can be obtained by sampling at higher rates than minimally required under the sampling theorem, as long as the noise is additive and white.

Simulation studies in which the properties of movement signals are evaluated in relation to measurement error properties and to model limitations are beginning to provide more confidence in the transformations needed for bridging the gap between what we can measure and what we need to know. For reviews and recent models on smoothing and differentiation of noisy data in the biomechanics field, see Hatze (1981, 1990), Wood (1982), Woltring (1985, 1986, 1990), Fioretti and Jetto (1989), Dohrmann & Busby (1990), and D'Amico and Ferrigno (1990, 1992). In the statistics and signal processing literature, see Kohn and Ansley (1987, 1989), Fessler (1991), and Vaughan (1982).

## GENERAL PROBLEMS WITH FIRST AND HIGHER DERIVATIVES

The noise-amplification character of the differentiation process is well known. If we have a time signal $x(t)$ consisting of a sum of sinusoidal harmonics,

$$x(t) = \sum_{i=0}^{n} a_i \sin(\omega_i t + \phi_i), \tag{5.1}$$

the amplitudes in $x(t)$ are defined by the $\{a_i\}$, but they are $\{\omega_i a_i\}$ in the first time derivative of $x(t)$, and $\{\omega_i^2 a_1\}$ in the second time derivative of $x(t)$ (Figure 5.1). Thus, the higher the frequency content in a position signal, for example, the more this will dominate in the velocity, acceleration, and higher derivatives.

Because of this high-frequency amplification effect, one should not choose a cut-off frequency on the basis of the relative power contribution (defined in terms of squared amplitudes) of the rejected signal components. Suppose, for example, a signal consists of a 1-Hz component with amplitude 1, plus a 10-Hz component with amplitude 0.01. The relative power of the 10-Hz component then amounts to a meager 0.01%, whereas the amplitudes on the second derivative level are equal.

Above a certain frequency, the spectral content of natural movement data should decay with frequency such that high-frequency signal components on

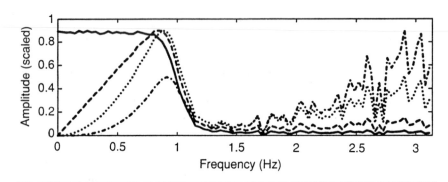

**Figure 5.1**  Example of a noisy signal spectrum (solid line) and of its first (- - -), second (· · ·), and third (- · - · - · - ·) time derivatives.

the derivative level do not produce excessive accelerations and inertial forces (Woltring, 1985). In many measurement devices, the white noise added by the equipment has a much higher bandwidth before it begins to decay with frequency (cf. Teulings & Maarse, 1984). Thus, there is usually a cross-over frequency beyond which the signal-to-noise ratio (SNR) is less than 1, for each frequency, and this provides a realistic cut-off frequency for low-pass smoothing. This cut-off frequency is the same for each derivative under the SNR criterion.

Signals in 3-D movement analysis are typically multidimensional, for example, expressed as xyz coordinates of multiple landmarks over time. The question then arises whether these signal components should be processed independently or not. For example, in rigid-body movement analysis, known and constant distances between landmark pairs might be used as constraints during the smoothing and differentiation process (e.g., Jennings & Wood, 1988), and they might be estimated simultaneously if they are not known a priori. Also, the fundamental frequencies in the sagittal plane during regular gait are twice those in the mediolateral direction, and heel-strike transients are hardly present in the lateral direction. Thus, one might perform optimal smoothing and differentiation *independently* from imposing constraints (Lange, Huiskes, & Kauer, 1990; Woltring, Lange, Kauer, & Huiskes, 1987). It is currently unknown which approach is better, but independent processing is much easier and usually faster. However, the preceding example indicates a pitfall: If independent processing is performed, crosstalk of high-frequency terms into low-frequency signal components cannot be avoided, whether constraints are imposed before or after smoothing.

It has been proposed that the problems of smoothing and differentiation can be avoided by iteratively integrating derivative estimates until the integrals fit the raw data within a certain tolerance. This argument is fallacious, because the approach is tantamount to derivative estimation. Effectively, it involves building a feedback system with an integrator in the feedback loop. As is known from classical analog computing, this is a standard way to build a differentiator. Whether any high-frequency components are found in the final solution depends on the strategy adopted for adjusting the estimated derivatives. If no high frequencies are used, then a low-pass filtering strategy has, in effect, been selected.

This fallacy does not occur only in purely kinematic models. It also occurs in comprehensive, dynamic models if force estimates are iteratively adjusted until a particular displacement criterion is met. Here, the adjustment criterion used to modify current force estimates will determine whether the final, dynamic solution contains low-frequency components. If high frequencies are used, one is likely to find that the criterion function is very shallow, possibly with multiple optima, so uniqueness and robustness of the solution under small measurement or model disturbances will be in doubt.

## Sampling Frequency Choice

The sampling frequency must be selected correctly if the signal is to be reliably reconstructed from digitized samples. The Nyquist criterion prescribes that the sampling frequency should be at least twice the highest signal frequency. This does not predict the utility of higher sampling frequencies. Sampling of a continuous signal will "map" all frequencies above the Nyquist frequency, $\omega_N = \pi/\tau$ (i.e., half the sampling frequency), into the Nyquist band, $-\omega_N \le \omega \le \omega_N$. The Nyquist criterion assures that this mapping does not occur for the *signal* component in measured data, but high-frequency noise will continue to be mapped into the Nyquist band. This is apparent from the height of the noise pedestal at the Nyquist frequency in Figure 5.2, a and b. Furthermore, the spectrum of the sampled data (signal plus noise) is repeated around the frequencies $2k\omega_N$, with $k = \ldots, -2, -1, 0, 1, 2, \ldots$.

This frequency repetition may seem strange until one realizes its relation to stroboscopic effects. Whenever a periodic phenomenon is sampled, it is impossible to determine its original frequency range. Between the sampled data points, harmonic components with low and high frequencies may be interpolated; at the sampling points, only the lowest frequencies (within the Nyquist band) contribute, but all other interpolating signals are zero. Mathematically, digital signal processing, both in the time and frequency domains, occurs along a *circular* time or frequency scale. Continuous signal processing occurs along a *linear* scale.

As long as the measurement noise above a chosen Nyquist frequency is white, it is useful to increase the sampling frequency, because the amount of noise mapped into the Nyquist band will be smaller (cf. Teulings & Maarse, 1984). Gustafsson (1977) and Lanshammar (1982a, 1982b) have derived the following relation between the noise level $\sigma_k^2$ of each sample after optimally smoothing and differentiating band-limited, sampled data:

$$\sigma_k^2 = \sigma^2 \tau \frac{\omega^{2k+1}}{(2k+1)\,\pi},\tag{5.2}$$

where $k > 0$ is the derivative order ($k = 0$ refers to smoothing without differentiation); $\omega$, the low-pass filter's cut-off frequency; $\sigma$, the standard deviation of the

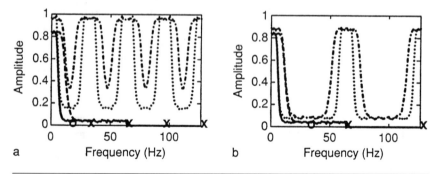

a                                          b

**Figure 5.2**   Aliasing effects shown for (a) low sampling rate and (b) high sampling rate in sampling of wide-band data, with narrow-band (solid line before sampling, • • • after sampling) and wide-band (- - - before sampling, - · - · - · - · after sampling) signal components. The sampling frequency and its multiples are marked with "x"; the Nyquist frequency ($\omega_N$) is marked "o."

noise in the raw samples; and $\tau$, the sampling interval (i.e., the inverse of the sampling frequency).

The factoring of Equation 5.2 in terms of an equipment-related, spatiotemporal resolution term, $\sigma^2\tau$, and of a process-related component, $\omega^{2k+1}/(2k+1)\,\pi$ (Furnée, 1989, 1990; Woltring, 1984), suggests that, for given $\sigma$, $\tau$ should be made as small as possible (or rather, as small as necessary). However, increasing the bandwidth of a digital system usually implies increasing the bandwidth of the preceding analog system, so reduction of $\tau$ may entail an increase in $\sigma$. How these opposing factors will affect the product, $\sigma^2\tau$, remains to be seen in practical situations; at any rate, a combined spatiotemporal resolution criterion is a meaningful entity for comparing given systems, each with its own $\sigma_i^2\tau_i$.

In the above discussion, signals were thought to be composed of sums of sinusoids, with an implicit assumption of independence between the components, except that the amplitudes decrease with frequency beyond some (usually unknown) frequency. Practical experiments usually produce this kind of information, which makes frequency-based smoothing procedures meaningful and attractive.

If we know more about the data (e.g., we know the frequency content embodied in the amplitudes $\{a_i\}$ of Equation 5.1, and also the phases $\{\Phi_i\}$, a different situation follows. We then know the full shape of the signal, except for a scaling factor $S$ and the time $T$ of its occurrence. In such a situation, only these two parameters have to be found from the noisy data, for which various parametric procedures can be used, including one known as synchronous detection (see pp. 83-84).

These various methods can all be handled in some optimal manner; for such optimality to apply, certain properties are assumed. One assumption is that deviations between the optimal model and the measured data meet certain stochastic conditions. Zero mean and whiteness of the residual errors are common assumptions. An important question is to what extent any data processing is *robust*, that is, to what extent the proper functioning of the optimal model can be assumed if the functional or stochastic assumptions do not hold. For example, to what extent is an optimal low-pass filter with automatic cut-off frequency selection (based on an assumption of white noise) insensitive to nonwhite noise? Some illustrations are provided later in this chapter.

## Smoothing, Prior Knowledge, and Nonlinear Transformations

The properties of the measurement noise may determine whether smoothing and differentiating should be done on the raw data, or after 3-D reconstruction. If the measurement errors are white, zero mean, stationary, and additive to some low-pass signal, any nonlinear transformation will destroy these properties to some extent: High-frequency, white noise will be warped into the low-frequency domain, and low-frequency signal components will be mapped onto the high-frequency range. The example in Figure 5.3, a and b shows where a noisy spectrum was transformed into the time domain, squared at each time instant, and transformed back into the frequency domain.

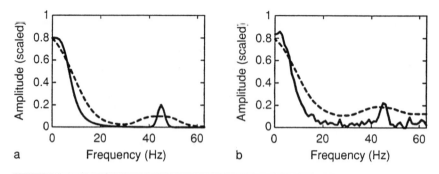

**Figure 5.3**   Nonlinearity effects in the time domain and spectrum distortion: spectrum of (a) a noisefree and (b) a noisy signal before (solid line) and after (- - -) the signal has been squared in the time domain.

In this situation, separation of signal and noise by low-pass filtering becomes suboptimal, and this effect will be more pronounced with greater nonlinearities in the transformation. In 3-D reconstruction of landmark coordinates from camera observations, only mild nonlinearities occur. However, the mapping from 3-D landmark coordinates to rigid-body data (and even more, that from rigid-body data to helical-axis entities, Woltring, 1985; Woltring et al., 1987) is highly nonlinear, so separation of low-pass signal components from wide-band, additive noise should be carried out first. Once smoothing has been done, differentiation can occur by finite differences and interpolations, even after strongly nonlinear transformations. However, the interpolation and differencing intervals should be sufficiently small (albeit sufficiently wide to avoid rounding errors in digital arithmetic) to allow linear approximations for first- and higher order derivatives.

One should note, however, that independent processing of raw data may not allow full use of available information. For example, the lines of sight from cameras to an observed marker point should ideally intersect at a single point. The residual errors encountered in this situation may help determine where stochastic or systematic errors must be assumed, especially if more than two cameras are involved. This is a consideration in favor of nonlinear, 3-D reconstruction before low-pass smoothing. Similarly, prior knowledge of the distances between landmarks on a rigid cluster might be exploited when directly estimating position and attitude of the rigid body from projected camera observations. An example is the direct linear transformation used for 3-D camera calibration: Position and attitude of one object with respect to another with certain additional parameters can be uniquely obtained from the 11 DLT-parameters that are assessed during calibration.

Theoretically, the best approach is to do all processing at the same time, using the tools of contemporary system identification, state and parameter estimation (Eykhoff, 1974; Söderström & Stoica, 1989). In this way, one can estimate a rigid body's position vector and attitude matrix and its velocity and acceleration vectors (both linear and rotational) from all observed landmark data on a cluster, even if some landmarks are not always observed by all cameras. In addition, one can merge position and acceleration measurements. The position measurements are particularly useful for the low-frequency range, whereas the acceleration data contribute at higher frequencies (Ladin & Wu, 1991).

The following sections discuss some intermediate approaches between the extremes of fully independent and fully simultaneous processing.

# PARAMETRIC METHODS: MODEL FITTING

The distinction between nonparametric and parametric models dates from the time of analog computing and continuous systems, where nonparametric models are "described by a curve, function or table" (Söderström & Stoica, 1989, p. 9), whereas parametric models have a given *structure* but a limited number of unknown *parameters*. In a digital signal-processing context, the distinction is somewhat artificial because both approaches involve a certain amount of parameter estimation. It is still a useful distinction in terms of the number of parameters relative to the amount of data available; in other words, in terms of the number of degrees of freedom. Thus, fitting a parabola to a set of noisy data points is a parametric procedure, whereas fitting a quintic spline to the data is nonparametric (even though a parabola is a special case of a quintic spline). Some procedures accommodate mixtures of parametric and nonparametric models in the time or frequency domains (see D'Amico & Ferrigno, 1990, 1992).

In a typical parameter estimation approach, a mathematical model is formulated with a small number of unknown, or partially known, parameters (e.g., they might be nonnegative), and the problem is how to estimate these parameters from a sufficiently redundant set of noisy data. For example, if a zero-mean signal of known shape $x(t)$ is hidden in a noisy signal $y(t)$, with additive, white, uncorrelated, stationary noise (zero mean and possibly unknown standard deviation $\sigma$), a procedure called *synchronous detection* in radar technology is based on finding the maximum value of the cross-correlation function

$$\rho(T) = \int x(t)\, y(t + T)\, dt. \tag{5.3}$$

The maximum value can be found efficiently in the frequency domain, because convolution in the time domain is equivalent to multiplication in the frequency domain. Using fast fourier transform methods, the amount of computational savings can be dramatic; see any general handbook on digital signal processing (e.g., Oppenheim & Schafer, 1989).

Generally, parameter estimation under conditions of additive noise can be translated into the problem of solving a system of linear(-ized) equations:

$$A\vec{a} = \vec{y}, \tag{5.4}$$

where $A$ is a known $m*n$ matrix of rank $m$, usually with $n >> m$; $\vec{y} = (y_1, \ldots, y_n)'$ the sum of the true, unknown signal and of the noise $\vec{\varepsilon} = (\varepsilon_1, \ldots, \varepsilon_n)'$ of which only certain stochastic properties are known, and where $\vec{a} = (a_1, \ldots, a_m)'$ contains the unknown parameters to be estimated.

A simple example is the fitting of a parabola $x(t) = a_0 + a_1t + a_2t^2$ to a set of $m \geq 3$ noisy data $\{y_i\}$ stored in an $n$-vector $\vec{y}$. Defining $n$ row vectors $[1, t_i, t_i^2]$, one for each measurement time point $(t_i)$, these rows can be adjoined into an $n*3$ matrix $A$. If the noise is zero mean, that is, $\xi(\vec{\varepsilon}) = 0$, with known, nonsingular noise covariance matrix $\xi(\vec{\varepsilon}\,\vec{\varepsilon}') = N$, the *best linear unbiased estimate* (*blue*) for $\vec{a}$ follows as

$$\hat{\vec{a}}_{\text{blue}} = N_{\text{blue}} \, A'/N^{-1} \, \vec{y} \tag{5.5}$$

with covariance matrix

$$N_{\text{blue}} = (A'N^{-1}A)^{-1}. \tag{5.6}$$

In many cases, $N$ is known up to a scaling factor, which can be obtained from the *residual variance* $\sigma_{\text{res}}^2$, such that

$$\hat{\sigma}_{\text{res}}^2 = \frac{\hat{\vec{\varepsilon}}\,'\,\hat{\vec{\varepsilon}}}{(n-m)}, \tag{5.7}$$

where $\hat{\vec{\varepsilon}}$ is the fitting error between the data and the model, or

$$\hat{\vec{\varepsilon}} = (I - AN_{\text{blue}}A'N^{-1})\vec{y}, \tag{5.8}$$

and where the estimate's covariance matrix $N_{\text{blue}}$ can be estimated as

$$\hat{N}_{\text{blue}} = \hat{\sigma}_{\text{res}}^2 \, (A'N^{-1}A)^{-1}. \tag{5.9}$$

Though conventional (weighted) least-squares fitting of a suitable model to noisy data may result in estimates that are optimal from an unbiased, minimum-variance point of view, the variances of these estimates may be very large if the data matrix $A$ is nearly singular. Differentiation of noisy data is an example of near singularity. Therefore, it may be useful to allow a small bias in the estimate, which will reduce the estimate's variance considerably (see Figure 5.4).

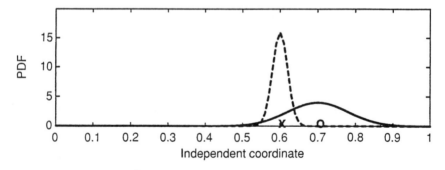

**Figure 5.4**   Example of bias and variance trade-off in parameter estimation. The probability density function (PDF) of a best linear unbiased estimate (*blue*; solid line) and a biased estimator (- - -) are plotted. The true mean is marked with "o" and the biased mean is marked "x."

Low-pass filtering in differentiation is an example of imposing such a bias. In order to reduce the stochastic error level, certain components are suppressed. This involves both noise and signal removal; the latter is, in fact, a form of biasing.

One way of imposing a small bias is by imposing a number of constraint equations on the linear system of Equation 5.2,

$$C\vec{a} = \vec{0}, \tag{5.10}$$

where $C$ is a $k*m$ matrix. To determine the amount of weight attributed to this bias by an additional weight factor, the regularization parameter $x$ must be greater than zero. In differentiation, for example, the constraint could be imposed on the amplitudes of the derivative(s) of interest. The resulting estimate then is

$$\hat{\vec{a}}_{\text{biased}} = (A'N^{-1}A + \alpha\ C'C)^{-1}A'N^{-1}\vec{y}. \tag{5.11}$$

Similar models exist in the field of medical imaging (CT, MRI). The use of weighted constraints in Equation 5.10 has the effect of imposing *correlations* between the elements of $\vec{a}$ and thus of reducing the effective dimensionality of $\vec{a}$. In fact, spline smoothing can be described in terms of these models, but where $m \approx n$, unlike the $m < < n$ situation of the parametric model.

The question remains how much bias should be imposed so as to reduce the noise level in the estimate $\vec{a}_{\text{biased}}$ sufficiently without distorting the signal too much. This issue of *optimal* regularization, by choosing suitable values for $\alpha$, is addressed in the section on optimal data processing.

## NONPARAMETRIC METHODS

Spline fitting in the time domain, and fast Fourier techniques in the frequency domain, are often viewed as nonparametric signal processing methods. Yet these methods are also parametric, despite the fact that the model to be fitted to the data is quite general and thus incorporates many unknown parameters. There is substantial literature on both topics, though this chapter does not review them. Rather, the reader is referred to the original literature, such as the 1989 book by Oppenheim and Schafer, and archive journals, such as the *IEEE Transactions on Acoustics, Speech, and Signal Processing*. Furthermore, there are many software packages that allow assessment of filter parameters from specified time- or frequency-domain requirements. These can be found in the published literature, on electronic fileservers (such as NETLIB@RESEARCH.ATT.COM, which returns an index in response to the one-line message, *send index*), and through commercial channels.

## Time-Domain Methods

The term *digital filtering* has a number of meanings, which are sometimes confused in biomechanics. Formally, the term may refer to any form of digital signal processing, whether low-pass smoothing, differentiating, or any other kind of transformation. In a time-domain context, the term *filtering* is often interpreted in a more narrow context, unlike *smoothing* and *prediction.* Given a certain amount of noisy measurements over a time interval, $T_{min}$ $T_{max}$, and where some transformation from these data has to be estimated for time $T$, *smoothing* refers to the case $T_{min} < T < T_{max}$, *filtering* to $T = T_{max}$, and *prediction* to $T > T_{max}$. For example, in Kalman filtering, past and present measurements are used to estimate current values for positions, velocities, and accelerations from noisy data. These may be positions, bearings, distances or derivatives thereof. When future data are also used, the term *smoothing* applies (e.g., Fioretti & Jetto, 1989). However, in the case of (in)finite impulse response filters or (non)recursive filters, the term *filter* is being used in the general sense. Such filters may estimate past, current, or future data from a batch of measurements. The same applies to Wiener filtering.

Many of the better known time-domain methods work well if the raw data can be assumed to be stationary. This includes the standard case of a low-pass signal with additive, zero-mean, white measurement noise. It is important, however, to realize that this assumption may not be valid for many biomechanical experiments. Events such as heel strike in gait analysis are not stationary but *transient* phenomena, and signal processing methods that do not take this into account may distort the signal sufficiently to render its interpretation doubtful, particularly after differentiation. A proper frequency analysis may reveal to what extent low-pass properties may be assumed for the signal component in measured data.

For standard digital filters, reference is again made to the literature. In biomechanics, the use of spline functions has become popular since the publications of Soudan and Dierckx (1979) and Wood (1982). However, splines are hardly referenced in the digital signal processing literature. This may be because their underlying, analytical basis and their popularity in statistics were rather remote from the more deterministic time-frequency domain approach in the field of engineering-biased signal analysis. Recently, the link between these two fields has become apparent (Craven & Wahba, 1979; cf. Woltring et al., 1987) when the equivalence between the classic Butterworth filter from electrical engineering and spline fitting to noisy data in statistics was noted. Generalization of spline fitting to multivariate data with correlated noise may further reduce the lack of familiarity between these fields (Fessler, 1991). Further links are being made to the dynamic modeling and state-space literature (Dohrmann & Busby, 1990; Kohn & Ansley, 1987, 1989).

## Frequency-Domain Methods

In the previous subsection, reference was made to *Wiener filtering*. Here, smoothing, filtering, prediction, differentiation, and other operations are formally conducted in the frequency domain. However, because of real-time applications in aerospace navigation and gunnery, efforts have been made to approximate these operations in the time domain. If the signal and noise are uncorrelated, with known spectral power densities $S(\omega)$ and $N(\omega)$, respectively, the Wiener filter is the optimal filter that minimizes the expected mean-squared error, and it has a transfer function (e.g., Woltring, 1990)

$$H(\omega) = \frac{S(\omega)}{S(\omega) + N(\omega)} . \qquad (5.12)$$

For example, the optimal, $k$th-order differentiator follows from the product $(j\omega)^k H(\omega)$, where $(j\omega)^k$ is the ideal, $k$th-order differentiator's frequency response, and $j$ the imaginary unit $\sqrt{-1}$. For any type of linear transformation, the ideal transfer function in the frequency domain is weighted with $H(\omega)$ in Equation 5.12. This is a consequence of the chosen optimality criterion; other procedures may have optimality criteria that depend on the order of the derivative sought.

Though the Wiener filter may be optimal from a frequency-domain point of view, it exhibits a number of adverse effects in the time domain. Excessive "ringing" of the filter when the data or noise are nonstationary is one example: If transient signals are modeled by a finite number of low-frequency, stationary signals, the result is likely to become less transient than the original.

Based on the work of Anderssen and Bloomfield (1974), Hatze (1981) described a fast Fourier transform procedure for estimating low-pass filtered signals and their derivatives. The problem for nonstationary data became evident because of the cyclic nature of the FFT-procedure. The raw data were modified by subtracting a linear trend $a_0 + a_1 t$ in such a way that the first and last data points became zero. Then the data were made cyclically continuous in the first derivative by negative mirroring, that is

$$x_i = y_i, \, x_{2n+1-i} = -y_i; \quad i = 1, \ldots, n \qquad (5.13)$$

and smoothing and differentiation were performed on the extended $\{x_1\}$ dataset in the frequency domain, following a Fourier transformation. Unfortunately, this approach requires imposing a second derivative discontinuity at the record boundaries ($i = 1$ and $i = 2n$) if the true second derivative is nonzero at these points. Low-pass filtering such data in the frequency domain resulted in considerable ringing in the second derivative of the signal, not only at the record boundaries but throughout the data range (Woltring, 1985). Similar problems occur if there are genuine transients within the data record. Alternative methods that avoid such artifacts should be used under these conditions.

Again, handbooks in the signal processing field provide many frequency-domain methods for the design of low-pass filters and differentiators, and software packages implementing them are available in the public literature and in the commercial domain. Also, there are methods in which a mixture of time- and frequency-domain criteria can be chosen.

## OPTIMAL DATA PROCESSING

Many standard software toolboxes provide the user with almost complete control over the desired time or frequency responses. Then the problem becomes how to choose such responses. This brings us to the problem of optimally estimating desired filter responses from measured data. A classic example is the Wiener filter discussed previously, where the signal and noise spectra were estimated from the data or assumed, in some fashion. Recently, optimal state-space methods have been investigated, especially in control engineering and in aerospace navigation.

Typically, generic filter responses are defined and a number of parameters must be estimated from the data. For example, one may decide to use a symmetric, forward-reverse Butterworth filter of the $m$th order, with transfer function

$$H_\alpha(\omega) = \frac{1}{1 + (\omega/\omega_\alpha)^{2m}} \tag{5.14}$$

and try to find a suitable cut-off frequency $\omega_\alpha$ from the data. Because the Butterworth filter is a so-called infinite impulse response (IIR) filter, usually implemented in a recursive fashion for direct processing in the time domain, its initial state (at the record boundaries) is important. This initial state may adversely influence the values of numerical derivatives subsequently obtained from the smoothed data. Thus, one might prefer finite impulse response (FIR) filters, whose initial and derivative behavior is better controlled, and estimate their length and frequency properties from the data (cf. D'Amico & Ferrigno, 1992). Alternatively, one may resort to splines of sufficiently high order, that is, the order of the selected spline should be higher than that of the highest derivative sought (Woltring, 1985). Then the regularization bias will have little effect on the derivatives of interest.

Hatze (1981, 1990), Woltring (1985, 1986, 1990), and Woltring et al. (1987) have described optimal methods for estimating smoothed data and their derivatives from noisy data records. Hatze applied such a method to an FFT-model, and Woltring (1985, 1986) and Dohrmann and Busby (1990) described such methods for splines. Recently, this work was generalized to correlated noise in vector measurements and published in the digital signal processing literature (Fessler, 1991). Busby and Trujillo (1985) and Kohn and Ansley (1987, 1989) have proposed state-space, dynamic programming and Kalman-filter-type approaches, which may be numerically more reliable than the finite difference B-splines of Lyche, Schumaker, and Sepehrnoori (1983) upon which Woltring (1986) based his method. This is especially true if the sampling interval in a measurement record is not constant.

Optimization presupposes the definition of realistic criteria to be optimized; these, in turn, imply certain generic assumptions about the data (signal *and* noise) to be processed by the optimal filters. Because models are always imperfect representations of reality, the question must be posed—How robust is the method if the model assumptions are not met? For example, low-frequency signals with additive, white noise are often assumed; what happens if the data contain nonwhite noise or high-frequency signals? For that matter, what is the significance of rigid-body-based estimates of joint moments and forces, considering the nonrigid nature of the human motor apparatus? (One example will be discussed in the next section). In essence, many of these methods can be reduced to the problem of optimally estimating regularization parameters such as $\alpha$ in Equation 5.11.

## Generalized Cross-validation

In the work of Craven and Wahba (1979), one approach to optimally determining $\alpha$ from the data is *generalized cross-validation* (GCV). In essence, this approach involves estimating a fitting function based on all data points, except the $k$th data point, for some given value of $\alpha$. The prediction error $\varepsilon_k$ at this point is then used to determine a root-mean-squared (RMS) fitting error over all $k$, where $k = 1, \ldots, m$. This results in a value of $\alpha$ for which this RMS fitting error is minimized. Thus, GCV implements the best prediction in the RMS sense. In contrast to ordinary cross-validation (CV, or OCV), GCV uses weight factors to accommodate unequal sampling intervals in the independent variable.

Interestingly, the topics of splines in statistics and of digital filtering in (electrical) engineering have approached each other in this area. It appears that the frequency-domain characteristics of regularized, *harmonic* splines (in which begin- and end-values of the spline and of its derivatives are equal) are more or less equivalent to those of the Butterworth filter (Craven & Wahba, 1979) and that the GCV-criterion has a simple frequency-domain interpretation. Other splines, such as *natural* and *complete* splines with different boundary conditions, are still very similar to the Butterworth filter (Woltring et al., 1987).

In essence, the GCV-function is determined by the mean residual error and by the effective stopband-width of the filter for sampled data. If $H_\alpha(\omega)$ is the filter's frequency response, and $X(\omega)$ the input spectrum, then the output spectrum $Y_\alpha(\omega)$ and the error spectrum $E_\alpha(\omega)$ follow as

$$Y_\alpha(\omega) = H_\alpha(\omega)\, X(\omega),$$

$$E_\alpha(\omega) = \{H_\alpha(\omega) - 1\}X(\omega).$$

(5.15)

The mean residual power per unit frequency in the stopband $\omega_{\text{stop}}$ follows as

$$\sigma_\alpha^2 = \frac{1}{\omega^{\text{stop}}} \int_0^{\pi/\tau} |E(\omega)|^2\, d\omega.$$

(5.16)

If the filter is an ideal low-pass filter, $H_\alpha = 1$ up to the filter's cut-off frequency $\omega_\alpha$, and 0 for higher frequencies, then $\omega_{\text{stop}} = \pi/\tau - \omega_\alpha$, and the mean residual power in the stopband follows as

$$\sigma_\alpha^2 = \frac{1}{(\pi/\tau - \omega_\alpha)} \int_{\omega_\alpha}^{\pi/\tau} |X(\omega)|^2\, d\omega,$$

(5.17)

where $\tau$ is the sampling interval. For white noise, $X(\omega)$ is essentially flat for any $\omega_\alpha$ beyond the signal's maximum frequency, so $\sigma^2$ will be constant, too. The GCV-function is now obtained by dividing once more by $\omega_{\text{stop}}$, and the optimal value follows by choosing that particular value for $\omega_\alpha$ at which the GCV-function is minimal.

For the ideal, digital low-pass filter, the effective number of filter parameters is simply the number of frequency components passed by the filtering procedure and thus a linear function of the stopband-width. In a digital filtering context, the number of different frequencies in the Nyquist band is finite. For practical, linear filters with a finite transition bandwidth between $H(\omega) \approx 1$ and $H(\omega) \approx 0$, this notion can be generalized by weighing each frequency component. It appears that the *Trace* of the low-pass filtering matrix $A_\alpha$ (see below) in the linear transformation

$$\vec{y}_\alpha = A_\alpha \vec{x}, \tag{5.18}$$

between the raw input data $\vec{x}$ and the smoothed output data $\vec{y}_\alpha$, can be used for that purpose (cf. Craven & Wahba, 1979). For the ideal filter, $n_\alpha = Trace\ \{A_\alpha\}$ is equal to the number of frequency components, and the stopband-width is proportional to $n - n_\alpha$. Because the total residual power can also be assessed in the time domain as the sum of squared differences between the raw and smoothed data, the residual variance per unit frequency and the GCV-function can be similarly assessed, without the need for a frequency-domain transformation.

For a truly low-pass signal and additive, wide-band, uncorrelated noise (i.e., a straight, horizontal line in the high-frequency range beyond the low-pass signal band), the result is similar to that shown in Figure 5.5a. Here, the raw data spectrum depicted by the continuous line is a function of frequency, whereas the residual standard deviation per unit frequency and the GCV-score, depicted by the dashed and dotted lines, respectively, are functions of the filter's *cut-off* frequency up to the Nyquist frequency. Because the high-frequency data contain white noise with a constant spectrum, the residual standard deviation coincides with the data. The GCV-function, by virtue of division by the stopband-width, goes to infinity for increasing

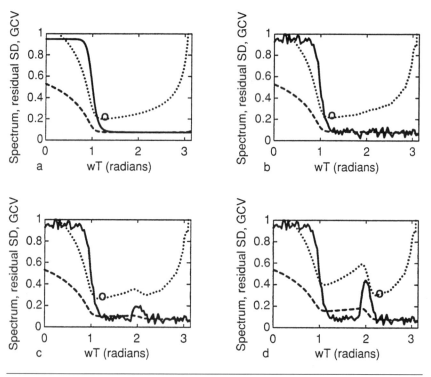

**Figure 5.5**  Optimal, GCV-based low-pass smoothing for an ideal low-pass filter, low-pass signal, and different noise types: (a) ideal data; (b) no hf peak; (c) weak hf peak; (d) strong hf peak. The symbol "o" indicates the absolute minimum of the GCV-function.

cut-off frequency. For decreasing cut-off frequency, the s.d. function starts to deviate from the signal spectrum as this spectrum begins to change with frequency.

Because real signals obtained in 3-D movement analysis will not perfectly meet the model assumptions, the question should be posed: How well does this procedure work in practice? This is illustrated in Figure 5.5, b-d, for the same signal spectrum and ideal low-pass filter, but for different noise types. In Figure 5.5b, the noise is still rather white, but in Figures 5.5c and 5.5d, a stronger peak has been added in the high-frequency range. For the former case, this peak is sufficiently small to be detected as noise by the algorithm, but this is not so in the latter case. However, the filter does not distinguish between what is signal and what is noise; ultimately, it is the filter user's interpretation of the data that should make the final choice.

## Spline Package GCVSPL

Some of these problems can be demonstrated and resolved using the GCVSPL-package (Woltring, 1986). This package finds local and global minima in the GCV-function by admitting user-defined starting values in the search for the optimal regularization parameter $\alpha$. This package can be retrieved by electronic mail by sending the request, *send gcvspl from gcv*, to NETLIB@RESEARCH.ATT.COM or to NETLIB@ORNL.GOV. For further details, include the request, *send index from gcv*, in the next line.

The package accommodates previously selected, even spline orders $2m$ (i.e., odd spline degrees $2m - 1$). It can calculate linear ($m = 1$), cubic ($m = 3$), quintic ($m = 5$), heptic ($m = 7$), and even nonic splines ($m = 9$), which, for double precision arithmetic, is usually the highest order that can be safely accommodated. It is advisable to select $m$ higher than the highest derivative being sought, in order to avoid boundary artifacts caused by the natural spline's property of imposing zero derivatives at the record boundaries from the $m$th up to the $(2m - 2)$ derivative. In this fashion, the regularization bias imposed on the derivatives of interest will be small.

For periodic, equidistantly sampled splines, the equivalence with the double Butterworth filter (Equation 5.14) can be demonstrated via a Fourier transformation and a variational argument. If the spline is defined as that linear transformation $h(\cdot)$, with Fourier transform $H(\omega)$, that minimizes the following cost function, for fixed regularization parameter $\alpha \geq 0$, such that

$$\frac{1}{n} \sum_{i=1}^{n} \mid h(x_i) - x_i \mid^2 + \alpha \int \mid \frac{\partial^m h[x(t)]}{\partial t^m} \mid^2 dt , \; x_i = x(t_i) . \qquad (5.19)$$

Equation 5.14 can be derived as the optimal transfer function $H_\alpha(\omega)$, with

$$\omega_\alpha = (n\tau\alpha)^{-1/2m}. \qquad (5.20)$$

Although this relation is strictly valid for harmonic splines only, the GCVSPL natural spline has been found to behave in approximately the same manner unless the amount of smoothing is very high. (In that case, periodicity does not apply because the spline then approximates a single polynomial over the whole data range.) Thus, the GCVSPL spline does not behave as the ideal low-pass filter of the preceding subsection. However, an effective low-pass bandwidth can be derived from $Trace\{A_\alpha\}$, and the effective stopband-width follows as proportional to $Trace\{I - A_\alpha\}$. As shown by Woltring (1986), $A_\alpha$ can be derived as

$$A_\alpha = B(B + \alpha E)^{-1}, \tag{5.21}$$

where the $m*m$ matrices $B$ and $E$ are band limited and related to the matrices $A$, $N$, and $C$ in Equation 5.11. $N$ is diagonal if the noise is uncorrelated. Because of the band-limited nature of these matrices, calculation of $\bar{y}_\alpha$ in Equation 5.18 and of $Trace\{A_\alpha\}$ is numerically efficient and does not require explicit matrix inversion in Equation 5.21. The residual standard deviation in the time domain (which is proportional to the residual standard deviation per unit frequency in the frequency domain) and the GCV-function follow simply as

$$\sigma_\alpha^2 = \frac{|\{I - A_\alpha\}\bar{x}|^2}{Trace\{I - A_\alpha\}} \tag{5.22}$$

and

$$\text{GVC}_\alpha = \frac{|\{I - A_\alpha\}\bar{x}|^2}{|Trace\{I - A_\alpha\}|^2}. \tag{5.23}$$

The similarity of Equations 5.7, 5.8, and 5.22, with $Trace\{I\} = n$ suggests that $Trace\{I - A_\alpha\}$ should be interpreted as the *effective number of degrees of freedom* of the smoothing problem, and $Trace\{A_\alpha\}$ as the *effective number of parameters*, $n_\alpha$, even though the total number, $m$, of spline parameters being estimated is approximately equal to the number of data points, $n$, for any value of $\alpha$.

The significance of this interpretation can be demonstrated for the limiting case $\alpha \to \infty$ of GCVSPL's smoothing, natural splines. The spline then reduces to a polynomial of order $m$ (i.e., of degree $m - 1$), and $Trace\{A_\alpha\}$ has been empirically found to reach this value. Because of this interpretation of $Trace\{A_\alpha\}$, it is one of the control parameters used for optimization in the GCVSPL-package.

In order to find multiple optima in the GCV-function, it may be useful to plot both the residual standard deviation and the GCV-score as a function of $\alpha$. The spline's cut-off frequency $\omega_\alpha$ in the equivalent Butterworth filter is approximately related to $\alpha$ by the relationship given in Equation 5.19, where $m$ is the order of the spline and of the equivalent Butterworth filters. Therefore, it is advisable to assess these functions with $\alpha$ varied logarithmically.

Experimentally, the following linear relation has been found between $Trace\{A_\alpha\}$ and $\omega_\alpha$,

$$Trace\{A_\alpha\} \approx n/2 + mk_n\omega_\alpha\tau \tag{5.24}$$

and

$$k_m = \int_0^\infty \frac{dx}{1 + x^{2m}} \downarrow \frac{1}{\pi} \text{ for } m \to \infty. \tag{5.25}$$

This relation has been found to hold except for extremely high values of the regularization parameter $\alpha$. Then the spline degenerates to a pure polynomial (see the $BW_{fre}$ graph in Figure 5.6). This means that the natural spline needs approximately $m/2$ samples to accommodate boundary effects. For the remaining points, the natural spline behaves approximately like a periodic spline up to the $(m - 1)$ derivative. The boundary artifacts will propagate throughout the higher derivatives of the data records if the higher derivatives of the signal are non-zero at the record boundaries (Woltring, 1985).

When the spline is calculated for a value of $\alpha$ chosen by the user, the GCVSPL-package is used in a nonrecursive fashion. Alternatively, the package can be used in a recursive mode, where the optimal values or a required value of $Trace\{A_\alpha\}$ are found automatically. This package includes a number of other options, such as processing vectorial and nonequidistantly sampled data and allowing weight factors, if some knowledge about relative noise distributions is available.

The question to be posed is, how optimal are any derivatives obtained from such smoothed data? The spline procedure is a compromise between fitting at

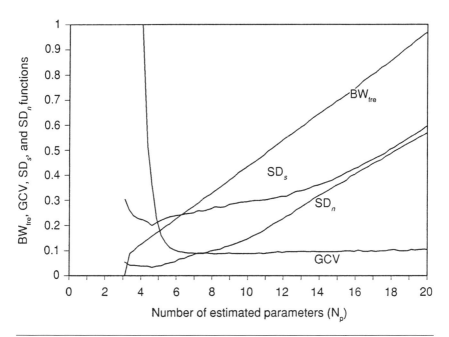

**Figure 5.6**  Smoothing results for the instantaneous helical axis from noisy landmark data (Woltring et al., 1987). The position errors $SD_s$ and direction errors $SD_n$ of the IHA, the equivalent Butterworth filter's cut-off frequency $BW_{fre}$, and the GCV-function are plotted as a function of the effective number $n_\alpha = Trace\ \{A_\alpha\}$ of estimated spline parameters.

the level of the data proper (0-th derivative) and biasing in the *m*th derivative. Hatze (1981) has proposed using different smoothing criteria for each required derivative, but the spline packages do not immediately accommodate this. Indeed, inspection of the results of Woltring et al. (1987) and Lange et al. (1990) indicates that the amount of smoothing was less than optimally required (see Figure 5.6). In these studies, the instantaneous helical axis (IHA) was estimated for a movement about a fixed axis. Therefore, the dispersion of the estimated IHA could be used as an overall criterion for the quality of the data acquisition and processing chain. Woltring et al. (1987) showed that the IHA is a first-derivative entity, involving the position and velocity of a reference point on a moving body and the rotation velocity vector of that body. Though there was impressive improvement due to GCV-optimal smoothing of the noisy landmark position data, stronger smoothing would, in fact, have reduced the amount of noise in the estimated IHA data by a factor 2 to 3 (see the $SD_s$ and $SD_n$ graphs in Figure 5.6 for the position and direction dispersions, respectively,

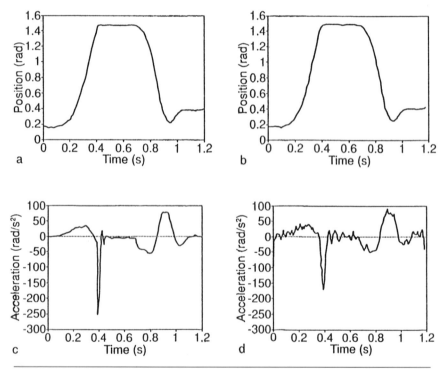

**Figure 5.7**   (a and b) Position data and (c and d) acceleration data derived from Dowling (1985). Both graphs on the left are from unprocessed recordings. The upper-right graph shows the position data smoothed by GCV-regularized natural splines (3rd, 5th, or 7th order). The lower-right graph shows the second derivative of the smoothed position data.
*Note.* From ''Model and Measurement Error Influences in Data Processing'' by H. Woltring. In *Biomechanics of Human Movement: Applications to Ergonomics, Sports, and Rehabilitation* (pp. 203-237) by A. Cappozzo & N. Berme (Eds.), 1990, Worthington, OH: Bertec. Copyright 1990 by Bertec Corporation. Adapted by permission.

of the IHA). This finding demonstrates that generalized cross-validation can be quite conservative, rejecting too little noise rather than too much signal (if the number of samples is sufficiently large—180, in this case).

The extent to which optimizing each derivative will substantially improve the smoothing and differentiation process merits further investigation. One example is given in Woltring (1990), where data containing a strong transient in the second derivative were processed with natural GCV-splines of different orders. In Figure 5.7, a-d, measured position and acceleration data are compared with natural GCV-spline estimates. Significant ringing is apparent in the estimated second derivatives. This results from imposing a low-pass filter on an essentially wide-band signal. Even though the signal's second derivatives vanished at the record boundaries, the cubic GCV-spline showed a strong high-frequency component not seen in the higher-order GCV-splines. Possibly, this was a noise effect at the record boundaries.

## ACKNOWLEDGMENTS

This work was supported under the CAMARC project (for Computer-Aided Movement Analysis in a Rehabilitation Context), within the *Advanced Informatics in Medicine* (AIM) action of the Commission of the European Communities (1989-1991). The author is grateful to Philips Medical Systems Nederland B.V. for granting leave of absence during that period and to the Universities of Nijmegen and Eindhoven, The Netherlands, for their hospitality. The Mathworks, Inc. of South Naticks, MA, consented to the incidental use of an AT-Matlab review copy for generating most of the illustrations in this chapter.

## REFERENCES

Anderssen, R.S. & Bloomfield, P. (1974). Numerical differentiation procedures for nonexact data. *Numerische Mathematik, 22*, 157-182.

Busby, H.R., & Trujillo, D.M. (1985). Numerical experiments with a new differentiating filter. *Journal of Biomechanical Engineering, 107*, 293-299.

Cappozzo, A., & Berme, N. (Eds.) (1986). *Biomechanics of human movement—Applications to ergonomics, sports and rehabilitation.* (Proceedings of a Study Institute and Summer Conference on Biomechanics of Human Movement at Formia, Italy). Worthington, OH: Bertec.

Craven, P., & Wahba, G. (1979). Smoothing noisy data with spline functions. *Numerische Mathematik, 31*, 377-403.

D'Amico, M., & Ferrigno, G. (1992). Comparison between the most recent techniques for smoothing and derivative assessment in biomechanics. *Medical and Biological Engineering and Computing, 30*, 193-204.

D'Amico, M., & Ferrigno, G. (1990). A technique for the evaluation of derivatives from noisy biomechanical displacement data by a model-based bandwidth-selection procedure. *Medical and Biological Engineering and Computing, 28*, 407-415.

Dohrmann, C.R., & Busby, H.R. (1990). A dynamic programming approach to smoothing and differentiating noisy data with spline functions. In A. Cappozzo & N. Berme (Eds.), *Biomechanics of human movement—Applications to ergonomics, sports, and rehabilitation* (pp. 248-262). Worthington, OH: Bertec.

Dowling, J.J. (1985). A modeling strategy for the smoothing of biomechanical data. In B. Johnsson (Ed.), *Biomechanics X-B* (pp. 1163-1167). Champaign, IL: Human Kinetics.

Eykhoff, P. (1974). *System identification—Parameter and state estimation.* London: Wiley.

Fessler, J.A. (1991). Nonparametric fixed-interval smoothing with splines. *IEEE Transactions on Signal Processing,* **39**(4), 852-859.

Fioretti, S., & Jetto, L. (1989). Accurate derivative estimation from noisy data: A state-space approach. *International Journal of Systems Science* **20**(1), 33-52.

Furnée, E.H. (1989). *TV-computer motion analysis systems—The first two decades.* Doctoral thesis, Delft University of Technology, The Netherlands. ISBN 90-9003095-6.

Furnée, E.H. (1990). PRIMAS: Real-time image-based motion measurement system. In J.S. Walton (Ed.), *Proceedings Mini-Symposium on Image-Based Motion Measurement,* **1356**, (pp. 56-62). First World Congress of Biomechanics, San Diego, CA. Bellingham, WA: Society of Photo-Optical Engineers.

Gustafsson, L. (1977). *ENOCH—An integrated system for measurement and analysis of human gait.* Doctoral thesis, Teknikum, Uppsala University, UPTEC 77 23R, Uppsala, Sweden.

Hatze, H. (1981). The use of optimally regularized Fourier series for estimating higher order derivatives of noisy biomechanical data. *Journal of Biomechanics,* **14**(1), 13-18.

Hatze, H. (1990). Data conditioning and differentiation techniques. In A. Cappozzo & N. Berme (Eds.), *Biomechanics of human movement—Applications to ergonomics, sports, and rehabilitation* (pp. 237-248). Worthington, OH: Bertec.

Jennings, L.S., & Wood, G.A. (1988). Co-joint data smoothing using splines. In A.E. Goodship & L.E. Lanyon (Eds.), *ESB 1988—European Biomechanics* (Proc. 6th Meeting of the European Society of Biomechanics). London: Butterworths.

Kohn, R., & Ansley, C.F. (1987). A new algorithm for spline smoothing based on smoothing a stochastic process. *SIAM Journal on Scientific and Statistical Computing,* **8**(1), 33-48.

Kohn, R., & Ansley, C.F. (1989). A fast algorithm for signal extraction, influence, and cross-validation in state-space models. *Biometrika,* **76**(1), 65-79.

Ladin, Z., & Wu, G. (1991). Combining position and acceleration measurements for joint force estimation. *Journal of Biomechanics,* **24**(12), 1173-1187.

Lange, A. de, Huiskes, R., & Kauer, J.M.G. (1990). Effects of data smoothing on the reconstruction of helical axis parameters in human joint kinematics. *Journal of Biomechanical Engineering,* **112**, 107-113.

Lanshammar, H. (1982a). On practical evaluation of differentiation techniques for human gait analysis. *Journal of Biomechanics,* **15**(2), 99-105.

Lanshammar, H. (1982b). On precision limits for derivatives numerically calculated from noisy data. *Journal of Biomechanics,* **15**, 459-470.

Lyche, T., Schumaker, L.L., & Sepehrnoori, K. (1983). FORTRAN subroutines for computing smoothing and interpolating natural spline. *Advances in Engineering Software,* **5**, 2-5.

Oppenheim, A.V., & Schafer, R.W. (1989). *Discrete time signal processing.* Englewood Cliffs, NJ: Prentice Hall.

Söderström, T., & Stoica, P. (1989). *System identification.* London: Prentice Hall International (UK).

Soudan, K., & Dierckx, P. (1979). Calculation of derivatives and Fourier coefficients of human motion data, while using spline functions. *Journal of Biomechanics,* **12**(1), 21-26.

Teulings, H-L. & Maarse, F.J. (1984). Digital recording and processing of handwriting movement. *Human Movement Science,* **3**(1/2), 193-217.

Vaughan, C.L. (1982). Smoothing and differentiation of displacement-time data. An application of splines and digital filtering. *International Journal of Biomedical Computing,* **13**, 345-386.

Woltring, H.J. (1984). On methodology in the study of human movement. In H. Whiting (Ed.), *Human motor actions—Bernstein reassessed* (pp. 35-73). Amsterdam: North-Holland.

Woltring, H.J. (1985). On optimal smoothing and derivative estimation from noisy displacement data in biomechanics. *Human Movement Science,* **4**(3), 229-245.

Woltring, H.J. (1986). A FORTRAN package for generalized cross-validatory spline smoothing and differentiation. *Advances in Engineering Software,* **8**(2), 104-113.

Woltring, H.J. (1990). Model and measurement error influences in data processing (pp. 203-237). In A. Cappozzo & N. Berme (Eds.), *Biomechanics of human movement—Applications to ergonomics, sports, and rehabilitation* (pp. 203-237). Worthington, OH: Bertec.

Woltring, H.J. (1992). One hundred years of photogrammetry in biolocomotion. In A. Cappozzo, M. Marchetti, & V. Tosi (Eds.), *Biolocomotion: A century of research using moving pictures* (pp. 199-225). Rome: Promograph.

Woltring, H.J., de Lange, A., Kauer, J.M.G., & Huiskes, R. (1987). Instantaneous helical axis estimation via natural cross-validated splines. In G. Bergmann, A. Kölbel, & A. Rohlmann (Eds.), *Biomechanics: Basic and applied research* (pp. 121-128). Dordrecht: Martinus Nijhoff.

Wood, G.A. (1982). Data smoothing and differentiation procedures in biomechanics. *Exercise and Sport Sciences Review,* **10**, 308-362.

---

This chapter was written shortly before Dr. Woltring died on November 3, 1992, in an automobile accident.

# Chapter 6

# Computer Graphics for Visualization and Animation

*T.W. Calvert*
**Armin Bruderlin**

This overview of the use of computer graphics systems in human movement studies includes an introduction to computer graphics as a tool for visualization, a description of the hardware and software of high-performance graphics workstations, and a discussion of specific techniques for representing and displaying the human body and for animating human movement. A number of excellent introductory texts on computer graphics are available with more information on the topics introduced here (see Foley, van Dam, Feiner, & Hughes, 1990; Magnenat-Thalmann & Thalmann, 1985; Newman & Sproull, 1979; Rogers, 1985).

The purpose of any computer graphics system is to allow visualization. This can be as simple as plotting data in the form of a graph or as complex as the realistic animation of human figures. Whatever the application, a satisfactory graphics system must achieve effective communication and supply the user with sufficient information to assess results effectively and quickly. In most cases the system should also allow input, so that the user can interact with the data and the display in real time.

Animation is a particular form of visualization in which a series of images of a changing situation are displayed rapidly, one after another, to give the user the illusion of movement. This is exactly the same process used in film or television. However, in computer animation the moving objects are frequently the result of computation. Thus, the calculations must be performed quickly if the frames are to be displayed at a speed fast enough to avoid flicker (in North America, standard display rates are 24 frames/s for film and 30 frames/s for video).

# COMPUTER GRAPHICS HARDWARE
# AND SOFTWARE

Any computer system is capable of producing some kind of graphic output. This can be patterns in alphanumeric output on a screen or printer, or of dots (pixels) on a screen or printer (dot matrix), the output of a plotter, or images on a high-quality display screen. Generally, computer graphics implies images of reasonable definition and resolution on a screen or printer. These can be produced by the smallest personal computer or the largest supercomputer. A typical graphics workstation is shown in Figure 6.1, and Figure 6.2 is a diagram of the functional components.

The most common output device is the screen of the cathode-ray tube (CRT). Like television, a beam of electrons causes a moving spot of light to be produced by the phosphor on the surface of the screen. In vector graphic systems, the beam of electrons traces out the lines of the shape being drawn in the same way that signals are displayed on an oscilloscope. Though these systems give good performance when displaying line drawings, they are becoming less common. Today, most systems use a raster display similar to television, in which the electron beam traces a pattern of horizontal scan lines on the screen and the intensity is modulated to produce an image. The resolution of a display is determined by the number of lines and the number of distinct points (or *pixels*) that can be drawn along each line (Figure 6.3). Common definitions include 320 × 240, 640 × 480, or 1,024 × 768 points × lines.

**Figure 6.1**   A typical high-performance graphics workstation.
*Note*. Photo courtesy of Silicon Graphics, Inc.

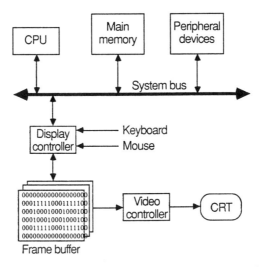

**Figure 6.2**    The functional components of a graphics workstation.

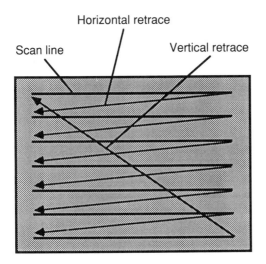

**Figure 6.3**    A raster display.

A simple form of black-and-white display uses a "bit map" in which each bit turns a corresponding pixel on or off on the screen depending on the state (1 or 0) of the bit in memory. Thus, for the Apple Macintosh Classic, the bit-mapped display with a resolution of 320 × 240 requires 320 × 240 or 76,800 bits of memory, or 9,600 bytes. The memory set aside to store the displayed image is frequently called the *frame buffer*. It is defined by its resolution in pixels and the depth in bits/pixel. The simple monochrome bit-mapped display has 1 bit per pixel.

With 2 bits/pixel, $2^2$, or 4, levels of gray can be defined. Because of the limited resolution of the human eye, it is generally agreed that 8 bits/pixel (with $2^8$, or 256, levels of gray) is a useful maximum. Color is usually generated by a display tube with three primary color phosphors on the screen: red, green, and blue (RGB). The intensity of each color is controlled independently from a minimum of 1 bit/color up to a useful maximum of 8 bits/color in the frame buffer. Thus a full-color frame buffer would have $3 \times 8$, or 24, bits/pixel.

The simplest image is made up of lines or vectors. For example, we can represent a box with a line drawing, as shown in Figure 6.4a. This representation is incomplete because it only shows the edges of the object, but it can be drawn very quickly. This is important for real-time interaction. A hidden-line removal algorithm can be used to produce a more realistic image (Figure 6.4b), but the additional computation slows down the display. Figure 6.5a shows a vector image of two human bodies without hidden-line removal, and Figure 6.5b shows the same bodies with hidden lines removed. Note that line drawings of complex objects can also be improved by the use of color.

Objects are more realistically displayed and easier to recognize if they are represented by surfaces rather than just the lines of their edges. The display of the surface of an object depends on its color, surface characteristics (shiny, matte, etc.) and illumination. Algorithms have been developed that give reasonably realistic representations if the color, surface characteristics, and illumination are known. Generally, the computing time needed to produce the image is proportional to the quality of the representation.

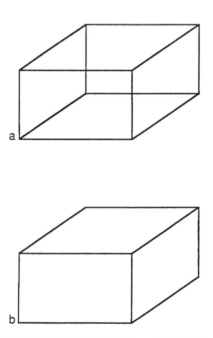

**Figure 6.4**    (a) Line drawing of a box, and (b) the box with hidden lines removed.

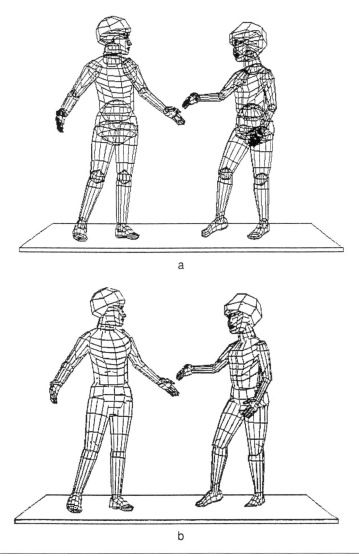

**Figure 6.5**  (a) A vector image of the human body with hidden lines visible, and (b) the same image with hidden lines removed. (Produced using Paracomp *Swivel 3D* on a Macintosh II.)

The simplest display of surfaces is flat-shaded, as illustrated in Figure 6.6a. In this case, no account is taken of the illumination or the surface characteristics; each surface has uniform shade and illumination. A shaded version of the same image is shown in Figure 6.6b. Here the surfaces have differential illumination, depending on the orientation of the surface relative to the light sources, and specular reflections (highlights) are added to shiny surfaces. Figure 6.6c shows the effect of adding shadows, which further increases realism.

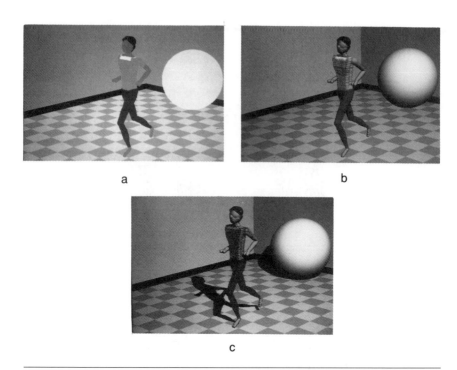

a                                                    b

c

**Figure 6.6**   (a) A flat, shaded image of a human figure, (b) shading that takes account of light source and body orientation, and (c) shading that also takes account of shadows and specular reflection (highlights).

## Graphics Workstations

The high-performance graphics workstations now available from a number of manufacturers have evolved certain common design features. In this section, the generic high-performance graphics workstation is described and compared to low-end workstations based on personal computers. Individual products may have particular features, such as special purpose hardware to improve performance.

## Hardware Characteristics

### Resolution of the Image

Generic workstations have a screen resolution of at least 768 vertical by 1,024 horizontal and often 1,024 by 1,280, or more. The underlying frame buffer is typically at least 1,024 × 1,024. The display is usually refreshed at least 30 times/s. If the display is interlaced, like television, then the odd-numbered lines of each frame are drawn in the first 1/60th of a second and the even lines in the second 1/60th. This gives less flicker than a simple 30-frames/s display but is not as good as a 60-frames/s, noninterlaced display. Some higher quality systems have a frame rate higher than 60 frames/s.

PC-based workstations have resolutions that start at $320 \times 240$ (black-and-white Macintosh) or $640 \times 480$ (4 bits/pixel VGA on IBM-PC family). Most systems can accept special plug-in cards that increase the resolution to that of the high-performance workstations; however, writing speed with such cards can be very slow.

## Color and Gray Levels—Bits/pixel

The minimum useful number of bits/pixel for a generic workstation is 8; this allows the display of up to 256 distinct color shades or gray levels. A full-color frame buffer normally has 8 bits/color or 24 bits/pixel. Because computer memory has become cheaper, users often request additional storage for each pixel. Smoother animation is possible if double-buffering is used; this involves splitting the frame buffer into two equivalent sections. The first stores the image currently being displayed while the next image to be displayed is built up in the second section. When the new image is complete, the roles are reversed, the new image is displayed, and computation starts on the next. Other reasons to have additional memory associated with each pixel include the use of additional bit-planes for overlays to the main display and the use of frame-buffer memory for z-buffering (this is a technique for fast, hidden surface removal; Foley et al., 1990).

PC-based workstations can produce useful black-and-white graphics with 1 bit/pixel. Though add-on cards can support at least 24 bits/pixel, a typical maximum is 8 bits/pixel.

## Drawing Speed—Pixels/s and Polygons/s

The speed at which an image can be drawn is generally measured either in pixels/second, vectors/second, or polygons/second. The latter is most useful because rendering software normally depends on the surfaces of objects in the image being broken down into many small polygons. Typical numbers for a generic workstation are 200,000 vectors/s and 24,000 polygons/s, whereas for a PC-based workstation the numbers would be one order of magnitude less.

## Computing Speed

Though the speed of the specialized graphic functions is important, the overall performance depends on having a fast central processor. In 1992, competitive low-end generic workstations claim to perform 1 to 2 MIPS (million instructions per second), whereas high-end systems claim 10 to 20 MIPS. This does not tell the whole story, however, because the effective speed depends on how many simple instructions the computer uses to perform complex operations. Another useful measure of speed is megaFlops, or millions of floating point operations per second (MFlops). Low-end generic workstations claim 0.5 MFlops, whereas high-end systems claim 1.5 MFlops. PC-based workstations can approach the speeds of the low-end generic systems.

## Memory and Disk

The total workstation memory (excluding the frame buffer) limits system performance in several ways. Current generic systems range from 8 Mb to 64 Mb (megabytes). A workstation's disk is typically at least 350 Mb, but in a networked environment, a much larger capacity (e.g., 2,000 Mb) may be available on a fileserver. PC-based workstations typically have from 1.0 Mb to 4 Mb of memory and a disk of 20 Mb to 200 Mb.

## Network Capability

High-performance generic workstations are designed to be connected to a high-capacity local area network, such as an ethernet. This network, which is typically based on a single coaxial cable or a shielded twisted pair, allows the workstation to communicate with other workstations at up to 10 Mb/s. One or more workstations on the network may be configured as fileservers, which are systems with very large disks. The programs, data, and images of each user can then be stored on a fileserver so that the workstation disk can be devoted to systems software and the programs, data, and images currently being used. Other facilities on a network can include powerful workstations used as "compute servers," printers, plotters, and specialized input devices, such as scanners.

Modern network software is designed to be transparent to the user, so that no matter which workstation on the network is being used, the same environment is presented. Networks can also be bridged to other networks in adjacent locations, across the continent, or around the world. The interaction capabilities are only limited by the speed of the links. With national networks now available with a 2-Mb/s transmission rate, it becomes feasible for researchers to interact in interesting ways with colleagues at other locations.

PC-based workstations can also be connected to networks. Typical transmission speeds are one order of magnitude less than for the ethernet. The Appletalk network built into all Macintosh computers, for example, has a transmission speed of 230.4 Kb/s.

# Operating Systems and Graphics Software

## Operating System

All high-performance workstations now use a time-shared (multitasking, multiprocessing) operating system with powerful networking capabilities. The most common choice is some version of Unix, originally developed at Bell Labs. This system is widely used in the research community and has the advantage of reasonable compatibility between the different versions being used by different suppliers.

A customized operating environment for a workstation provides a window interface and a window manager. This allows the user to open up a number of windows on the screen. Different programs can be run simultaneously, each from its own window. Separate windows can be used for different forms of output (numeric and graphic, for example). The window software is provided as part of the system and enhances the power and flexibility available to the user.

Although time-shared operating systems are available for PC-based workstations, they are less common because the operating system overhead results in poor performance on all except the most powerful systems. Window software is now widely used on these workstations; for example, it comes as an integral part of the Macintosh environment and is available through Microsoft Windows for IBM-PCs that run DOS.

## Standard 2-D Graphics Software

The standard features of software for generating 2-D graphics allow the programmer to build up complex images from simple primitives. The simplest functions allow lines or objects to be drawn on the screen using screen coordinates, possibly with a color or linetype specified. Examples of these functions include

> Point (x,y, color),
>
> Line (x1,y1,x2,y2, color, linetype), and
>
> Polygon (x1,y1,x2,y2, . . . . . . x$n$,y$n$, color).

Curves can usually be specified in much the same way as a polygon, but a spline will be fitted to give a smooth result. Standard functions are provided for circles, ellipses, and their arcs.

## 3-D Graphics Software

For 3-D graphics, the display is defined in the object space rather than the screen space. Objects are specified in terms of their natural three-dimensional coordinates and a transformation is defined to map a projection of the 3-D object to the 2-D screen. The transformation in its simplest form is

$$\mathbf{x}_s = T \cdot \mathbf{x}_b \, , \qquad (6.1)$$

where $\mathbf{x}_s$ is a 2-D vector of the horizontal and vertical coordinates on the screen, $\mathbf{x}_b$ is a 3-D vector of the 3-space coordinates of a point in object space, and $T$ is a transformation matrix that performs the appropriate projection.

If the object is to be displayed as a line drawing, then the graphics functions are essentially the same as for the 2-D screen coordinate system, except that the coordinates are all in 3-D. The transformation $T$ can perform rotation, scaling (zoom) and translation (pan) to provide the required view. If the transformation is driven by a user-controlled input device (e.g., a joy stick or a mouse), the rotation, zoom, and pan operations are performed under direct user control in real time. The clarity of the displayed images can be substantially improved if hidden lines are removed with an appropriate algorithm (this can be done in real time for all except the most complex objects).

If the object is to be displayed using colored and shaded surfaces, then a rendering algorithm must be chosen. The Gouraud shading algorithm provides reasonable shading with highlights. This distinguishes between most common surface finishes. With several light sources, it can provide real-time images of simple objects (e.g., 100 polygons) on low-end workstations and more complex objects on high-end workstations. More accurate rendering requires advanced techniques, such as ray-tracing and radiosity. The ray-tracing algorithm takes account of reflections, transparency, and shadows and provides excellent images of shiny surfaces. Simple objects can be rendered in near-real time (0.5-2 s), whereas complex objects may require many minutes or even hours. The radiosity algorithm handles diffuse light and flat surfaces well, but also requires significant computing time. For most research purposes, algorithms such as Gouraud are quite adequate, and real-time or close to real-time performance is possible (for a more extensive discussion, see Foley et al., 1990).

# Interactive Devices

## The Mouse

This has become the preferred input device for selecting items on the screen, providing analog input, and other functions. The user moves the mouse around on a flat surface and thus can input two dimensions simultaneously. A mouse typically has up to three buttons, which allow the user to signal when an object is selected. Three buttons provide sophisticated control; with any more buttons, the user may forget the meaning of each.

## Other Pointers—Trackballs, Joysticks, Tablets

Trackballs and joysticks are functionally equivalent to the mouse and have the advantage that they do not move around on a surface. Some users prefer them to the mouse, and they are widely available. A tablet is a flat surface with an underlying electrosensitive grid. A drawing or photograph can be placed on the tablet, and a special stylus allows points to be selected. This is useful for digitizing drawings or other images.

## Three-Dimensional Devices

A variety of devices are available to assist in the input of 3-D data. The simplest is a 3-D device for which the position and orientation in space (six parameters in all) can be determined using magnetic fields (e.g., those devices produced by Polhemus Inc., Colchester, VT, or Ascension Technology Corporation, Burlington, VT). Other devices include the following:

- the DataGlove (VPL Research Inc., Redwood City, CA)—This glove uses a Polhemus sensor to signal the position of the wrist while other optical sensors signal the bend in each finger. This allows the user to capture the shape of the hand and gestures made by the hand in real time.
- the BodySuit—Research continues on fully instrumented suits, which will extend the capability of the glove to the whole body. The system is complex, and there are obvious calibration problems.

# Video Versus Graphic Images

Although the technology underlying video (television) and computer displays is essentially the same, there are differences in the standards used, which often make it difficult to interchange images between the two. In the future, the differences may become less significant as the world moves towards new standards for high-definition television and digital video. However, for some time, those who use computer-generated images will have to convert them to the existing standards of NTSC (North America), PAL (Europe, Brazil, etc.), and SECAM (France and the former Soviet Union).

## Characteristics of NTSC Video

The North American standard for video is called NTSC. The original version was developed in the 1930s and was later modified to handle color. Adding a color capability that would be compatible with the existing black-and-white standard resulted in a rather complex system.

The NTSC standard provides 525 horizontal lines of which only 480 are visible. There are 30 frames per second, but these are displayed in two fields of 262.5 lines each; the first field displays the odd lines, and the second the even lines. Thus, in any area of the screen, lines are being rewritten 60 times each second, and this interlacing reduces flicker. The resolution along each horizontal line is equivalent to approximately 600 distinct pixels. However, the average domestic TV set can only display a resolution equivalent to 300 lines and 300 pixels a line.

## Characteristics of PAL and SECAM Video

Although some countries outside North America use NTSC, most use PAL (Europe, except France, and Brazil) or SECAM (France and the former Soviet Union). Both of these systems are based on 625 lines and 50 frames/s. They have slightly different ways of encoding the color signal onto alternate lines.

## Digitizing Film and Video

Digitizing modules are available that can accept a video signal and digitize it in real time. Capturing one frame is called *frame grabbing*. Unless expensive synchronization equipment is available, it is not possible to specify exactly which frame is digitized. The resolution of the digitized image may be less than that of the video, but once the digitized image has been stored, it can be displayed like any other digital image.

Film can be digitized either directly or by first transferring it to video. Direct digitization involves either scanning each frame with a flying-spot scanner or capturing the image with a digital camera. In either case, the resolution of the film is normally higher than that of the digitized result and depends on the equipment used.

## Transferring Images to Video

If a computer graphics system has a video-compatible output (available either as a standard or an option on generic workstations), this output can be fed to further video equipment for display, mixing, or recording. As long as new frames are generated at video rates, animation can be recorded on video tape. Unless expensive time-base correction and synchronization equipment is used, the resulting signals will not meet the standards for broadcast video. However, they may be good enough for research purposes.

Frequently a computer graphics system produces images at less (often much less) than the video rate of 30 frame/s. In this case, each image must be written onto video tape one frame at a time for later display at video rates. This cannot be done with standard videorecorders. Although several methods are available, the simplest is to use a high-quality videorecorder with single-frame capability. Such systems cost up to 10 times as much as a standard recorder.

## Video in a Graphics System

In a high-performance workstation, digitized video can be viewed interchangeably with computer-generated images. The system can be configured so that "live" video is displayed in one of the workstation windows. Video images can be mixed with others and modified. This allows comparison of synthesized results or plots of data with videos of live action.

# VISUALIZATION— REPRESENTING THE HUMAN BODY

For display purposes, an appropriate representation of the body is one that shows the information needed—not necessarily the one that is most realistic. Many different kinds of display are possible for the human body, ranging from a simple stick figure to a fully fleshed-out and clothed figure. Almost all models start with an underlying articulated structure that approximates the human skeleton.

There is no difficulty in devising an articulated structure to represent the human skeleton. The only question is the level of detail required. Figure 6.7 shows the topology of a minimal stick figure with 17 independent limb segments. A more elaborate version with 22 segments is shown in Figure 6.8a (see p. 114). More realistic figures will have fingers, toes, and facial characteristics, but there are no fundamental differences. If the number of segments in the back and neck is increased to approach the number of vertebrae, then smooth flexion of the back is possible, but the large number of segments is difficult to control. This can be simplified if the system fits a spline curve to the segments or if the vertebrae are grouped so that certain sets can be moved together (e.g., upper back, midback, and lower back).

Though the stick figure represents the skeleton very well, it does not show accurate contact with support surfaces or other figures, and it cannot show the rotation of a limb about its axis. Furthermore, a stick figure frequently produces ambiguous body postures (e.g., inability to distinguish whether an arm is in front of or behind the body).

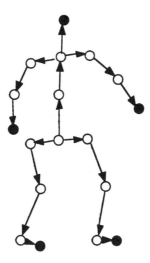

**Figure 6.7**    The topology of a simple stick figure.

## Putting Flesh On the Bones—Solid Models of the Human Body

Standard computer-aided design (CAD) techniques can be used to define solid models for each limb segment. The solid model of a limb segment is built around the skeletal or stick-figure segment for that limb. Although an accurate solid model of the human body can be defined for any given stance, there is no simple way to adjust this solid model as the joints flex, as tissue deforms with acceleration, or as clothing moves. Because of this, many applications represent the individual limb segments by rigid models and make no attempt to model the deformation of tissue at joints. The figures shown in Figure 6.6 were built up from some 800 spheres; using spheres to approximate limb segments describes the joints fairly well. A more advanced approach to handling the joints was proposed by Chung (1987), who interpolated between stored prototypes; later, Forsey (1988) developed a technique for a hierarchical definition of spline-based surfaces that gives good results.

The deformation of tissue due to acceleration presents a further problem, which could in theory be predicted by biomechanical models. Other problems with realistic models involve the need to represent surface characteristics of skin and clothing. Skin is particularly difficult to represent convincingly. Clothing presents its own problems, because if it is not skin tight, it will move independently of the body. For this reason, many displays of human figures assume that the subjects wear bodysuits or other skintight clothing.

Other data-capture systems provide information on the internal structure of the human body. These include X-ray images, axial tomography images (CT scans), magnetic resonance images (MRI), and positron emission tomography (PET) images. The CT, MRI, and PET scans give digital images that show a cross-section of a part of the body (e.g., the arm, the torso, or the head). Each gives a rather different picture of the same physical elements; the CT scan shows structures detected by a conventional X-ray, principally bone and higher density tissues. The MRI image shows the concentration of selected nuclei with certain magnetic resonance properties, whereas the PET image shows structures in which a radioactive substance has concentrated. In all cases, the result of the scan is a series of cross-sections, or slices, of the body segment being scanned. Computer techniques can be used to combine these slices into a solid model, which can then be viewed on a 3-D graphics display. These cross-sections can be used to improve models of limb segments. In addition to the exterior contours, CT scans show how the bones relate to the surface of the body segment involved and MRI scans can show the pathways followed by muscles and tendons.

## Displaying Solid Models of the Human Body

Whatever approach is used to produce a solid model of the human body, many different approaches provide useful displays. These include the following:

- stick figure displays—This is the fastest and simplest display, but it can be difficult to interpret (Figure 6.8a).
- outline displays—Easier to interpret than a simple stickfigure when drawn efficiently, but like the stick figure, it may be ambiguous in some orientations (Figure 6.8b).

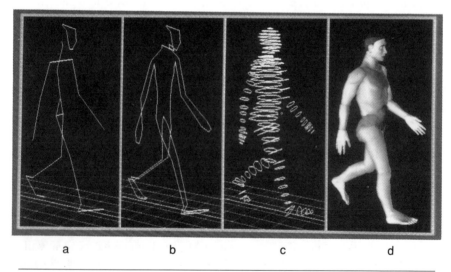

a          b          c          d

**Figure 6.8**   Four examples of different ways to display a human figure: (a) stick figure display, (b) outline figure display, (c) contour figure display, and (d) a fully shaded display of a solid model. This example has over 3,000 polygons and is based on a digitized model of an articulated male body. (Obtained from Datapoint Corporation, Provo, Utah.)

- contour display—This is more realistic and is a useful compromise between clarity of display and efficient drawing (Figure 6.8c).
- simple, shaded displays—Even simple, shaded displays, like the monochrome example shown in Figure 6.8d, require extensive computation compared to simpler figures.

# ANIMATION AS A TOOL
# TO VISUALIZE HUMAN MOVEMENT

Although it has limitations, computer animation of human figures is a very useful tool. Besides the obvious applications in the production of cartoons for entertainment and education, interest is developing in scientific visualization for sports and biomechanics, in simulation of microworlds (Zeltzer, 1989), and in the ergonomic evaluation of workplaces (Phillips, Zhao, & Badler, 1990). Animation is also a medium of artistic expression, and animated dance is a growing artform (Schiphorst, Calvert, Lee, Welman, & Gaudet, 1990; Van Baerle, 1990). The computer animator of human figures needs, among other things, a high-level way to specify natural, goal-oriented, primary and secondary movement. Though photo realism is currently unattainable for clothed and fleshed-out figures, an animation should at least provide movement that is accurate and does not distract the viewer from the main educational or entertainment goal; at best, it should allow the specification of subtleties of artistic expression.

## Animation Techniques

Three-dimensional computer animation can be described as the specification and display of moving objects. Generally, this process consists of object modeling, motion specification, and image rendering. Although all three phases contribute to making a computer-animated sequence, the second is the essence of animation, for without it, the problem reduces simply to generating still images. Whereas the animator specifies the idea of a motion, the computer has to translate it into the actual positions and orientations for each time step (frame). This aspect of an animation is termed *motion control*. In simple animation, where the objects to be moved are independent, rigid bodies like boxes or abstract symbols, motion control is straightforward, and the paths of ''flying logos'' can be fairly readily expressed by simple concatenations of matrix transformations. As the complexity of both the objects and their movements increases, motion control becomes the principal issue. The animation of articulated bodies, such as humans and animals, is especially challenging.

An articulated body is represented by a hierarchical structure of rotational joints, each of which has up to three rotational degrees of freedom. The human body, for example, possesses over 200 degrees of freedom (DOF) and is capable of complex movements that research is still trying to measure, analyze, and represent. The ''DOF problem'' (Zeltzer, 1985) indicates the nontrivial task of coordinating and controlling the limbs of an articulated body to achieve a desired motion out of the vast range of possibilities. For example, consider the very simple skeletal model of a human character in Figure 6.8a. This has 22 segments, and, to specify all of the joint angles and a reference point in space for the body, 69 parameters must be specified for each frame of animation. For 1 min of animation at the North American video rate of 30 frames/s, 124,200 numbers are required to determine the motion of the model. The animation of a realistic human model with "flesh" requires many more parameters; problems such as facial expressions, clothing, and the adjustment of tissue around the joints have yet to be resolved (Calvert, 1988).

A practical motion control system should offer an easy, natural means of specifying a motion and then should generate a realistic animation. Most current animation systems make a trade-off between these two demands. For example, in traditional keyframing (in which gross motion is specified by key frames and intermediate positions are interpolated), the quality of a motion is usually directly proportional to the number of key positions that are specified. Although the computer calculates the in-between frames, the animator is still left to work out a lot of noncreative, tedious detail (i.e., *all* the joint rotations) at each key frame. If the desired movements are complicated, the animator, rather than the system, does more of the motion control.

The oldest and, until recently, most successful approach to the animation of human movement involves copying the actual movement of live subjects. Called *rotoscoping* in the animation industry, this technique is the standard approach to data collection in biomechanics (Vaughan, Davis, & O'Connor, 1992); movement is recorded on film or video from at least two orthogonal directions, and then digitizing and filtering are applied to obtain the joint coordinates of all body segments. Live action can also be captured using special instrumentation, such as goniometers (Calvert, Chapman, & Patla, 1980), and, more recently, using cameras to track light-emitting diodes, or even by having the subject wear a completely instrumented data-suit. The input of these approaches to animation is limited in that they lack generality: Each different movement pattern must be captured separately. However, they are important in acquiring data for biomechanics research, so animation of this kind of data has a special place in human movement studies.

# Motion Control Systems

Motion control can be interactive or scripted, high level or low level, and based on dynamics or kinematics.

## Interactive vs. Scripted Motion Control

Direct motion control methods can be classified as either interactive or scripted. Keyframing is an example of an interactive technique. Here the animator works at an interactive graphics workstation to specify a series of key body positions, or key frames. Then the computer mathematically interpolates intermediate frames to produce smooth movement. This can give good results, but is extremely tedious because each of the joint rotation angles must be specified at each keyframe. In scripted animation, the motion is described as a formal script by the user and interpreted by the computer. Examples are the systems developed to interpret dance notation scores, such as Labanotation (Calvert, Chapman, & Patla, 1982) and Benesh notation (Politis, 1987). Script-based systems can be quite powerful, but without a knowledge base and some built-in intelligence, they cannot specify subtle movement.

## High-Level vs. Low-Level Control

Motion control techniques can also be scaled from low level to high level, depending on the amount of detail needed to define motions or, conversely, the amount of knowledge the system has about general types of movements. Keyframing is placed at the low end of the scale, because every joint angle has to be meticulously specified. Using techniques from artificial intelligence, those high-level, knowledge-based approaches recently developed (Morawetz & Calvert, 1990; Zeltzer, 1985) relieve the animator of this tedious task. Knowledge about primary, goal-oriented movements is stored in well-structured knowledge bases. An inference system then deduces appropriate movement patterns for the limbs to carry out high-level tasks such as walking to the door or fetching the cup from the table. Though this method has proved to be very powerful and provides ease of movement specification, the movement patterns it produces will lack detail or individuality unless the knowledge bases are very richly detailed. Also, this method of representing movement requires a deep understanding of movement itself (Badler, 1986).

## Kinematic vs. Dynamic Control

A third aspect of classifying motion control techniques distinguishes between kinematic and dynamic approaches. Kinematics refers to motion specification in terms of positions, velocities, and acceleration over time, neglecting the forces and torques that actually cause the motion. It may produce movements that are somewhat unrealistic in appearance. In particular, when the whole body is in motion, figures often look as if they are being pulled by strings. Whereas all the above techniques are kinematic, a dynamic approach describes a system in terms of the underlying physical laws of motion (Wilhelms, 1985). (This approach is often referred to as kinetics in biomechanics.) Its main advantage over kinematic systems is that the motion looks natural, because bodies have mass and move under the influence of forces and torques just as they do in the real world. The difficulty here is to find appropriate joint torque patterns to achieve a particular complex movement. Also, solving the equations of motion (forward dynamics problem) might cause numerical instabilities, because a system having $n$ equations for $n$ degrees of freedom usually has to be solved using numerical integration.

The two most important goals we strive for in human figure animation are the convenient specification of movement and the realistic appearance of that motion. A high-level motion control technique satisfies the first goal, whereas dynamics produces natural animation. The following sections introduce a hybrid approach for animating human gait and a system for the kinematic animation of multiple figures. These two examples illustrate how dynamics and kinematics can be combined in systems that allow high-level control.

# A Hybrid Approach to the Animation of Human Gait

Certain types of movement, like walking or running, are conceptually well understood, yet a complete dynamic simulation of such actions is difficult because generating the proper torques for a locomotion cycle is complicated by problems of maintaining balance and coordinating leg motion (Van den Bogert, Schamhardt, & Crowe, 1989). On the other hand, kinematic approaches are inflexible and often produce a "weightless," unrealistic animation.

At Simon Fraser University, Armin Bruderlin has adopted a hybrid kinematic-dynamic approach (Bruderlin & Calvert,1989). Much like Zeltzer's (1982) process, knowledge of the locomotion cycle is incorporated into a hierarchical motion control process. The key idea is that the system "knows" about certain classes of motion and provides the animator with a set of movement commands or parameters, which completely control the figure. The animator no longer has to specify how joint angles change over time because the system, once initiated, is able autonomously to execute a desired motion. This approach parallels our understanding from neurophysiology of how complex articulated movements are controlled in mammals (see Figure 6.9).

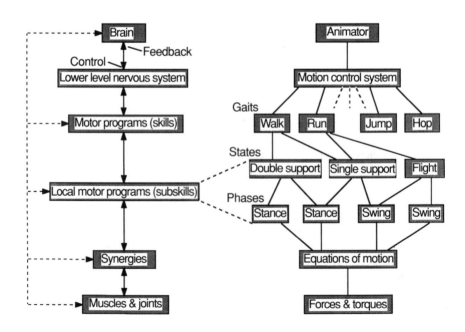

In order to animate bipedal locomotion, the animator takes over the tasks of the brain in the neurophysiological model. At the top level, the researcher initializes the desired motion by specifying parameters such as step length, frequency, or velocity. The lower level nervous system is represented by the motion control system, which contains knowledge about locomotion. Depending on the animator's specification, the control system selects a proper gait that is then decomposed by the (local) motor programs into its underlying components. Every type of gait (e.g., walking) is made up of several different states (e.g., the single-support state, in which one foot is off the ground). These states, in turn, are broken up into the individual leg phases (stance and swing phases). Thus, a gradual reduction in the number of degrees of freedom, along with a decrease in the levels of coordination, is achieved by the control system which parallels that of its natural, biological counterpart. At the bottom level, knowledge is incorporated in the form of dynamic equations of motion for the legs. These equations are tailored to perform a specific application. For example, a compound pendulum simulates the swing leg. The dynamic equations can be considered the low-level manager that produces natural movements, while being guided and coordinated from higher levels.

Our experience suggests that this hybrid of dynamic and goal-directed techniques provides a useful method of animating walking and running under a range of conditions. The hierarchical decomposition of a task supports a convenient motion specification. The dynamic equations of motion produce realistic locomotion patterns (Bruderlin & Calvert, 1991). Figure 6.10 shows an example of the display used to evaluate this model of locomotion and to check the quality of the overall animation. Note that the display combines a contour-based body display with a vector display showing leg angles over time.

# Life Forms—A System for Kinematic Animation of Multiple Figures

In 1984, the Graphics Research Group at Simon Fraser University began to develop a set of tools to support the choreographer in the compositional process. This effort grew out of several years of experience working with computer-based systems to edit and interpret the dance notation Labanotation. These efforts have now evolved to the *Life Forms* system, which is implemented in slightly different forms on Silicon Graphics IRIS and Apple Macintosh workstations (Schiphorst et al., 1990). *Life Forms* is an example of a computer-based system for human animation that has proved useful in practice.

The *Life Forms* system has evolved in the context of a developing understanding of the composition of human movement sequences in relation to hierarchical process, alternate views, use of knowledge, and visualization of motion from concept to final realization. For example, when using a computer to compose a dance, the spatial interrelationships of the dancers can be studied and refined at any point. Alternatively, a representation analogous to a musical score allows the composer to review and edit developments over time. A third view allows

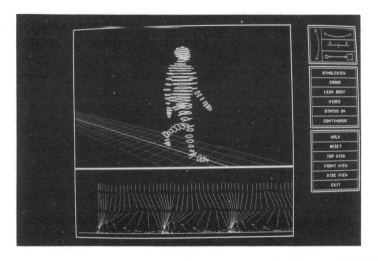

**Figure 6.10**   A display used to evaluate a model of locomotion; the top half shows a typical step, whereas the lower half summarizes lower limb movement.

the movement paths of the dancers on the stage to be visualized. All of these components contribute to the final physical realization.

Using the *Life Forms* system, the choreographer can visually compose the key scenes for the dance piece in space and time. A key scene consists of a number of dancers represented by human figures of various colors placed appropriately on a stage, each having a posture chosen from a menu (Figure 6.11a). Key scenes are stored as they are composed, and additional, intermediate scenes are created as composition proceeds. (This is analogous to the development of a storyboard for a movie.) Once the key scenes have been sketched out, the choreography can be refined to a more detailed level, and an animation of the resulting motion can be produced.

The key scenes are built up from menus of prestored body stances or movement sequences. If none of the existing stances or sequences in the menu are suitable for a new composition, then a new stance or sequence can be built up using the interactive sequence editor (Figure 6.11b). The sequence editor allows individual stances to be adjusted and sequences to be edited.

A preliminary version of the IRIS implementation of *Life Forms* is being tested by New York choreographer Merce Cunningham. He used the system in the composition of his new piece, ''Trackers,'' which premiered in New York on March 19, 1991, to very favorable reviews. The Apple Macintosh implementation is being tested by other choreographers and has been used in teaching dance at several universities and colleges, including Ball State University, Swarthmore College, and Connecticut College. Both versions are licensed to Kinetic Effects, Incorporated (Suite 310, 100 Stewart Street, Seattle, WA 98109), and the Macintosh version is distributed by Macromedia Corporation (600 Townsend Street, San Francisco, CA 94103).

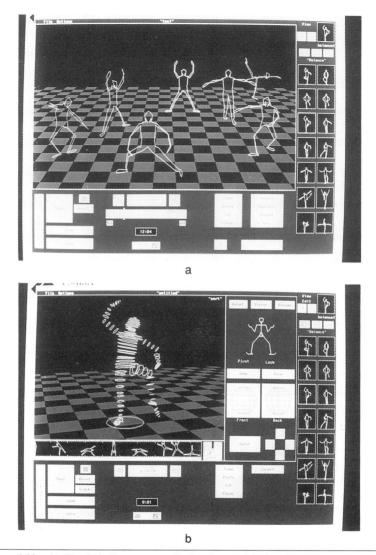

a

b

**Figure 6.11**   (a) The *Life Forms* stage display for spatial planning. (b) The Sequence Editor component of *Life Forms* used for editing stances and temporal planning.

## SUMMARY AND CONCLUSIONS

In the use of 3-D computer graphics to animate human body movements, the two most important components are efficient (and effective) methods for representing and displaying the human body and easy-to-use techniques for describing human movement. Computer-based animation of human figures is an important activity in its own right. It has much in common with the simulation of human movement carried out in biomechanics research and with the development of control strategies for bipedal robots. However, the goals are different: In animation, the most important objective is to generate movement that is accurate, no matter how it is produced. An important secondary objective is to provide the animator with high-level tools for movement specification.

# REFERENCES

Badler, N.J. (1986). Animating human figures: Perspectives and directions. In *Proceedings of Graphics Interface '85* (pp. 115-120). Palo Alto, CA: Morgan Kaufmann.

Bruderlin, A., & Calvert, T.W. (1989). Goal-directed, dynamic animation of human walking. *Computer Graphics*, **23**(3), 233-242.

Bruderlin, A. & Calvert, T.W. (1991). Animation of human gait. In A. Patla (Ed.), *Adaptability of human gait* (pp. 305-330). New York: Elsevier Science Publishers.

Calvert, T.W. (1988). The challenge of human figure animation. In *Proceedings of Graphics Interface '88* (pp. 203-210). Palo Alto, CA: Morgan Kaufmann.

Calvert, T.W., Chapman, J., & Patla, A. (1980). The integration of subjective and objective data in animation of human movement. *Computer Graphics*, **14**(3), 198-203.

Calvert, T.W., Chapman, J., & Patla, A. (1982). Aspects of the kinematic simulation of human movement. *IEEE Computer Graphics and Applications*, **2**(9), 41-50.

Chung, T. (1987). *An approach to human surface modelling using cardinal splines*. Unpublished master's thesis, Simon Fraser University, Burnaby, BC.

Foley, J.D., van Dam, A., Feiner, S.K., & Hughes, J.F. (1990). *Computer graphics: Principles and practice*. Reading, MA: Addison-Wesley.

Forsey, D.R. (1988). Hierarchical B-spline refinement. *Computer Graphics*, **22**(3), 205-212.

Magnenat-Thalmann, N., & Thalmann, D. (1985). *Computer animation: Theory and practice*. Tokyo: Springer-Verlag.

Morawetz, C.L. & Calvert, T. (1990). A framework for goal-directed human animation with secondary movement. In *Proceedings of Graphics Interface '90* (pp. 60-67). Palo Alto, CA: Morgan Kaufmann.

Newman, W., & Sproull, R. (1979). *Principles of interactive computer graphics*. New York: McGraw-Hill.

Phillips, C.B., Zhao, J., & Badler, N.J. (1990). Interactive real-time articulated figure manipulation using multiple kinematic constraints. *Computer Graphics*, **24**(2), 115-120.

Politis, G. (1987). *A computer graphics interpreter for Benesh movement notation*. Unpublished doctoral dissertation, University of Sydney.

Rogers, D.F. (1985). *Procedural elements for computer graphics*. New York: McGraw-Hill.

Schiphorst, T., Calvert, T.W., Lee, C., Welman, C., & Gaudet, S. (1990). Tools for interaction with the creative process of composition. In *Proceedings of ACM Computer Human Interface Conference, 1990.* (pp. 167-174). Seattle, WA.

Van Baerle, S. (1990). A case study of flexible figure animation. *SIGGRAPH Course notes: Human Figure Animation*, pp. 52-69. New York: ACM.

Van den Bogert, A.J., Schamhardt, H.C., & Crowe, A. (1989). Simulation of quadrapedal locomotion using rigid body dynamics. *Journal of Biomechanics*, **22**, 33-41.

Vaughan, C.L., Davis, B.L., & O'Connor, J. (1992). *Gait analysis laboratory*. Champaign, IL: Human Kinetics Publishers.

Wilhelms, J. (1985). Using dynamic analysis to animate articulated bodies such as humans and robots. In *Proceedings of Graphics Interface '85* (pp. 97-104). Palo Alto, CA: Morgan Kaufmann.

Zeltzer, D. (1982). Motor control techniques for figure animation. *IEEE Computer Graphics and Applications*, **2**(9), 53-59.

Zeltzer, D. (1985). Towards an integrated view of 3-D computer character animation. In *Proceedings of Graphics Interface '85* (pp. 105-115). Palo Alto, CA: Morgan Kaufmann.

Zeltzer, D. (1989). Direct manipulation of virtual worlds. In B. Barsky, N. Badler, & D. Zeltzer (Eds.), *Making them move: Mechanics, control and animation of articulated figures* (pp. 35-50). Palo Alto, CA: Morgan Kaufmann.

# Chapter 7

# X-ray Photogrammetry

*Ian A.F. Stokes*

Human motion can be measured by using X-rays in a way analogous to visible-light photogrammetry. Two or more views of a subject in a sequence of positions permits three-dimensional measurement of shape and motion, or photogrammetry. In normal practice, for dosage and other reasons, only two image projections are used. This chapter outlines the general methods used and practical considerations, and surveys previous uses of these techniques.

The following definitions are used in this chapter:

*Stereo X-ray photogrammetry or radiography*: Procedure in which a single film plane is used to provide two or more images from different projections (X-ray tube positions; see Figure 7.1a)

*Biplanar X-ray photogrammetry or radiography*: Procedure in which two film planes (often at 90° angles to each other) are used to produce pairs of images (see Figure 7.1b)

*Object point*: The point on an object that produces an identifiable image and whose spatial position can be measured

*Image point*: A feature visible in an X-ray image that corresponds to an object point

*Object space*: The physical space occupied by object points in a photogrammetric study

## PECULIARITIES OF RADIOGRAPHIC SYSTEMS

Radiographic methods represent a special case of photogrammetry in which X-rays are used in place of visible light. Because visible light provides an image corresponding to light emission or reflection from a surface, and X-rays provide an image corresponding to the transmission of electromagnetic radiation through the object field, a number of significant practical differences arise. The absence of lenses simplifies the X-ray technique. Also, because the recording film and the source of X-rays lie on opposite sides of the object field, recorded images are always larger than the objects being recorded. Despite these differences, there

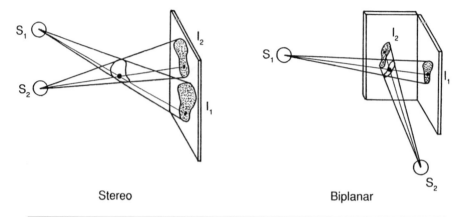

Stereo                              Biplanar

**Figure 7.1**   Arrangement of X-ray equipment for (a) stereo and (b) biplanar X-ray photogrammetry studies. Two X-ray sources (tubes) or two positions of the same tube, identified as $S_1$ and $S_2$, are used to create images $I_1$ and $I_2$. The position of a landmark point in each image corresponds to an object point O, located at the intersection of the rays that join the sources to the images.

are many similarities with visible light photogrammetry. X-ray photogrammetry can also be used both to record the shape of objects and to track their motion.

## How X-ray Images Are Produced

X-rays can pass through biological tissues but tend to be absorbed by denser elements, especially those that mineralize bone. The degree of absorption depends on the spectrum of X-ray wavelengths. The spectrum is controlled by the voltage applied to the X-ray tube, by the filtration added to the exit port, and to some extent by the air gap between the tube and the patient. X-rays are absorbed and scattered by tissues within the subject. After exiting the patient, they may be further scattered or filtered by the air gap between the subject and the film. Also, a Potter-Bucky grid may be placed in front of film to reduce the amount of scattered radiation reaching the film. This is a device consisting of strips of lead placed edgewise to the X-ray tube. It is designed to absorb scattered radiation by permitting only rays coming directly from the X-ray tube to pass through it to the X-ray film.

The image on the film is produced by light emitted from scintillation screens placed on either side of it. Therefore, the quality of the image also depends on the properties of the scintillation screen. For instance, highly sensitive screens tend to have larger grain size and a nonlinear response to the intensity of irradiation. The same principles also apply to newer techniques (image intensifiers, digital radiography, and computerized tomography) in which an electronic detector is used in place of the conventional intensifying screen and silver emulsion film.

The resulting image therefore can be considered to be a gray scale showing the accumulated attenuation of X-rays by the tissues and air gap through which they have passed, superimposed over a "noise," or background intensity created by scattered X-rays. A major advantage of computerized tomography is that it generates an image from a uniformly thin "slice" of the patient and has a more uniform geometric scaling. In the case of plane radiography, the divergent beam from the X-ray source causes geometric magnification. Parts of the body close to the X-ray film are magnified less than those at a greater distance from the film. Greater distance from the film also causes image degradation because of image scattering and the finite size of the X-ray source within the tube.

## Principles and History of X-ray Photogrammetry

Three-dimensional X-ray techniques imply two or more images. A stereo pair of images with a narrow convergence angle can be placed in a stereocomparator instrument to provide the illusion of a 3-D image to the operator. However, because the images show radiodensity rather than object surfaces, it is difficult to obtain a good 3-D image and to make measurements using this technique. Most applications involving 3-D measurement rely on landmarks located in each image and subjected to a mathematical reconstruction of their 3-D spatial location.

Special considerations are required in the selection and identification of image points (landmarks). X-ray images depend on the geometry of the object being radiographed as well as the distribution of radiodensity within it. Bones provide strong images, but the selection of consistent landmarks on these bones can present significant problems. The most accurate landmarks are provided by using small metallic markers either attached to the body or inserted into (or attached to) selected internal structures or organs. Although radiographic images can be digitized by a number of techniques, including the detector arrays used in computerized X-ray tomography and the two-dimensional detectors used in digital radiography, most practical systems use films, which are subsequently marked and digitized manually. Therefore, the principal components of a practical system for X-ray photogrammetry include the imaging equipment, calibration equipment, film-marking and digitization equipment, and the software used for analysis of the digitized data.

X-ray photogrammetry has a long history. In the first decade after the discovery of X-rays, Davidson (1898) reported a method for localizing the positions of foreign bodies. Radiographs were made with two X-ray tubes, separated by a distance. After the photographic plates had been developed, they were replaced in their original positions and light, silk threads were used to join the two images of the foreign body in these plates to their respective X-ray sources (see Figure 7.2). The intersections of these threads gave the original location of the foreign body at the time of radiography, but this physical reconstruction technique never gained popularity.

In 1970, Hallert published practical methods and the geometric principles by which X-ray photogrammetry could take advantage of the computational methods then available for optical photogrammetry. These methods were exploited by Selvik (1989), who developed the now widely used techniques that employ

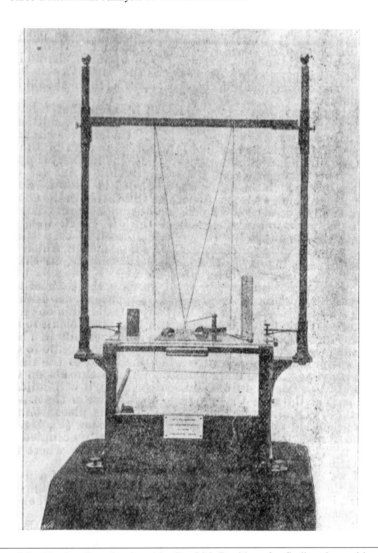

**Figure 7.2**   The localizer developed by Dr. J.M. Davidson for finding the positions of foreign bodies. After developing the X-ray plates, silk threads were used to reconstruct the paths of the X-rays from tubes to image points, and hence locate the positions of object points at the intersections of the threads.
*Note.* From "Roentgen Rays and Localisation" by J.M. Davidson, 1898, *British Medical Journal*, p. 11. Reprinted by permission of BMJ Publishing Group.

spherical tantalum markers. Selvik (1989) listed 92 papers he had coauthored on the application of roentgen stereophotogrammetry to a wide variety of medical problems, including quantification of bone growth, tumor development, joint kinematics, and morphologic effects of surgical procedures. This methodology has been widely used by others (Karrholm, 1989).

# PRINCIPAL COMPONENTS OF AN X-RAY PHOTOGRAMMETRY SYSTEM

The major equipment in an X-ray room includes the X-ray tube, which provides the source of X-rays, the generator that drives it, and film holders (see Figure 7.3). Also common are devices for enhancing image quality and controlling dose, which include Potter-Bucky grids and image-intensifying screens. Three-dimensional measurement from X-ray films requires calibration equipment and procedures. Other components, such as automatic exposure systems, film-developing equipment, and mechanical systems such as gantries, will not be considered here except inasmuch as rigid mounting of tubes, maintenance of film dimensions and flatness, and high-quality images are all essential for accurate geometric reconstructions.

## Imaging Procedures and Dose

X-ray tubes provide an effective point source of X-rays of a certain spectrum of energy or wavelength. In the tube, electrons bombard the anode, and X-rays result from the energy dissipated in this collision. The spectrum of energy depends mainly on the energy of impact, which is determined by the voltage applied to the tube. The energy flux also depends on the current flow. Subsequent collimation and filtering of the emitted X-rays can be used respectively to limit the field of exposure and modify the energy spectrum. The selection of high-quality tubes with a small focal spot, the appropriate selection of kilovoltage and exposure factors, and the filtration and collimation of the beam are all crucial to good image quality. Additionally, the use of Potter-Bucky grids and the selection of suitable film and screen combinations affect both image quality and the X-ray

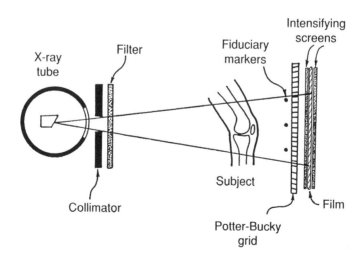

**Figure 7.3**   Major components of an X-ray imaging system.

dose required to produce the image. The Potter-Bucky grid is focused in one direction so as to transmit rays that originate at the X-ray tube and to absorb any scattered X-rays that strike the grid at an angle. Therefore, the orientation of the Potter-Bucky grid is important in stereo systems where the same film plane and Potter-Bucky grid are used with two different tubes or tube positions. The axis of the Potter-Bucky grid must be parallel to the stereo base to permit transmission of both images.

Dosage is of major importance. X-rays are harmful, and the damage they do depends on the volume of the body irradiated, the intensity of the radiation, and the susceptibility of the tissues. In general, fast-dividing cells, such as epithelial cells, are the most susceptible, but the consequences of damage to cells of the reproductive system can be more profound. It is still not certain whether there is a lower limit of dose below which no damage is done or cannot be repaired. Limits of safe, or acceptable, dose are generally set by reference to the whole-body background dose from the environment. The dose to a particular organ can be estimated from measurements of entry and exit doses on the tube and film sides of the body, respectively. The entry dose from a typical chest film is on the order of 20% of annual background dose, but the exit dose is much less, about 1% of annual background. Suitable positioning of subjects can exploit this difference. Less sensitive organs, such as muscles and bones, can be used to shield more sensitive ones, such as the breast, thyroid gland, and gonads.

## Calibration and Fiduciary Marks

In addition to the conventional X-ray equipment, special components are required to perform 3-D photogrammetric measurements. In effect, the relative positions of X-ray sources and X-ray films must be recorded, although this may not necessarily be done directly. A standard calibration object is normally used in this process (Figure 7.4). In general, one of two approaches may be used. In the first, all the control and calibration points are included in each image of the subject. The alternative is to maintain rigid control over the tube positions and film planes once the calibration procedure has been completed; the calibration object can then be removed. This second approach is particularly useful *if* the equipment can be repositioned precisely. In either case, at least two control (fiduciary) point images on each X-ray film are required in order to establish the film's position and relative rotation compared to the fixed position of the X-ray tubes.

The exact form of a calibration object depends on the reconstruction technique being employed. In general, an object with several defined landmarks in known positions is required. Each landmark provides a sharp image in the calibration films. The calibration points should be distributed throughout the object space volume in which reconstruction is required. A typical calibration object consists of a framework constructed of radiolucent material with markers, such as spherical metallic balls, mounted in known or measurable positions. The design and construction of such calibration objects is often easier than for optical systems, in which control points may become obscured by parts of their supporting structure.

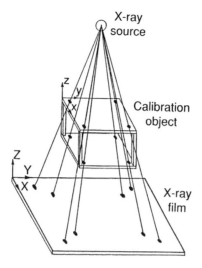

**Figure 7.4**    The image of a calibration object, having markers in known position, is used to calculate the relationship between the local xyz coordinate system of the calibration object and the global XYZ coordinate system. The X-ray source is considered to be the intersection of a bundle of rays that pass through corresponding object and image points.

## Digitization

The final component of a stereo-photogrammetric system is the digitizing instrument by which coordinates of image points are measured and recorded. In most cases, manual digitization has been used, but an entire X-ray image can be gray-scale digitized for subsequent automatic or semiautomatic detection of image points. Mechanical, acoustic, and magnetic systems can all be used in the construction of manual digitizing tablets. If the digitizing tablet can be back-lit, then image points can be selected and digitized at the same time. However, in general it is better to select and mark image points in advance on a light box, so the ability to back-illuminate during digitization is not essential. Manual digitization has a precision on the order of 0.5 mm (depending on the care taken), which is about 1 part in 500 of a typical image.

## LANDMARKS: ANATOMIC AND METALLIC

The selection of landmarks provides the greatest practical problem in most X-ray photogrammetry. Image points should be invariant with respect to the direction of projection and the orientation of the image plane. Because an X-ray image represents the accumulated radiodensity of material between the tube and the film, even distinct landmarks such as metallic spheres can appear as distorted images in the presence of other radiodense objects in the path of the X-ray beam.

Anatomic landmarks are particularly problematic. There are few anatomical features that provide a consistent image point. However, sometimes image points can be reconstructed from a geometric feature of the skeleton, which does show clearly on a radiograph. For instance, if the femoral head is considered to be a part of a sphere, then the center of its circular image will always represent an image point at the center of the sphere.

A point in the middle of the femoral shaft and half-way along it could also make a reasonable landmark, but this depends on consistent location of the most distal and the most proximal points of the bone (Figure 7.5). Images of the vertebral endplates can be considered to represent a planar, elliptical shape, so the center of area of their images would be an invariant point corresponding to the center of area of the actual endplate surfaces. Points on the pedicles, articular facets, and spinous processes of vertebrae have also been used as landmarks (Figure 7.6, a-c).

Metallic markers are generally much more accurate. However, if applied superficially, they have the disadvantage of moving relative to each other and relative to skeletal features. If they are inserted into bone or attached to bones or other structures this requires an invasive procedure. Aronson, Hansson, and Selvik (1978) investigated the use of spherical tantalum markers (0.1 mm in diameter) and found that in rabbits they caused no biologic reaction and there was no evidence of migration over time. He also developed an instrument for inserting these markers percutaneously. However, a possible disadvantage of such markers is that their positions are generally arbitrary, and one needs to establish their

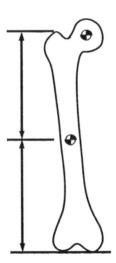

**Figure 7.5**   Possible landmarks on the femur. If the femoral head is assumed to be spherical, then the center of its image always lies at the center of the sphere. Depending on the shapes of the ends of the bone, and the range of X-ray projection used, a point half-way along the femoral shaft is also invariant.

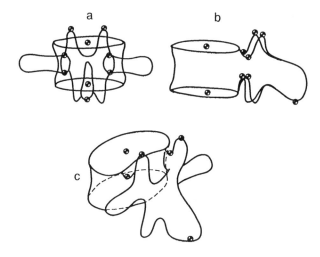

**Figure 7.6**   Landmarks on a vertebra. Tracing of (a) frontal and (b) lateral X-ray images, showing landmarks at the vertebral endplate centers, tips of superior articular facets, bases of the pedicles, and inferior point of the spinous processes. (c) A perspective view and the landmarks visible in this view.

relationship to an anatomically significant coordinate system. Thus, for kinematic studies, additional anatomic markers must be identified and the coordinate transformation system, which relates them to the frame of reference identified by the applied metallic markers, must be well defined.

## CALIBRATION AND
## RECONSTRUCTION PROCEDURES

In principle, 3-D reconstruction of image points can be performed by simple vector algebra. If the coordinates of two image points are known, along with the coordinates of the X-ray sources, then the object point is reconstructed as the intersection of, or the closest approach between, the two vectors joining the image points to their respective X-ray sources. Early calibration and reconstruction procedures were based on this principle. Geometric principles are outlined by Hallert (1970), and a practical solution method is outlined in Brown, Burstein, Nash, and Schock (1976). The disadvantage of this approach is that the film planes, the calibration object, and the X-ray sources must all be fixed in the same global coordinate system in the calibration and reconstruction. This means that the calibration object must be positioned in a known location and orientation relative to the film coordinate systems during calibration.

The direct linear transformation (DLT) approach to photogrammetry (originally outlined by Abdel-Aziz & Karara, 1971) has a number of practical advantages. The global coordinate system is established by the local coordinate system of the calibration object. During calibration, this object can be placed in any position relative to the X-ray equipment and films. This means that very little special equipment and very few special precautions are required to perform X-ray photogrammetry. Providing that the X-ray equipment can be positioned in stable locations, only the calibration object and at least two fiduciary marks are required to localize the position and orientation of each film. A FORTRAN program implementing this method was given by Marzan (1975; see chapter 2).

## Choice of X-ray Projections

Three-dimensional photogrammetry requires at least two different image projections. In practice, a number of factors must be taken into account in selecting these projections. These factors include the resulting X-ray dose, compatibility with standard X-ray projections (in the case of clinical studies), visibility of landmarks, and the accuracy of reconstruction. These sometimes competing requirements may necessitate a compromise: For instance, in their studies of lumbar spine motion, Stokes, Wilder, Frymoyer, and Pope (1981) used anteroposterior and lateral projections in a biplanar radiography configuration because this provided films compatible with clinical use, good landmark visibility, and accurate reconstruction. On the other hand, in reconstruction of the entire thoracic and lumbar spine for studies of scoliosis deformity, a stereoconfiguration (the same film plane for both views) was used with a posteroanterior projection and subsequent movement of the tube to a 15°-lateral projection (Stokes, Bigalow, & Moreland, 1987). Later, this technique was modified by raising the second tube to an angle of 20° (Stokes, Dansereau, & Moreland, 1989) to permit reconstruction of the rib cage in addition to the spine.

In general, a living subject must maintain a static position while two or more X-ray exposures are made. This may take more than 1 minute if film changes or tube repositioning is required. (Changing the film is usually required in stereo studies and is also often necessary in biplanar studies because scattered X-rays would otherwise result in the exposure of both films.) Subject-positioning devices should be used to maintain static position during such film changes.

It is desirable to use a different X-ray tube for each exposure. However, a single tube can be used if it can be repositioned accurately. If this is not possible, calibration can be performed for each image pair by incorporating calibration points into each image. Typically, the human subject is positioned within a box or frame that supports the calibration points.

## Line Reconstructions

Conventional photogrammetry involves the matching of image points for reconstruction of object points. However, image line reconstructions are also possible under certain special circumstances. If each image line is considered as a series of points, then the reconstruction starts with the problem of finding matching image points on the two lines. If the image lines are oriented perpendicular to the stereo base (the line joining the two X-ray tube positions) then this matching problem is unique and can be solved by minimizing the error term associated with the reconstruction of a pair of image points. Iteration is required to find the best matching points. This approach has been exploited in the reconstruction of vertebral body geometry (Szirtes, 1981) and in reconstruction of rib- cage geometry (Dansereau & Stokes, 1988). A similar approach has been used for calculating the shape of blood vessels and constrictions in them from arteriograms (Reiber, Gerbrands, Troost, Kooijman, & Slump, 1983; Siebes, D'Agenio, & Selzer, 1985; Wu, 1991).

## Evaluation of Accuracy and Precision

In X-ray studies, the major factors affecting accuracy are the dimensional stability of the X-ray equipment and the quality of the image as it affects the ability to identify landmark positions accurately. Because the image is generally of similar size to the object being studied, a stereo reconstruction typically has errors in the plane parallel to the film of the same order of magnitude as errors in identifying and measuring landmark positions. However, errors in the direction *perpendicular* to the film are magnified by a trigonometric function of the angle between the X-ray projections. Accuracy and precision of reconstructions can be determined experimentally. In a typical experiment, a rigid object with a number of landmarks is displaced and rotated by known amounts and radiographed in each position. After reconstruction, the precision of landmarks can be estimated by examining the constancy of the linear distances between pairs of landmarks. The precision for measuring the relative motion of a body can be estimated by grouping landmarks to represent bodies and calculating the apparent (though actually zero) relative motion between them. The accuracy of the kinematic study can be examined by measuring the whole-body motion from the landmarks and comparing it with the known displacements in rotation introduced at the time of radiography (Lange, Huiskes, & Kauer, 1990a).

It may be possible to improve the accuracy of a reconstruction in a kinematic study by use of a redundant number of landmarks (>3) on each moving body and the rigid-body assumption (i.e., because the linear distances between all pairs of landmarks on a rigid body should remain constant, landmark reconstruction errors can be detected). In calculating body motion, function minimization procedures can be used to find the combination of translations and rotations that, when applied to the initial position coordinates of given landmark points, transforms their coordinates to positions corresponding to a best match to the measured subsequent-position coordinates. This approach has been used by Selvik (1989) and by Plamondon and Gagnon (1990) in radiographic studies. For a sequence of more than two positions of a body, Stokes (1981) developed an ad hoc method for adjusting landmark coordinates to force them to conform to the rigid-body assumption. This gave modest improvements in measurement precision. The accuracy with which body motion can be deduced from landmark coordinates is the same as that achieved in visible light photogrammetry and is discussed in chapter 2.

# SURVEY OF APPLICATIONS

In the study of human motion, X-ray stereophotogrammetry has been used in conventional kinematic studies of the musculoskeletal system, as well as in a number of applications specific to X-ray techniques. The latter include measurement of bony growth with an accuracy of 30 μm (Aronson, Hansson, & Selvik, 1978), studies of craniofacial development (Rune, Sarnäs, Selvik, & Jacobsson, 1986), measurement of the deformation of soft tissue structures (Brown, Jevsevar, & Rowell, 1991), measurement of the relative motion of joint prostheses (Ryd, Albrechtsson, Herberts, Lindstrand, & Selvik, 1988), and the study of cardiac motion (Philips, Prenis, Santamore, & Bove, 1983).

## Applications in the Upper Extremity

Functional anatomical studies of the elbow joint (Chao & Morrey, 1978) and the wrist (Lange, Kauer, & Huiskes, 1986) have been conducted with the help of stereo X-ray measurement techniques, using metallic markers applied to cadaveric specimens. Smith, An, Cooney, Linscheid, and Chao (1989) extended their studies of normal wrist kinematics to establish the effects of certain scaphoid bone osteotomies on the motion occurring between extreme positions of the wrist. Lange, Huiskes, and Kauer (1990b) deduced wrist ligament length changes from kinematic studies.

## Studies of the Spine

Lysell (1969) and White (1969) described early applications of stereoradiography to the measurement of spinal kinematic properties, including the ''coupling'' of motion that occurs in spinal motion segment specimens. Pope, Wilder, Matteri, and Frymoyer (1977) used biplanar X-ray techniques with whole cadavers subjected to motion of the trunk in an attempt to determine whether ''coupling'' of spinal motion also occurred in the intact spine. The spine has been the focus of several in vivo studies, facilitated by the existence of anatomic features of vertebrae that serve as radiographic landmarks. The accuracy and precision of these landmarks have been studied by Rab and Chao (1977), Stokes, Medlicott, and Wilder (1980), and by Plamondon and Gagnon (1990).

### Spinal Deformity

Serial stereoradiographic studies of the scoliotic spine have been reported (DeSmet, Tarlton, Cook, Berridge, & Asher, 1983; Hindmarsh, Larsson, & Mattsson, 1980; Stokes, Shuma-Hartswick, & Moreland, 1988) with a view to more precise detection of progression of these deformities. Reuben, Brown, Nash, and Brower (1982) reported a change in the shape of scoliosis deformities under the influence of axial loading and compared these measurements with those in their earlier report (1979) on the effects of traction on a healthy, undeformed spine. Allard,

Dansereau, Duhaime, and Geoffroy (1984) and Hierholzer and Lüxmann (1982) both exploited 3-D measurements of spinal deformity by using invariant measurements of the spinal shape. These are independent of the position of the patient relative to the X-ray equipment at the time of radiography. Saraste and Österman (1986) used stereophotogrammetry to evaluate vertebral rotation changes after scoliosis surgery.

## Motion in the Painful Spine

Several investigations of subjects with low-back pain have been conducted using biplanar X-ray methods in an attempt to establish correlations between lumbar spinal motion and pain (Pearcy, 1985; Stokes & Frymoyer, 1987; Stokes et al., 1981). These studies are complicated by the fact that subjects with low-back pain tend to have a reduced range of motion close to the precision of the measurement technique. Also, subjects must hold the spine in a selected position for the duration of the radiographic study. It is probable that a more continuous, more precise measurement of spinal motion would be more successful in providing information of diagnostic value. Olsson, Selvik, and Willner (1977) used markers implanted at the time of surgery for accurate measurement of the motion of the lumbosacral spine after operations that fused this part of the spine. Even in fusions that appeared sound, motion of up to 4° was found to occur. The precision of the method was better than 0.5°.

# Studies of the Lower Extremity

Detailed functional anatomical studies of knee and ankle joints have been conducted in cadaver specimens using stereophotogrammetry. In the knee, the tracking motion of the patella relative to the femur has been reported by Veress, Lippert, Hon, and Takamoto (1979) and by Kampen and Huiskes (1990). Van Dijk, Huiskes, and Selvik (1979) used measurements of relative motion of the knee joint, combined with an anatomical model, to deduce the changes in length of the cruciate ligament under various conditions.

A stereo X-ray method of determining tarsal bone motion was used together with anatomic measurements of ligament geometry in a study of the ligaments' functional role by Allard, Thiry, and Duhaime (1985). Studies of motion of the ankle-joint complex and the instantaneous axes of rotation have been performed in anatomic specimens by van Langelaan (1983) and in volunteers by Lundberg (1989) and Lundberg, Svensson, Nemeth, and Selvik (1989).

Very precise X-ray photogrammetric measurements, employing implanted metallic markers, have been used for detecting micromotion of implanted prostheses (Mjoberg, 1991). This idea was reported by Lippert et al. in 1982; more recently, Ryd, Albrechtsson, Herberts, Linstrand, and Selvik (1988) used these techniques for detecting both migration over time and acute motion under load of total knee replacements, as did Ryd, Linstrand, Rosenquist, and Selvik in 1986. Turner-Smith (1990) and Turner-Smith, White, and Bulstrode (1990) adapted these techniques to the study of hip joint prostheses.

# CONCLUSION AND FURTHER APPLICATIONS AND DEVELOPMENTS

X-ray photogrammetric techniques have a number of potential advantages over other methods, particularly for the study of internal structures and direct measurement of skeletal kinematics. However, there are a number of disadvantages to using X-rays, including dosage and the need to freeze motion in order to obtain at least two good-quality images. Digital radiography (which uses digital sensors in place of the conventional photographic film) would overcome many of these difficulties by allowing lower dose techniques and permitting more rapid analysis of the images. Automated image processing, permitting automatic recognition of landmarks, could also provide substantial benefits. This approach is being investigated in the field of angiography for measurement of motion of the heart (Philips et al., 1983; Smith & Quarendon, 1985).

Computerized tomographic methods provide an interesting comparison with the stereo and biplanar methods outlined here. In principle, 3-D models of anatomic structures constructed from serial tomographic images could be used to measure the motion of these structures. However, the problems of 3-D modeling (surface detection, etc.) and of fitting a coordinate system to such models for the measurement of motion are extremely complex and substantially more difficult than the 2-D analog in which two images are overlaid to establish their relative positions. Overall, stereo and biplanar X-ray techniques hold a special position in the study of 3-D human motion, which is currently dictated by the technology of image acquisition and analysis. No doubt, the range of applications of these techniques will be expanded as technological enhancements are made.

# REFERENCES

Abdel-Aziz, Y.I., & Karara, H.M. (1971). Direct linear transformation from comparator coordinates into object space coordinates in close-range photogrammetry. *Proceedings of the American Society of Photogrammetry VIth Symposium on Close-Range Photogrammetry* (pp. 1-18). Falls Church, VA: American Society of Photogrammetry.

Allard, P., Dansereau, J., Duhaime, M., & Geoffroy, G. (1984). Scoliosis assessment in Friedreich's ataxia by means of intrinsic parameters. *Canadian Journal of Neurological Science, 11,* 582-587.

Allard, P., Thiry, P.S., & Duhaime, M. (1985). Estimation of the ligament's role in maintaining foot stability using a kinematic model. *Medical & Biological Engineering & Computing, 23,* 237-242.

Aronson, A.S., Hansson, L-I., & Selvik, G. (1978). Roentgen stereophotogrammetry for determination of daily longitudinal bone growth in the rabbit. *Acta Radiologica (Diagnosis), 19,* 97-105.

Brown, G.A., Jevsevar, D.S., & Rowell, D. (1991). X-ray stereophotogrammetric measurement of meniscal circumferential strains. *Proceedings of the 37th Annual Orthopaedic Research Society, Paper No. 295.*

Brown, R.H., Burstein, A.H., Nash, C.L., & Schock, C.C. (1976). Spinal analysis using a three-dimensional radiographic technique. *Journal of Biomechanics,* **9**, 355-365.

Chao, E.Y.S., & Morrey, B.F. (1978). Three-dimensional rotation of the elbow. *Journal of Biomechanics,* **11**, 57-73.

Dansereau, J., & Stokes, I.A.F. (1988). Measurements of the three-dimensional shape of the rib cage. *Journal of Biomechanics,* **21**, 893-901.

Davidson, J.M. (1898). Roentgen rays & localisation: An apparatus for exact measurement and localisation by means of roentgen rays. *British Medical Journal,* **1**, 10-13.

DeSmet, A.A., Tarlton, M.A., Cook, L.T., Berridge, A.S., & Asher, M.A. (1983). The top view for analysis of scoliosis progression. *Radiology,* **147**, 369-372.

Hallert, B. (1970). X-ray photogrammetry. *Basic Geometry and Quality.* Amsterdam: Elsevier.

Hierholzer, E., & Lüxmann, G. (1982). Three-dimensional shape analysis of the scoliotic spine using invariant shape parameters. *Journal of Biomechanics,* **15**(8), 583-598.

Hindmarsh, J., Larsson, J., & Mattsson, O. (1980). Analysis of changes in the scoliotic spine using a three-dimensional radiographic technique. *Journal of Biomechanics,* **13**, 279-290.

Kampen, A. van, & Huiskes, R. (1990). The three-dimensional tracking pattern of the human patella. *Journal of Orthopaedic Research,* **8**, 372-382.

Karrholm, J. (1989). Roentgen stereophotogrammetry: Review of orthopedic applications. *Acta Orthopaedica Scandinavica,* **60**, 491-503.

Lange, A. de, Huiskes, R., & Kauer, J.M.G. (1990a). Measurement errors in roentgen stereophotogrammetric joint-motion analysis. *Journal of Biomechanics,* **23**, 259-269.

Lange, A. de, Huiskes R., & Kauer, J.M.G. (1990b). Wrist joint ligament length changes in flexion and deviation of the hand: An experimental study. *Journal of Orthopaedic Research,* **8**, 722-730.

Lange, A. de, Kauer, J.M.G., & Huiskes, R. (1986). Kinematic behavior of the human wrist joint: A roentgen stereophotogrammetric analysis. *Journal of Orthopaedic Research,* **3**, 56-64.

Lippert, F.G., III, Harrington, R.M., Veress, S.A., Fraser, C., Green, D., & Bahniuk, E. (1982). A comparison of convergent and biplane X-ray photogrammetry systems used to detect total joint loosening. *Journal of Biomechanics,* **15**, 677-682.

Lundberg, A. (1989). Kinematics of the ankle and foot: In vivo roentgen stereophotogrammetry. *Acta Orthopaedica Scandinavica Supplementum,* **233**, 1-24.

Lundberg, A., Svensson, O.K., Nemeth, G., & Selvik, G. (1989). The axis of rotation of the ankle joint. *Journal of Bone & Joint Surgery [Br],* **71**, 94-99.

Lysell, E. (1969). Motion in the cervical spine. *Acta Orthopaedica Scandinavica* (Suppl), **123**, 1-61.

Marzan, G.T. (1975). Rotational design for close-range photogrammetry. (Ph.D. thesis, Univ. of Illinois, 1976). Ann Arbor: Xerox University Microfilms.

Mjoberg, B. (1991). Fixation and loosening of hip prostheses. A review. *Acta Orthopaedica Scandinavica,* **62**, 500-508.

Olsson, T.H., Selvik, G., & Willner, S. (1977). Mobility in the lumbosacral spine after fusion studied with the aid of roentgen stereophotogrammetry. *Clinical Orthopaedics,* **129,** 181-190.

Pearcy, M.J. (1985). Stereo radiography of lumbar spine motion. *Acta Orthopaedica Scandinavica* (Suppl), **212,** 1-46.

Philips, C.M., Prenis, J., Santamore, W.P., & Bove, A.A. (1983). Recognition and storage of metal heart marker position from biplane X-ray images at video rates. *IEEE Transactions on Bio-Medical Engineering,* **30**(1), 10-17.

Plamondon, A., & Gagnon, M. (1990). Evaluation of Euler's angles with a least squares method for the study of lumbar spine motion. *Journal of Biomedical Engineering,* **12,** 143-149.

Pope, M.H., Wilder, D.G., Matteri, R.E., & Frymoyer, J.W. (1977). Experimental measurements of vertebral motion under load. *Orthopedic Clinics of North America,* **8,** 155-167.

Rab, G.T., & Chao, E.Y.S. (1977). Verification of roentgenographic landmarks in the lumbar spine. *Spine,* **2,** 287-293.

Reiber, J.H.C., Gerbrands, J.J., Troost, G.J., Kooijman, C.J., & Slump, C.H. (1983). 3-D reconstruction of coronary arterial segments from two projections. In P.H. Heinman and R. Brennecke (Eds.), *Digital imaging in cardiovascular radiology* (pp. 151-163). New York: G. Thieme Verlag.

Reuben, J.D., Brown, R.H., Nash, C.L., & Brower, E.M. (1979). In vivo effects of axial loading on healthy, adolescent spines. *Clinical Orthopaedics,* **139,** 17-27.

Reuben, J.D., Brown, R.H., Nash, C.L., & Brower, E.M. (1982). In vivo effects of axial loading on double-curve scoliotic spines. *Spine,* **7,** 440-447.

Rune, B., Sarnäs, K.V., Selvik, G., & Jacobsson, S. (1986). Roentgen stereometry in the study of craniofacial anomalies: The state of the art in Sweden. *British Journal of Orthodontics,* **13,** 151-157.

Ryd, L., Albrechtsson, B.E.J., Herberts, P., Linstrand, A., & Selvik, G. (1988). Micromotion of noncemented Freeman-Samuelson knee prostheses in gonarthrosis: A roentgen stereophotogrammetric analysis of eight successful cases. *Clinical Orthopaedics,* **229,** 205-212.

Ryd, L., Linstrand, A., Rosenquist, R., & Selvik, G. (1986). Tibial component fixation in knee arthroplasty. *Clinical Orthopaedics,* **213,** 141-149.

Saraste, H., & Östeman, A. (1986). The effect of a device for transverse traction on vertebral rotation in surgery for scoliosis as studied by X-ray stereophotogrammetry. *International Orthopaedics,* **10,** 131-133.

Selvik, G. (1989). Roentgen stereophotogrammetry: A method for the study of the kinematics of the skeletal system. *Acta Orthopaedica Scandinavica* (Suppl), **232,** 1-51.

Siebes, M., D'Agenio, D.Z., & Selzer, R.H. (1985). Computer assessment of hemodynamic severity of coronary artery stenosis from arteriograms. *Computer Methods and Programs in Biomedicine,* **21,** 143-152.

Smith, D.K., An, K.N., Cooney, W.P., 3d, Linscheid, R.L., & Chao, E.Y.S. (1989). Effects of a scaphoid wrist osteotomy on carpal kinematics. *Journal of Orthopaedic Research,* **7,** 590-598.

Smith, L.D.R., & Quarendon, P. (1985). Four-dimensional cardiac imaging. *Society of Photo-optical Instrumentation Engineers,* **593,** 74-76.

Stokes, I.A.F. (1981). Computational technique for optimizing accuracy of radiographic measurements of intervertebral joint motion. In W. Welkowitz (Ed.), *Bioengineering: Proceedings of the 9th Northeast Conference* (pp. 229-233). New York: Pergamon.

Stokes, I.A.F., Bigalow, L.C., & Moreland, M.S. (1987). Three-dimensional spinal curvature in idiopathic scoliosis. *Journal of Orthopaedic Research,* **5**, 102-113.

Stokes, I.A.F., Dansereau, J., & Moreland, M.S. (1989). Rib cage asymmetry in idiopathic scoliosis. *Journal of Orthopaedic Research, 7*, 599-606.

Stokes, I.A.F., & Frymoyer, J.W. (1987). Segmental motion and instability. *Spine,* **12**, 688-691.

Stokes, I.A.F., Medlicott, P.A., & Wilder, D.G. (1980). Measurement of movement in painful intervertebral joints. *Medical & Biological Engineering & Computing,* **18**, 694-700.

Stokes, I.A.F., Shuma-Hartswick, D., & Moreland, M.S. (1988). Measurement of spinal and back shape changes in patients with scoliosis. *Acta Orthopaedica Scandinavica,* **59**, 128-133.

Stokes, I.A.F., Wilder, D.G., Frymoyer, J.W., & Pope, M.H. (1981). Assessment of patients with low back pain by biplanar radiographic measurement of intervertebral motion. *Spine,* **6**, 233-240.

Szirtes, T. (1981). *Development and application of contour radiography.* Unpublished doctoral dissertation, McGill University, Montreal.

Turner-Smith, A.R. (1990). X-ray photogrammetry of artificial joints. *The Photogrammetric Record,* **13**(75), 347-366.

Turner-Smith, A.R., White, S.P., & Bulstrode, C. (1990). X-ray photogrammetry of artificial hip joints: Close-range photogrammetry meets machine vision. In A. Gruen & E.P. Baltsavias (Eds.), *Proceedings of the Society of Photo-optical Instrumentation Engineers,* **1395**, 587-594.

van Dijk, R., Huiskes, R., & Selvik, G. (1979). Roentgen stereophotogrammetric methods for the evaluation of the three-dimensional kinematic behavior of the cruciate ligament length patterns of the human knee joint. *Journal of Biomechanics,* **12**, 727-731.

van Langelaan, E.J. (1983). A kinematic analysis of the tarsal joints: An X-ray photogrammetric study. *Acta Orthopaedica Scandinavica Supplementum,* **204**, 1-269.

Veress, S.A., Lippert, F.G., Hon, M.C.Y., & Takamoto, T. (1979). Patella tracking patterns measurement by analytical X-ray photogrammetry. *Journal of Biomechanics,* **12**, 639-650.

White, A.A. (1969). Analysis of the mechanics of the thoracic spine in man. *Acta Orthopaedica Scandinavica,* (Suppl.), **127**.

Wu, J. (1991). *Three-dimensional reconstruction of coronary arteries from multiple projections.* Doctoral dissertation, University of Utah, Salt Lake City. (University Microfilms No. AD 691-14782).

# Part II

## Mechanical and Neuromuscular Modeling

# Chapter 8

# Euler's and Lagrange's Equations for Linked Rigid-Body Models of Three-Dimensional Human Motion

*James G. Andrews*

Biomechanical investigations often involve a simplified mechanical model of a human performing an activity, the derivation and solution of the equations that govern model behavior, and subsequent validation of the model. One of the most commonly used biomechanical models represents the human body (or some portion thereof) as a collection of interconnected rigid segments subject to applied external forces and moments (e.g., the effects of gravity, muscles, constraints, etc.). The equations governing the three-dimensional (3-D) motion of such models may be obtained by a variety of methods and are referred to by several names (e.g., Euler's equations, McGill & King, 1989; Lagrange's equations, Greenwood, 1988; Kane's equations, Kane & Levinson, 1985; variational equations, Haug, 1992; etc.).

Because of the relative simplicity of most biomechanical system models and the relative ease with which Euler's and Lagrange's equations can be generated for such simplified models, these equations have been and probably will continue to be the equations most frequently used in biomechanical investigations. This chapter describes and illustrates the process of generating Euler's and Lagrange's equations for mechanical models of the human body consisting of interconnected rigid segments with finite mass that can exhibit 3-D motion subject to a general external force-couple system and typical system constraints.

# RESTRICTIONS AND ASSUMPTIONS

Before proceeding, it is useful to list the particular restrictions and assumptions that apply to the analysis that follows.

## System Model

The physical system of interest (whether the entire human body or some portion thereof) is represented by a simplified mechanical model composed of a collection of massive rigid segments interconnected by smooth, spherical (ball-and-socket) joints. The inclusion of deformable system elements and the representation of segment articulations by more elaborate joint models (e.g., allowing for relative segmental translations in the joint neighborhood that are resisted by springs and dashpots, etc.) are beyond the scope of this analysis.

## System Coordinates and Constraints

General 3-D motion of the linkage system model is possible, although not required. System motion is constrained primarily by the intersegmental joints, but also by whatever additional constraints act on the model during the activity of interest (e.g., foot contact with the floor, etc.). Thus to describe the configuration history (i.e., the variable location and orientation) of a typical rigid segment, $S$, requires, in general,

- the specification of three independent linear coordinates (e.g., Cartesian coordinates, $X$, $Y$, $Z$) to locate a characteristic point of $S$ (e.g., the mass-center point $G$) relative to the origin $O$ of some global (inertial) Cartesian coordinate system or reference frame $B$: $OXYZ$ (with the Z-axis directed vertically upward), and
- the specification of three independent ordered angular coordinates (e.g., Cardan angles, $\phi$, $\theta$, $\psi$; see Appendix A) to orient a local Cartesian coordinate system embedded in $S$ (e.g., a reference frame, $R$: $Gxyz$, with origin at $G$) relative to the global system $B$.

The six independent generalized coordinates ($X^G$, $Y^G$, $Z^G$, $\phi$, $\theta$, and $\psi$) are used to specify the variable location and orientation of $S$ (or the local system $R$ embedded in $S$ at $G$) relative to the global system $B$ (Figure 8.1).

## External Force-Couple System

Each rigid system segment $S$ has an identifiable proximal joint $P$ and distal joint $D$ at which the joint resultant forces and joint resultant moments act on $S$ (i.e., $\mathbf{F}^P$ and $\mathbf{M}^P$ at $P$; $\mathbf{F}^D$ and $\mathbf{M}^D$ at $D$; see Figure 8.2). These joint resultants are due to the presence of adjacent segments and represent the combined kinetic effect of the muscles, ligaments, and bones that transmit forces to $S$ in the joint neighborhood (Crowninshield & Brand, 1981). In addition, $S$ is subject to a resultant

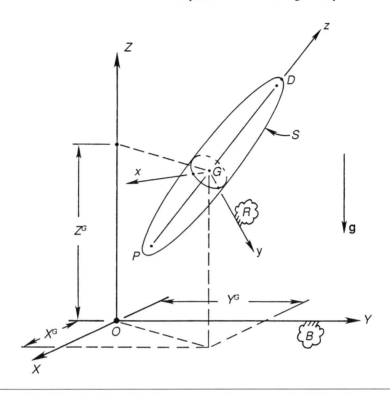

**Figure 8.1**   The global (inertial) Cartesian coordinate system $B$: $OXYZ$ with its origin at Point $O$ and its $Z$-axis directed vertically upward, and a typical rigid segment $S$ with its local Cartesian coordinate system $R$: $Gxyz$ fixed in $S$ at the mass center, Point $G$, and its $z$-axis along the longitudinal axis of geometric and mass symmetry connecting the proximal $P$ and distal $D$ joints.

gravity force $m\mathbf{g}$ acting at $G$ and to an arbitrary external force $\mathbf{F}^A$ and moment $\mathbf{M}^A$ acting at point $A$ (Figure 8.2). Thus, the external force-couple system acting on a typical segment $S$ consists of

(a) $\mathbf{F}^P$ and $\mathbf{M}^P$ at $P$, where $P$ is located relative to $G$ by $\mathbf{r}^{P/G}$,
(b) $\mathbf{F}^D$ and $\mathbf{M}^D$ at $D$, where $D$ is located relative to $G$ by $\mathbf{r}^{D/G}$,
(c) $m\mathbf{g}$ at $G$, where $G$ is located relative to the origin $O$ of the global system $B$ by $\mathbf{r}^{G/O}$, and
(d) $\mathbf{F}^A$ and $\mathbf{M}^A$ at $A$, where $A$ is located relative to $G$ by $\mathbf{r}^{A/G}$.

The resultant force $\mathbf{F}$, and the resultant moment $\mathbf{M}^G$ about $G$ associated with this external force-couple system are therefore given by

$$\mathbf{F} = \mathbf{F}^P + \mathbf{F}^D + \mathbf{F}^A + m\mathbf{g} \; ; \tag{8.1}$$

$$\mathbf{M}^G = (\mathbf{r}^{P/G} \times \mathbf{F}^P) + (\mathbf{r}^{D/G} \times \mathbf{F}^D) + (\mathbf{r}^{A/G} \times \mathbf{F}^A) + \mathbf{M}^P + \mathbf{M}^D + \mathbf{M}^A , \tag{8.2}$$

where $\times$ denotes the vector product.

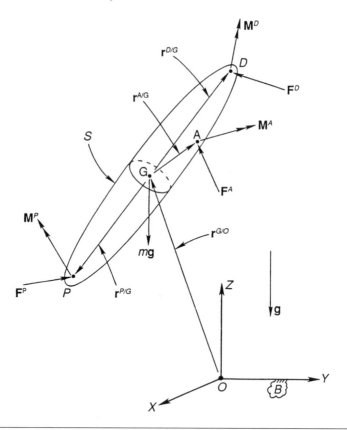

**Figure 8.2**   The components of the general external force-couple system acting on a typical rigid segment $S$.

## Orientation of $R$: $Gxyz$ in $S$

For simplicity and convenience, each rigid segment $S$ in the system model will be assumed to have a longitudinal axis of geometric and mass symmetry that passes through $P$ and $D$, and on which therefore its mass center $G$ lies. The local system $R$: $Gxyz$ embedded in $S$ at $G$ will be oriented and fixed in $S$ such that its $x$, $y$, and $z$ axes are principal axes of inertia for $S$ at $G$ (see subsequent discussion of inertia properties), with the $z$-axis coincident with the segment's longitudinal axis of geometric and mass symmetry.

## Inverse Dynamics Problem

The mathematical problem of interest here, and one that is frequently encountered in biomechanical investigations, is commonly called the inverse dynamics problem (Crowninshield & Brand, 1981). In such problems and in what follows, the configuration history of the linkage system model is known (i.e., the time variation of the six independent generalized coordinates that describe the configuration of each system element $S$ in the global system $B$ have been determined), and the equations of motion that govern model behavior (e.g., Euler's or Lagrange's equations) are used to determine the unknown components of the external force-couple system that together produce this known motion.

## Governing Equations

Only the process of generating Euler's and Lagrange's equations of motion for the system model is considered here. Other methods for obtaining the system equations of motion (e.g., variational methods) are beyond the scope of this chapter.

## Theoretical Basis

It is assumed that the reader is familiar with the methods and procedures of mechanical analysis used to describe two-dimensional (2-D) rigid body motion (e.g., material normally covered in an undergraduate physics or engineering dynamics course and usually employing vector algebra, vector calculus, elementary ordinary differential equations, matrices, etc.). With such a background and some additional effort, one should be able to understand and use the extensions of 2-D methodology to analyze general 3-D rigid body motion. Note, however, that this presentation should not be regarded as a comprehensive, self-contained treatment, even for the greatly simplified situations that it considers.

# PRELIMINARY TASKS

Assuming that the system model, the global and local coordinate systems, and the segment generalized coordinates have been selected, it is necessary to complete some additional preliminary tasks before constructing either Euler's or Lagrange's equations for the system model.

## Inertia Properties (Body Segment Parameters)

For each segment $S$, it is necessary to know, by estimation or by direct measurement, the mass $m$, the location of the mass center $G$, and the six independent components of the $3 \times 3 = 9$ element, symmetric, mass-center inertia matrix $\mathbf{I}^G$. Expressed in terms of local $R$-system components, these six independent inertia matrix elements include the three moments of inertia that appear as diagonal elements,

$$I^G_{xx} = \int (y^2 + z^2)\, dm \, ,$$

$$I^G_{yy} = \int (z^2 + x^2)\, dm \, , \tag{8.3}$$

$$I^G_{zz} = \int (x^2 + y^2)\, dm \, ,$$

and the three products of inertia that appear symmetrically as off-diagonal elements (note the conventional negative algebraic signs in these definitions):

$$I^G_{xy} = -\int xy\, dm = I^G_{yx} \, ,$$

$$I^G_{yz} = -\int yz\, dm = I^G_{zy} \, , \tag{8.4}$$

$$I^G_{zx} = -\int zx\, dm = I^G_{xz} \, .$$

Thus, the mass-center inertia matrix $\mathbf{I}^G$ can be expressed, in local $R$-system components,

$$\mathbf{I}^G = \begin{bmatrix} I^G_{xx} & I^G_{xy} & I^G_{xz} \\ I^G_{yx} & I^G_{yy} & I^G_{yz} \\ I^G_{zx} & I^G_{zy} & I^G_{zz} \end{bmatrix} = (\mathbf{I}^G)^T \, , \tag{8.5}$$

where the superscript $T$ denotes the transpose and indicates that $\mathbf{I}^G$ is a symmetric matrix.

Note that because the local system $R$ is fixed in $S$ at $G$ and $S$ is rigid, the $R$ components of $\mathbf{I}^G$ are constants and do not change as $S$ moves in the global system $B$. Note also that for different orientations of $R$ fixed in $S$, the elements of $\mathbf{I}^G$ take on different constant values. It can be shown (McGill & King, 1989) that there always exists a particular special orientation for the axes of $R$: $Gxyz$ in $S$, called the principal axes of inertia orientation, such that (a) all three independent products of inertia simultaneously vanish, and (b) the three corresponding principal moments of inertia include the largest and the smallest moments of inertia for all possible axes passing through $G$. Recalling the previous assumption that each segment $S$ has a longitudinal $z$-axis of geometric and mass symmetry that passes through $P$ and $D$ and contains $G$ and denoting the segment's principal moments of inertia at $G$ by

$$I^G_{xx} = I^G_{yy} = I_t \, ; \; I^G_{zz} = I_l \, ,$$

the segment's mass-center inertia matrix $\mathbf{I}^G$ can therefore be expressed, in local $R$ components,

$$\mathbf{I}^G = \begin{bmatrix} I^G_{xx} & 0 & 0 \\ 0 & I^G_{yy} & 0 \\ 0 & 0 & I^G_{zz} \end{bmatrix} = \begin{bmatrix} I_t & 0 & 0 \\ 0 & I_t & 0 \\ 0 & 0 & I_l \end{bmatrix} = (\mathbf{I}^G)^T. \tag{8.6}$$

# Kinematics

Letting

$$\mathbf{q} = [X^G, Y^G, Z^G, \phi, \theta, \psi]^T = [q_1, \ldots, q_6]^T$$

denote the $6 \times 1$ vector of scalar generalized coordinates $q_k$ ($k = 1, \ldots, 6$) for a typical segment $S$, it is necessary to write all appropriate constraint equations that these six scalar coordinates must satisfy, for all system segments, during system motion. This task is problem specific (see following examples) and is regarded here as having resulted in a set of $m$ independent, scalar, holonomic (i.e., integrable) constraint equations. These constraint equations, when they exist, can always be expressed in the form

$$\mathbf{c}(\mathbf{q}, t) = \mathbf{0}, \tag{8.7}$$

where the $m \times 1$ vector $\mathbf{c}$ is a function of the system's generalized coordinates $q_k$ and possibly the time, $t$. For a system model composed of $N$ rigid segments, the generalized coordinate vector $\mathbf{q}$ in Equation 8.7 is to be regarded as a $6N \times 1$ vector composed of all $N$, $6 \times 1$ generalized coordinate vectors for the individual segments. Nonholonomic (i.e., differential but nonintegrable) constraints, and differential constraints for which integrals exist but cannot be found, are relatively rare and are not considered in this analysis.

The preliminary kinematic tasks are completed by deriving expressions for each segment's mass-center velocity $\mathbf{v}^G$ and angular velocity $\Omega$ as functions of the segment's six generalized coordinates and their time-derivatives. This process leads to an expression for $\mathbf{v}^G$ of the form

$$\mathbf{v}^G = d(\mathbf{r}^{G/O})/dt = d(X^G\mathbf{I} + Y^G\mathbf{J} + Z^G\mathbf{K})/dt = \dot{X}^G\mathbf{I} + \dot{Y}^G\mathbf{J} + \dot{Z}^G\mathbf{K}, \tag{8.8}$$

where the overdot denotes the time-derivative and $\mathbf{I}$, $\mathbf{J}$, and $\mathbf{K}$ are, respectively, the unit vectors associated with the positive $X$, $Y$, and $Z$ axes of the global $B$ system. Following a different and angular coordinate dependent procedure (see Appendix A), the angular velocity $\Omega$ can be expressed, in local $R$ components,

$$\Omega = \Omega_x \mathbf{i} + \Omega_y \mathbf{j} + \Omega_z \mathbf{k}, \tag{8.9}$$

where

$$\Omega_x = \dot{\phi}c\theta c\psi + \dot{\theta}s\psi,$$

$$\Omega_y = -\dot{\phi}c\theta s\psi + \dot{\theta}c\psi, \tag{8.10}$$

$$\Omega_z = \dot{\phi}s\theta + \dot{\psi},$$

with $\mathbf{i}$, $\mathbf{j}$, and $\mathbf{k}$ the unit vectors associated with the positive $x$, $y$, and $z$ axes, respectively, of the local $R$ system, and the symbols "$s$" and "$c$" denoting the sine and cosine functions, respectively.

## Angular Momentum

The final preliminary task involves writing an expression for each segment's angular momentum about its own mass center $G$. This vector quantity, denoted by $\mathbf{H}^G$, can be expressed, in local $R$ components,

$$\mathbf{H}^G = \mathbf{I}^G\ \Omega = H^G_x \mathbf{i} + H^G_y \mathbf{j} + H^G_z \mathbf{k}, \qquad (8.11)$$

where

$$H^G_x = I^G_{xx}\Omega_x,$$

$$H^G_y = I^G_{yy}\Omega_y, \qquad (8.12)$$

$$H^G_z = I^G_{zz}\Omega_z.$$

# EULER'S EQUATIONS

Euler's equations conveniently describe the motion of a single rigid body or, when collected together, the motion of a system of $N$ interconnected rigid bodies. These equations relate the resultant external force ($\mathbf{F}$) and the resultant external moment about the mass center $G$ ($\mathbf{M}^G$) that act on each rigid body $S$, to the motion of $S$ that these kinetic quantities produce. Euler's equations are extensions of Newton's second law for particle motion (i.e., $\mathbf{F} = m\mathbf{a}$), and are second-order, ordinary differential equations that express Euler's first and second laws of motion for multiparticle systems (McGill & King, 1989). When the multiparticle system is a single rigid body $S$ with mass center $G$, these laws can be expressed as follows.

## Euler's First Law

The principle of the motion of the mass center, or the principle of linear momentum, takes the form,

$$\mathbf{F} = m\mathbf{a}^G, \qquad (8.13)$$

where $\mathbf{F}$ is the resultant external force acting on $S$ (Equation 8.1), $m$ is the segment mass, and $\mathbf{a}^G$ is the acceleration of $G$ in the global (inertial) reference frame B as given by the time-derivative of $\mathbf{v}^G$ in Equation 8.8. Consequently,

$$\mathbf{a}^G = d(\mathbf{v}^G)/dt = d(\dot{X}^G\mathbf{I} + \dot{Y}^G\mathbf{J} + \dot{Z}^G\mathbf{K})/dt = \ddot{X}^G\mathbf{I} + \ddot{Y}^G\mathbf{J} + \ddot{Z}^G\mathbf{K}, \qquad (8.14)$$

where the double overdot indicates the second derivative with respect to time.

## Euler's Second Law

The principle of angular momentum takes the form

$$\mathbf{M}^G = \dot{\mathbf{H}}^G , \qquad (8.15)$$

where $\mathbf{M}^G$ is the resultant external moment acting on $S$ about $G$ as in Equation 8.1, and $\dot{\mathbf{H}}^G$ is the time-derivative of the segment's angular momentum about $G$ given by Equation 8.11. Noting that Equation 8.11 gives $\mathbf{H}^G$ in local $R$ components, its time-derivative can be expressed (Greenwood, 1988),

$$\dot{\mathbf{H}}^G = [I^G_{xx}\,\dot{\Omega}_x - (I^G_{yy} - I^G_{zz})\,\Omega_y\Omega_z]\mathbf{i}$$

$$+ [I^G_{yy}\dot{\Omega}_y - (I^G_{zz} - I^G_{xx})\Omega_z\Omega_x]\mathbf{j} \qquad (8.16)$$

$$+ [I^G_{zz}\dot{\Omega}_z - (I^G_{xx} - I^G_{yy})\Omega_x\Omega_y]\mathbf{k} .$$

The local $R$ components of $\Omega$ in Equation 8.16 are given in Equation 8.10, and the time-derivatives of these components can be expressed

$$\dot{\Omega}_x = \ddot{\phi}c\theta c\psi - \dot{\phi}\dot{\theta}s\theta c\psi - \dot{\phi}\dot{\psi}c\theta s\psi + \ddot{\theta}s\psi + \dot{\theta}\dot{\psi}c\psi ;$$

$$\dot{\Omega}_y = -\ddot{\phi}c\theta s\psi + \dot{\phi}\dot{\theta}s\theta s\psi - \dot{\phi}\dot{\psi}c\theta c\psi + \ddot{\theta}c\psi - \dot{\theta}\dot{\psi}s\psi ; \qquad (8.17)$$

$$\dot{\Omega}_z = \ddot{\phi}s\theta + \dot{\phi}\dot{\theta}c\theta + \ddot{\psi} .$$

Thus, for a single rigid segment $S$, Euler's equations are a system of two independent vector equations (Equations 8.13 and 8.15) or a system of six independent scalar equations. Writing the three scalar equations corresponding to Equation 8.13 in global $B$ components, and the three scalar equations corresponding to Equation 8.15 in local $R$ components, Euler's six scalar equations for the motion of $S$ in the global reference frame $B$ can be expressed

$$F_X = m\ddot{X}^G ,$$

$$F_Y = m\ddot{Y}^G , \qquad (8.18)$$

$$F_Z = m\ddot{Z}^G ,$$

$$M^G_x = I^G_{xx}\dot{\Omega}_x - (I^G_{yy} - I^G_{zz})\Omega_y\Omega_z ,$$

$$M^G_y = I^G_{yy}\dot{\Omega}_y - (I^G_{zz} - I^G_{xx})\Omega_z\Omega_x , \qquad (8.19)$$

$$M^G_z = I^G_{zz}\dot{\Omega}_z - (I^G_{xx} - I^G_{yy})\Omega_x\Omega_y ,$$

where $\mathbf{F}$ from Equation 8.1 is expressed in global $B$ components such that

$$\mathbf{F} = F_X\mathbf{I} + F_Y\mathbf{J} + F_Z\mathbf{K},$$

and $\mathbf{M}^G$ from Equation 8.2 is expressed in local $R$ components

$$\mathbf{M}^G = M^G_x \mathbf{i} + M^G_y \mathbf{j} + M^G_z \mathbf{k}.$$

Recalling that the $x$, $y$, and $z$ axes of $R$ are principal axes of inertia for $S$ at $G$ with

$$I^G_{xx} = I^G_{yy} = I_t \text{ and } I^G_{zz} = I_l ,$$

and using Equations 8.10 and 8.17, Equations 8.19 can be re-expressed, in terms of the three Cardan angles and their time-derivatives,

$$M^G_x = I_t (\ddot{\phi}c\theta c\psi + \ddot{\theta}s\psi + \dot{\phi}^2 s\theta c\theta s\psi - 2\dot{\phi}\dot{\theta}s\theta c\psi)$$
$$- I_t (\dot{\phi}^2 s\theta c\theta s\psi - \dot{\phi}\dot{\theta}s\theta c\psi + \dot{\phi}\dot{\psi}c\theta s\psi - \dot{\theta}\dot{\psi}c\psi);$$

$$M^G_y = I_t (-\ddot{\phi}c\theta s\psi + \ddot{\theta}c\psi + \dot{\phi}^2 s\theta c\theta c\psi + 2\dot{\phi}\dot{\theta}s\theta s\psi) \qquad (8.20)$$
$$- I_t (\dot{\phi}^2 s\theta c\theta c\psi + \dot{\phi}\dot{\theta}s\theta s\psi + \dot{\phi}\dot{\psi}c\theta c\psi + \dot{\theta}\dot{\psi}s\psi);$$

$$M^G_z = I_l (\ddot{\phi}s\theta + \ddot{\psi} + \dot{\phi}\dot{\theta}c\theta).$$

For a single rigid body, $S$, Euler's six independent differential equations (Equations 8.18 and 8.19), together with the $m$ independent algebraic constraint equations (Equation 8.7), form a system of $6 + m$ scalar differential-algebraic equations (Haug, 1992) that govern the motion of $S$ in the global reference frame $B$. When viewed in terms of the associated inverse dynamics problem, the $m$ constraint equations (Equation 8.7) can be used to reduce the number of points that must be digitized to determine the motion-dependent right-hand sides of Euler's equations, and these equations can then be solved for at most six unknown scalar components of the external force-couple system acting on $S$.

For a system model with $N$ interconnected rigid segments, the collection of $6N$ independent scalar Euler's equations, together with the $m$ independent scalar constraint equations (Equation 8.7), form a system of $6N + m$ scalar differential-algebraic equations that govern the motion of the system model in the global reference frame $B$. When viewed in terms of the associated inverse dynamics problem, the $m$ constraint equations (Equation 8.7) can be used to reduce the number of points that must be digitized to determine the motion-dependent right-hand sides of the collection of Euler's equations, and this collection of $6N$ scalar equations can then be solved for at most $6N$ scalar components of the associated external force-couple system.

# LAGRANGE'S EQUATIONS

Lagrange's equations may also be used to describe the motion of a single rigid body or the motion of a system of interconnected rigid bodies. These equations, which can be derived from Euler's equations for multiparticle systems (Greenwood, 1988), relate the motion of the system to the external forces and moments that do virtual work on the system during a kinematically admissible virtual displacement (i.e., an infinitesimal or first-order displacement of the system that satisfies all constraints on the system's generalized coordinates). Note that because the joint resultant forces acting on adjacent segments do zero net virtual work on the multibody system (i.e., the virtual work done on one joint segment is equal in magnitude but opposite in algebraic sign to the virtual work done on the other joint segment), these workless constraint forces do not appear in Lagrange's equations. Hence, Lagrange's equations cannot be used, within the context of the associated inverse dynamics problem, to determine the joint resultant forces acting on adjacent segments in multibody systems.

Before generating Lagrange's equations, some additional preliminary tasks must be completed. These include deriving expressions for the system's kinetic energy $T$, its gravitational potential energy $V$, its Lagrangian $L$, and its active generalized force $\mathbf{Q}'$.

## Kinetic Energy

For each rigid segment $S$, the kinetic energy $T^S$ can be expressed

$$T^S = T^S_v = T^S_\Omega , \tag{8.21}$$

where

$$T^S_v = (1/2)m(\mathbf{v}^G)^T\mathbf{v}^G = (1/2)m[(\dot{X}^G)^2 + (\dot{Y}^G)^2 + (\dot{Z}^G)^2] \tag{8.22}$$

is the translational kinetic energy, and

$$T^S_\Omega = (1/2)\,\mathbf{\Omega}^T\mathbf{H}^G$$

$$= (1/2)[I^G_{xx}(\Omega_x)^2 + I^G_{yy}(\Omega_y)^2 + I^G_{zz}(\Omega_z)^2] \tag{8.23}$$

is the rotational kinetic energy. For a system of $N$ interconnected rigid segments, the system's kinetic energy $T$ is simply the algebraic sum of the segmental kinetic energies:

$$T = \Sigma(T^S). \tag{8.24}$$

## Gravitational Potential Energy

For each segment $S$, the gravitational potential energy $V^S$ can be expressed

$$V^S = mgZ^G, \tag{8.25}$$

where $m$ is the segment mass, $g$ is the local acceleration due to gravity, and $Z^G$ denotes the vertical coordinate of $G$ above the horizontal $X, Y$ plane of the global system $B$. For a system of $N$ interconnected rigid segments, the system's gravitational potential energy $V$ is simply the algebraic sum of the segmental gravitational potential energies.

$$V = \Sigma(V^S).$$  (8.26)

## Lagrangian

For a mechanical system consisting of one or more rigid segments, the associated Lagrangian $L$ is defined

$$L = T - V ,$$  (8.27)

where $T$ and $V$ are given by Equations 8.24 and 8.26, respectively.

## Generalized Force

The generalized force $\mathbf{Q}$ for a mechanical system with an $n \times 1$ generalized coordinate vector $\mathbf{q}$ is an $n \times 1$ vector obtained by writing the virtual work, $\delta W$, done on the mechanical system by the associated external force-couple system during a kinematically admissible virtual displacement of the system. Although a detailed description of the virtual work concept is beyond the scope of this presentation, a brief outline of the process required for the construction of the active generalized force vector $\mathbf{Q}'$ appearing in Lagrange's equations is presented in Appendix B. It will be assumed here that the virtual work done on the system model during a kinematically admissible virtual change in the system's generalized coordinate vector $\mathbf{q}$ has been constructed in accordance with the procedures outlined in Appendix B. This process leads to an expression for the system's active generalized force $\mathbf{Q}'$ in the form:

$$\mathbf{Q}' = \Sigma \, [(\mathbf{D}^A)^T \, \mathbf{F}^A] + \Sigma(\mathbf{J}^T\mathbf{M}^A) - \mathbf{K}^T\boldsymbol{\mu}.$$  (8.28)

In Equation 8.28, the first righthand-side sum is carried out over all external forces $\mathbf{F}^A$ acting on the system at points $A$ (except for the gravity forces); the corresponding matrix $\mathbf{D}^A$ is determined by the velocity of $A$ as given by Equation B.1; the second righthand-side sum is carried out over all external moments $\mathbf{M}^A$ acting on the system at points $A$; the corresponding matrix $\mathbf{J}$ is determined by the angular velocity of the segment that contains $A$ as given by Equation B.4; the constraint matrix $\mathbf{K}$, if present, is determined by the holonomic constraint equation 8.7 as given by Equation B.7; and $\boldsymbol{\mu}$ is the Lagrange multiplier vector associated with the constraints on the system's generalized coordinates.

For a holonomically constrained mechanical system of interconnected rigid segments whose configuration is specified by a generalized coordinate vector $\mathbf{q}$ with $n$ components $q_k$ $(k = 1, \ldots , n)$, Lagrange's equations can be expressed

$$d(\partial L/\partial \dot{q}_k)/dt - \partial L/\partial q_k = Q'_k ; k = 1, \ldots , n$$  (8.29)

where the system's Lagrangian $L$ is defined by Equation 8.27, and the system's active generalized force vector $\mathbf{Q}'$ has $n$ components $Q'_k$ as given by Equation 8.28. Thus, Lagrange's equations are a set of $n$ independent, scalar, second-order, ordinary differential equations that relate system motion (expressed in terms of the system's $n$ generalized coordinates $q_k$ and their time-derivatives) to the external forces ($\mathbf{F}^A$), external moments ($\mathbf{M}^A$), and holonomic constraints ($\boldsymbol{\mu}$) that together produce that motion.

In the absence of constraints, the $n$ generalized coordinates $q_k$ are independent and Lagrange's equations (Equation 8.29) form a system of $n$ independent, scalar, differential equations that govern model behavior. When viewed in the context of the associated inverse dynamics problem, these $n$ scalar equations can be used to solve for, at most, $n$ unknown scalar components of the asociated external force-couple system that do virtual work on the system during a kinematically admissible virtual displacement. However, if the $n$ generalized coordinates $q_k$ are constrained to satisfy $m$ independent scalar holonomic constraint equations (Equation 8.7), these $m$ scalar equations must be appended to the $n$ scalar Lagrange's equations (Equation 8.29) to form the appropriate differential-algebraic equations that govern model behavior. When viewed in the context of the associated inverse dynamics problem, this system of $n + m$ scalar equations can be used to solve for, at most, $n + m$ unknown scalar components of the associated external force-couple system that do virtual work on the system during a kinematically admissible virtual displacement.

For a single rigid, axially-symmetric segment $S$ with generalized coordinate vector

$$\mathbf{q} = [X^G, Y^G, Z^G, \phi, \theta, \psi]^T = [q_1, \ldots, q_6]^T$$

that moves in the global system $B$ subject to an external force-couple system with resultant force ($\mathbf{F}$) and resultant moment about $G$ ($\mathbf{M}^G$) given by Equations 8.1 and 8.2, respectively, Langrange's equations (Equation 8.29) can be expressed (see Equation C.15)

$$m\ddot{X}^G = F_X \,,$$

$$m\ddot{Y}^G = F_Y \,, \tag{8.30}$$

$$m\ddot{Z}^G = F_Z \,,$$

$$I_t(\ddot{\phi}c^2\theta - 2\dot{\phi}\dot{\theta}s\theta c\theta) + I_t(\ddot{\phi}s^2\theta + \ddot{\psi}s\theta + 2\dot{\phi}\dot{\theta}s\theta c\theta + \dot{\theta}\dot{\psi}c\phi)$$

$$= M^G_x c\theta c\psi - M^G_y c\theta s\psi + M^G_z s\theta \,; \tag{8.31}$$

$$I_t(\ddot{\theta} + \dot{\phi}^2 s\theta c\theta) - I_l\,(\dot{\phi}^2 s\theta c\theta + \dot{\phi}\dot{\psi}c\theta) = M^G_x s\psi + M^G_y c\psi \,;$$

$$I_t(\ddot{\phi}s\theta + \ddot{\psi} + \dot{\phi}\dot{\theta}c\theta) = M^G_z \; ; \tag{8.32}$$

where **F** is expressed in global $B$ components, $\mathbf{M}^G$ is expressed in local $R$ components, and

$$I^G_{xx} = I^G_{yy} = I_t \; ; \quad I^G_{zz} = I_l \; . \tag{8.33}$$

# APPLICATIONS OF LAGRANGE'S EQUATIONS

Lagrange's equations may be used to describe the behavior of system models in a wide variety of biomechanical investigations. The examples that follow serve to illustrate their derivation in two commonly occurring situations.

## Example 1: Three-Segment, Open-Loop Model

To illustrate the application of Lagrange's equations to systems modeled as collections of massive rigid segments interconnected by smooth spherical joints, consider first a three-segment, open-loop model (Figure 8.3, a and b; e.g., the

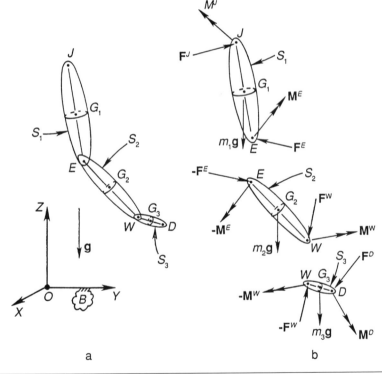

a

b

**Figure 8.3** (a) A three-segment, open-loop, mechanical model of the upper extremity moving in general 3-D motion relative to the global coordinate system $B$: $OXYZ$, with Points $J$, $E$, and $W$ representing the shoulder, elbow, and wrist joints, respectively. (b) Separate free-body diagrams of the three system segments.

upper extremity, with points $J$, $E$, and $W$ denoting the shoulder, elbow, and wrist, respectively). Let this system exhibit general 3-D motion such that the distal end $D$ of the distal segment $S_3$ (the hand) is subject to known joint resultants $\mathbf{F}^D$ and $\mathbf{M}^D$, and the proximal end $J$ of the proximal segment $S_1$ (the upper arm) is subject to unknown joint resultants $\mathbf{F}^J$ and $\mathbf{M}^J$.

Let the configuration of this system model be specified by the $12 \times 1$ generalized coordinate vector $\mathbf{q}$, where the first three components of $\mathbf{q}$ are the three independent global $B$ components $(X^J, Y^J, Z^J)$ of the vector $\mathbf{r}^{J/O}$ that locates the moving joint $J$ relative to $O$, and the remaining nine components of $\mathbf{q}$ are the three sets of three independent Cardan angles $(\phi_i, \theta_i, \psi_i ; i = 1, 2, 3)$ that orient each segment $S_i$ of mass $m_i$ (or $R_i : G_i x_i y_i z_i$ embedded in $S_i$ at $G_i$) relative to the global system $B$. Thus

$$\mathbf{q} = [X^J, Y^J, Z^J, \phi_1, \theta_1, \psi_1, \phi_2, \theta_2, \psi_2, \phi_3, \theta_3, \psi_3]^T$$

$$= [q_1, \ldots, q_{12}]^T . \tag{8.34}$$

The velocity difference equation for two points $P$ and $G$ fixed in a rigid segment $S$ takes the form (Greenwood, 1988)

$$\mathbf{v}^G = \mathbf{v}^P + ( \mathbf{\Omega} \times \mathbf{r}^{G/P}), \tag{8.35}$$

where $\mathbf{v}^G$ is the velocity of $G$, $\mathbf{\Omega}$ is the angular velocity of $S$ (given in terms of that segment's Cardan angles and their time-derivatives by Equations 8.9 and 8.10), $\mathbf{v}^P$ is the velocity of $P$, and $\mathbf{r}^{G/P}$ locates $G$ relative to $P$ (and also can be expressed in terms of that segment's Cardan angles). Hence, starting with segment $S_1$ and moving distally through the linkage system, Equation 8.35 can be used repeatedly to write $\mathbf{v}^G$ for each segment as a function of the components of $\mathbf{q}$ and their time-derivatives. Equations 8.21 through 8.24 can then be used to express the system's kinetic energy $T$ as a function of the components of $\mathbf{q}$ and their time-derivatives.

In a similar fashion, the vector addition law,

$$\mathbf{r}^{G/O} = \mathbf{r}^{G/P} + \mathbf{r}^{P/O}, \tag{8.36}$$

can be used repeatedly to relate $Z^J$ for each segment to $Z^J$ and to the nine Cardan angles that orient the three segments in the global system $B$. Hence, Equations 8.25 and 8.26 can be used to express the system's gravitational potential energy $V$ as a function of the components of $\mathbf{q}$. The system's Lagrangian $L$, given by Equation 8.27, can then be expressed as a function of the components of $\mathbf{q}$ and their time-derivatives, and the left-hand sides of Lagrange's equations (Equation 8.29) can be routinely constructed.

The system's $12 \times 1$ active generalized force vector,

$$\mathbf{Q}' = [Q'_1, \ldots, Q'_{12}]^T ,$$

is obtained by constructing the active virtual work $\delta W'_i$ for each segment $S_i$ using Equation B.12, summing these three algebraically-signed scalars to form the system's active virtual work $\delta W'$, and then using Equation B.13 to identify the scalar generalized forces $Q'_k$ corresponding to each scalar generalized coordinate $q_k$ ($k = 1, \ldots, 12$). Recalling that the gravity forces are ignored when calculating each segment's active virtual work, and that the joint resultant forces at the intermediate system joints $E$ and $W$ do zero net virtual work on the three-segment system, this process leads to the following results (see free-body diagrams in Figure 8.3).

$$\delta W' = \Sigma(\delta W'_i) = \Sigma(Q'_k \delta q_k) , \tag{8.37}$$

where

$$\delta W'_3 = (\mathbf{F}^D)^T \delta \mathbf{r}^D + (\mathbf{M}^D - \mathbf{M}^W)^T \delta \boldsymbol{\pi}_3 ,$$

$$\delta W'_2 = (\mathbf{M}^W - \mathbf{M}^E)^T \delta \boldsymbol{\pi}_2 , \tag{8.38}$$

$$\delta W'_1 = (\mathbf{F}^J)^T \delta \mathbf{r}^J + (\mathbf{M}^E + \mathbf{M}^J)^T \delta \boldsymbol{\pi}_1 ;$$

or

$$\Sigma(Q'_k \delta q_k) = (\mathbf{F}^D)^T \delta \mathbf{r}^D + (\mathbf{F}^J)^T \delta \mathbf{r}^J + (\mathbf{M}^D)^T \delta \boldsymbol{\pi}_3 + (\mathbf{M}^J)^T \delta \boldsymbol{\pi}_1$$

$$+ (\mathbf{M}^W)^T(\delta \boldsymbol{\pi}_2 - \delta \boldsymbol{\pi}_3) + (\mathbf{M}^E)^T(\delta \boldsymbol{\pi}_1 - \delta \boldsymbol{\pi}_2) . \tag{8.39}$$

Hence, only the joint resultant forces at the system's extreme joints $D$ and $J$ contribute to the scalar active generalized forces $Q'_k$, and the joint resultant moments at the system's intermediate joints $E$ and $W$ contribute to the $Q'_k$ quantities as a function of the virtual joint rotations (i.e., the differences between the adjacent segment's virtual rotations).

The system's active generalized forces $Q'_k$ on the right-hand sides of Lagrange's equations (Equation 8.29) are identified from Equation 8.39, and Lagrange's equations finally can then be constructed. When viewed in the context of the associated inverse dynamics problem, these 12 independent scalar equations can be used to determine at most 12 scalar kinetic unknowns. Because the joint resultants $\mathbf{F}^D$ and $\mathbf{M}^D$ at $D$ are regarded as known (either identically equal to zero or measured experimentally during system motion), Equation 8.39 indicates that Lagrange's equations can be used to determine the remaining 12 scalar kinetic unknowns (i.e., the three components of the unknown joint resultant force $\mathbf{F}^J$ at $J$, plus the $3 \times 3 = 9$ components of the unknown joint resultant moments $\mathbf{M}^J$ at $J$, $\mathbf{M}^E$ at $E$, and $\mathbf{M}^W$ at $W$).

# Example 2: Four-Segment, Closed-Loop Model

As a second illustration of the application of Lagrange's equations to multibody systems of massive rigid segments interconnected by smooth spherical joints, consider a four-segment, closed-loop model (Figure 8.4, a and b), for example, the lower extremity plus stationary bicycle system, with three moving segments—the

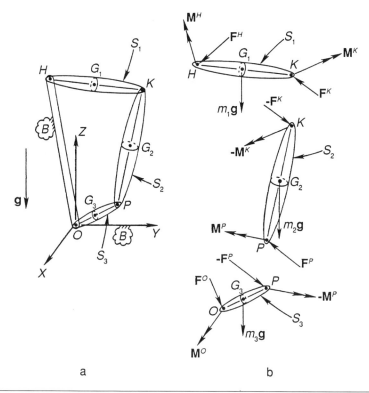

**Figure 8.4** (a) A four-segment, closed-loop, mechanical model of the lower extremity plus stationary bicycle system moving in general 3-D motion relative to the global coordinate system $B: OXYZ$, with points $H$, $K$, and $P$ representing the hip, knee, and pedal joints, respectively. (b) Separate free-body diagrams of the three moving system segments.

thigh, $S_1$; the shank-plus-foot, $S_2$; and the pedal crank, $S_3$—and one fixed segment—the extended bicycle frame connecting the sprocket center $O$ and the hip $H$—and with intermediate joints $K$ and $P$ denoting the knee and pedal joints, respectively. Note that this simplified model permits no motion of the foot relative to the shank at the ankle.

Let the configuration of this system model be specified by the $9 \times 1$ generalized coordinate vector $\mathbf{q}$, where the nine components $q_k$ ($k = 1, \ldots, 9$) are the three sets of three independent Cardan angles ($\phi_i, \theta_i, \psi_i; i = 1, 2, 3$) that orient each moving segment $S_i$ with mass $m_i$ (or $R_i$: $G_i x_i y_i z_i$ embedded in $S_i$ at $G_i$) relative to the global system $B$. In contrast to the previous example in which an open-loop multibody system model was considered, the components of the generalized coordinate vector $\mathbf{q}$ used to specify the configuration of this closed-loop system model must satisfy three independent scalar constraint equations. These constraint equations are the three scalar components of the vector equation that expresses the fact that the four interconnected system segments must always form a closed loop. This loop closure constraint can be expressed

$$\mathbf{r}^{H/O} = \mathbf{r}^{H/K} + \mathbf{r}^{K/P} + \mathbf{r}^{P/O} , \qquad (8.40)$$

where each relative location vector in Equation 8.40 can be written as a function of the components of $\mathbf{q}$. Thus, Equation 8.40 is a holonomic constraint that can be written in the form of Equation 8.7, where $\mathbf{c}$ is a $3 \times 1$ vector function of the components of $\mathbf{q}$, and the time $t$ does not appear explicitly.

Proceeding in a manner similar to that followed in the previous open-loop example, the individual segment's contributions to the system's kinetic energy $T$ and gravitational potential energy $V$ are first constructed in terms of the nine scalar generalized coordinates $q_k$ and their time-derivatives. Adding these segmental quantities to form the corresponding system energies, the system's Lagrangian $L$ is constructed, and the lefthand-side terms of the nine scalar Lagrange's equations (Equation 8.29) can then be routinely determined.

The components $Q'_k$ of the system's $9 \times 1$ active generalized force vector $\mathbf{Q}'$ are obtained, for this case of constrained generalized coordinates, by modifying the procedure used in the previous open-loop example where the generalized coordinates $q_k$ were unconstrained. This modification involves introducing an unknown $3 \times 1$ Lagrange multiplier vector $\boldsymbol{\mu}$ in the expression for $\mathbf{Q}'$ given in Equation 8.28, where the active virtual work done by the holonomic constraints is given by Equation B.15, and the corresponding active generalized force is given by Equation B.16. Thus, in a manner similar to that used in the previous open-loop example, the closed-loop system's active virtual work, $\delta W'$ (see free-body diagrams in Figure 8.4), can be expressed

$$\delta W' = \Sigma(\delta W'_i) + \delta W^c = \Sigma(Q'_k \delta q_k), \qquad (8.41)$$

where

$$\delta W'_3 = (\mathbf{M}^O - \mathbf{M}^P)^T \delta \boldsymbol{\pi}_3 ,$$

$$\delta W'_2 = (\mathbf{M}^P - \mathbf{M}^K)^T \delta \boldsymbol{\pi}_2 , \qquad (8.42)$$

$$\delta W'_1 = (\mathbf{M}^K + \mathbf{M}^H)^T \delta \boldsymbol{\pi}_1 ,$$

and

$$\delta W^c = -\boldsymbol{\mu}^T \mathbf{K} \delta \mathbf{q}$$

represents the active virtual work done by the constraints. Thus

$$\Sigma(Q'_k \delta q_k) = (\mathbf{M}^O)^T \delta \boldsymbol{\pi}_3 + (\mathbf{M}^H)^T \delta \boldsymbol{\pi}_1 + (\mathbf{M}^P)^T (\delta \boldsymbol{\pi}_2 - \delta \boldsymbol{\pi}_3)$$

$$+ (\mathbf{M}^K)^T (\delta \boldsymbol{\pi}_1 - \delta \boldsymbol{\pi}_2) - \boldsymbol{\mu}^T \mathbf{K} \delta \mathbf{q} . \qquad (8.43)$$

Note that none of the joint resultant forces contributes to the $Q'_k$ quantities, and the joint resultant moments at the extreme joints $O$ and $H$ contribute in proportion to the respective segment's virtual rotations, whereas the joint resultant moments at the intermediate joints $K$ and $P$ contribute in proportion to the respective joint's virtual rotations (i.e., the differences between the adjacent segment's virtual rotations).

The nine components $Q'_k$ of the active generalized force vector $\mathbf{Q}'$ that appear on the right-hand sides of Lagrange's equations (Equation 8.29) are identified from Equation 8.43, and Lagrange's equations can then be written. These nine independent scalar differential equations are augmented by the three independent scalar algebraic constraint equations (8.40) to form a system of 12 scalar differential-algebraic equations that govern the motion of the system. When viewed in the context of the associated inverse dynamics problem, these 12 scalar equations can be used to determine at most 12 unknown scalar components of the applied external forces and moments that do virtual work on the system during a kinematically admissible virtual displacement. Because Equation 8.43 involves 15 scalar kinetic quantities (i.e., four joint resultant moments, each with three scalar components, plus the three scalar components of the Lagrange multiplier vector $\boldsymbol{\mu}$ ), the associated inverse dynamics problem cannot be solved unless three of these 15 scalar kinetic quantities have been determined previously by some other means (e.g., by experimentally measuring one of the joint resultant moments). This difficulty is a characteristic phenomenon associated with the solution of all closed-loop inverse dynamics problems irrespective of whether Euler's equations or Lagrange's equations are used to describe system motion.

Before leaving this example, it is convenient to discuss the important concept of degrees of freedom and to indicate how simple constraints on joint motion can be introduced into the analysis. The number of degrees of freedom of a mechanical system model is simply the number of generalized coordinates used to specify its configuration ($n$), minus the number of independent constraint equations these generalized coordinates must satisfy during system motion ($m$). Thus, in the first example where 12 unconstrained generalized coordinates were used to describe the system configuration, $n = 12$, $m = 0$, and the number of degrees of freedom is equal therefore to 12. In contrast, the system model in the second example involved nine generalized coordinates that must satisfy three independent holonomic constraint equations (Equation 8.40). In this case, $n = 9$, $m = 3$, and therefore the number of degrees of freedom for this system is equal to 6. The number of degrees of freedom is an important concept in dynamics because

(a) it is a system property independent of the particular set of $n$ generalized coordinates used to describe system configuration,
(b) it indicates how many of these $n$ generalized coordinates are independent, and
(c) it therefore determines how many initial conditions can be arbitrarily and independently specified to solve the associated direct dynamics problem where the system's differential-algebraic equations must be integrated to obtain a unique solution for system motion (i.e., a number equal to twice the number of degrees of freedom, to uniquely specify both the system's initial configuration and its initial velocity state).

To illustrate how simple constraints on joint motion can be introduced conveniently into the analysis, consider a modification of the second example where the joint at point $O$ is now a simple hinge with its axis fixed to coincide with the $X$-axis of the global system $B$ (see Figure 8.4). In this situation, the orientation of $S_3$ is constrained such that both $\theta_3$ ($q_8$) and $\psi_3$ ($q_9$) must remain equal to zero during system motion, and only $\phi_3$ ($q_7$) can vary. This constraint can be imposed either by deleting $q_8$ and $q_9$ from the set of nine generalized coordinates used to describe the system configuration, or by retaining all nine generalized coordinates but adding, to the three independent holonomic closure constraint equations (Equation 8.40), the two independent holonomic constraint equations

$$q_8 = 0 \quad ; \quad q_9 = 0. \tag{8.44}$$

Hence, the unique number of degrees of freedom for this additionally constrained example is given either in the first instance by $(9 - 2) - 3 = 4$ or in the second instance by $9 - (3 + 2) = 4$. The favored choice between these two alternatives is the first, because it decreases the number of Lagrange's equations that must be generated to describe system motion. However, using holonomic constraint equations to reduce the number of generalized coordinates is not always the most convenient alternative from an analytical perspective. For example, due to the transcendental nature of the three constraint equations (Equation 8.40), it is quite cumbersome to use them to reduce the initial set of nine dependent generalized coordinates to a set of six independent generalized coordinates.

# DISCUSSION

This chapter presents a brief description of the process of constructing Euler's and Lagrange's equations for multibody mechanical models composed of massive rigid segments interconnected by smooth spherical joints and moving in three dimensions subject to a general external force-couple system and typical constraints. Such models are often used in biomechanical investigations, and Euler's and Lagrange's equations, although of special interest and importance, are not the only second-order, ordinary differential equations that can be generated and used to describe and analyze model behavior. Bear in mind that there exist related and important first-order differential equations representing integrals of the motion (e.g., the work-energy integral and the linear and angular impulse-momentum integrals) that may be derived and used to obtain valuable information about the system's dynamic behavior. And it is also important to note that there now exist a variety of conventional computer software packages that facilitate the generation and solution of the equations of motion and constraint for mechanical models of linked multibody systems.

The material presented in this chapter indicates that for the type of system models considered, Euler's equations are somewhat easier to generate than Lagrange's equations and that Euler's equations must be used whenever the workless joint resultant forces need to be included in the analysis. Lagrange's equations are somewhat more difficult to generate because they require, in addition to what is needed for Euler's equations, use of the concept of virtual work to determine the active generalized force $Q'$ that is responsible for system motion.

A comparison of Euler's first set of three equations (Equation 8.18) to Lagrange's first three equations (Equation 8.30) shows that they are identical. Likewise, Lagrange's sixth equation (Equation 8.32) is identical to the third of Euler's second set of three equations (Equation 8.20). Finally, it can be shown by algebraic manipulation that Lagrange's fourth and fifth equations (Equation 8.31) can be obtained as linear combinations of Euler's second set of three equations (Equation 8.20). Hence, in terms of the generalized coordinates used in this presentation to specify the configuration of $S$ in the global system $B$, Euler's and Lagrange's equations are nearly identical and completely equivalent systems of six independent, scalar, second-order, ordinary differential equations that govern the 3-D motion of $S$ in $B$.

Irrespective of whether Euler's equations or Lagrange's equations are to be constructed, it is always necessary to choose an appropriate set of generalized coordinates to describe the system configuration and to establish the independence of all constraint equations that these coordinates must satisfy during system motion. This process is perhaps the most challenging task facing even the experienced analyst, because it often involves the balancing of sometimes conflicting objectives (e.g., analytical simplicity vs. ease of physical interpretation). Of particular concern in this regard is the choice of angular coordinates used to specify segment orientation. Many possibilities exist (e.g., projection angles, direction angles, Euler angles, Cardan angles, Euler parameters, etc.). Depending on how the equations of motion are to be used (e.g., to solve the direct vs. the inverse dynamics problem), the selection of independent angular coordinates (e.g., the three Cardan angles used in this presentation vs. the nine dependent direction angles) can lead to severe problems (e.g., gimbal lock) that can make the integration of the equations of motion difficult, if not impossible.

Another important concern associated with the choice of angular coordinates is whether, for linked multibody systems, each segment's orientation should be specified relative to the same global reference frame $B$ (as was done here using independent Cardan angles for analytical simplicity and convenience) or relative to the adjacent proximal segment, except for one reference segment in the linkage system. The latter choice, although often analytically cumbersome, may nevertheless be justified based on such considerations as the manner in which joint motions are commonly described (e.g., flexion of the distal joint segment relative to the proximal joint segment), the associated ease of constraining admissible joint motions when that is appropriate (e.g., treating the knee as a joint that does not permit abduction or adduction), and the ability to relate joint resultant forces and moments to the individual joint structures that produce these kinetic quantities (e.g., muscles, ligaments, and bones).

Finally, it should be emphasized that both Euler's and Lagrange's equations can be used conveniently to analyze not only the 3-D motion of linked multibody systems of rigid segments, but also their 2-D motion and equilibrium states. In many cases, model validation requires the construction and solution of these 3-D equations to establish that the errors associated with the simplified model selected for use are in fact negligible and can therefore be safely ignored for the activity under investigation.

# APPENDIX A

# Rigid Body Orientation and Angular Velocity Using Cardan Angles

Let three ordered Cardan angles, $\phi$, $\theta$, and $\psi$ (Wittenberg, 1977), be used to specify the orientation of a rigid segment $S$ (or the orientation of a local Cartesian coordinate system $R$: $Gxyz$ embedded in $S$ at the mass center $G$, with unit vectors $\mathbf{i}$, $\mathbf{j}$, and $\mathbf{k}$) relative to a global (inertial) Cartesian coordinate system $B$: $OXYZ$ (with origin $O$, $Z$-axis directed vertically upward, and unit vectors $\mathbf{I}$, $\mathbf{J}$, and $\mathbf{K}$). Assume, for simplicity and convenience, that the respective axes of the local system $R$ and the global system $B$ initially coincide (i.e., $G$ lies at $O$), and the initial orientation of $R$ is denoted by $R_1$: $Gx_1y_1z_1$, with unit vectors

$$\mathbf{i}_1 = \mathbf{I};\ \mathbf{j}_1 = \mathbf{J};\ \mathbf{k}_1 = \mathbf{K}.$$

An arbitrary final orientation of the local system $R$ (or $S$) relative to the global system $B$ is obtained by the following ordered sequence of simple rotations:

1. An initial rotation of $R_1$ through the Cardan angle $\phi$ about $x_1 = X$ (Figure A.1), resulting in a new orientation of $R_1$ denoted by $R_2$: $Gx_2y_2z_2$ with unit vectors $\mathbf{i}_2$, $\mathbf{j}_2$, and $\mathbf{k}_2$, where

$$\mathbf{I} = \mathbf{i}_1 = \mathbf{i}_2,$$

$$\mathbf{J} = \mathbf{j}_1 = c\phi\mathbf{j}_2 - s\phi\mathbf{k}_2, \tag{A.1}$$

$$\mathbf{K} = \mathbf{k}_1 = s\phi\mathbf{j}_2 + c\phi\mathbf{k}_2.$$

(The symbols "$c$" and "$s$" denote the cosine and sine functions, respectively.) The angular velocity of $R_2$ relative to $R_1 = B$, denoted by $\mathbf{\Omega}_{2/B}$, can be expressed

$$\mathbf{\Omega}_{2/B} = \dot{\phi}\,\mathbf{i}_2. \tag{A.2}$$

2. An intermediate rotation of $R_2$ through the Cardan angle $\theta$ about $y_2$ (Figure A.2), resulting in a new orientation of $R_2$ denoted by $R_3$: $Gx_3y_3z_3$ with unit vectors $\mathbf{i}_3$, $\mathbf{j}_3$, and $\mathbf{k}_3$, where

$$\mathbf{i}_2 = c\theta\mathbf{i}_3 + s\theta\mathbf{k}_3,$$

$$\mathbf{j}_2 = \mathbf{j}_3, \tag{A.3}$$

$$\mathbf{k}_2 = -s\theta\mathbf{i}_3 + c\theta\mathbf{k}_3.$$

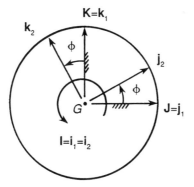

**Figure A.1**    The Cardan angle $\phi$ that orients the unit vectors $\mathbf{i}_2$, $\mathbf{j}_2$, $\mathbf{k}_2$ of the first intermediate coordinate system $R_2$: $Gx_2y_2z_2$ relative to the unit vectors $\mathbf{i}_1$, $\mathbf{j}_1$, $\mathbf{k}_1$ of the initial coordinate system $R_1$: $Gx_1y_1z_1 = B$: $OXYZ$ (with unit vectors $\mathbf{I}$, $\mathbf{J}$, $\mathbf{K}$).

The angular velocity of $R_3$ relative to $R_2$, denoted by $\boldsymbol{\Omega}_{3/2}$, can be expressed

$$\boldsymbol{\Omega}_{3/2} = \dot{\theta}\mathbf{j}_3 \,. \tag{A.4}$$

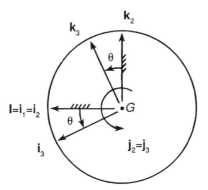

**Figure A.2**    The Cardan angle $\theta$ that orients the unit vectors $\mathbf{i}_3$, $\mathbf{j}_3$, $\mathbf{k}_3$ of the second intermediate coordinate system $R_3$: $Gx_3y_3z_3$ relative to the unit vectors $\mathbf{i}_2$, $\mathbf{j}_2$, $\mathbf{k}_2$ of the first intermediate coordinate system $R_2$: $Gx_2y_2z_2$.

3. A final rotation of $R_3$ through the Cardan angle $\psi$ about $z_3$ (Figure A.3), resulting in a new orientation of $R_3$ denoted by $R$: $Gxyz$ with unit vectors $\mathbf{i}$, $\mathbf{j}$, and $\mathbf{k}$, where

$$\mathbf{i}_3 = c\psi\,\mathbf{i} - s\psi\,\mathbf{j} \,,$$

$$\mathbf{j}_3 = s\psi\,\mathbf{i} + c\psi\,\mathbf{j} \,, \tag{A.5}$$

$$\mathbf{k}_3 = \mathbf{k} \,.$$

The angular velocity of $R$ (or $S$) relative to $R_3$, denoted by $\mathbf{\Omega}_{R/3}$, can be expressed

$$\mathbf{\Omega}_{R/3} = \dot{\psi}\mathbf{k} . \tag{A.6}$$

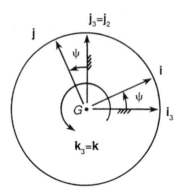

**Figure A.3**   The Cardan rotation angle $\psi$ that orients the unit vectors $\mathbf{i}$, $\mathbf{j}$, $\mathbf{k}$ of the local coordinate system $R$: $Gxyz$ relative to the unit vectors $\mathbf{i}_3$, $\mathbf{j}_3$, $\mathbf{k}_3$, of the second intermediate coordinate system $R_3$: $Gx_3y_2z_3$.

Letting $\mathbf{\Omega} = \mathbf{\Omega}_{R/B}$ denote the angular velocity of $S$ (or local system $R$ embedded in $S$ at $G$) relative to global system $B$, and using the vector addition law, $\mathbf{\Omega}$ can be expressed

$$\mathbf{\Omega} = \mathbf{\Omega}_{R/B} = \mathbf{\Omega}_{R/3} + \mathbf{\Omega}_{3/2} + \mathbf{\Omega}_{2/B} = \dot{\psi}\mathbf{k} + \dot{\theta}\mathbf{j}_3 + \dot{\phi}\mathbf{i}_2 . \tag{A.7}$$

Using the unit vector transformation equations (A.3) and (A.5), $\mathbf{\Omega}$ in Equation A.7 can be re-expressed, in local $R$ components,

$$\mathbf{\Omega} = \Omega_x \mathbf{i} + \Omega_y \mathbf{j} + \Omega_z \mathbf{k} , \tag{A.8}$$

where

$$\Omega_x = \dot{\phi}c\theta c\psi + \dot{\theta}s\psi; \quad \Omega_y = -\dot{\phi}c\theta s\psi + \dot{\theta}c\psi; \quad \Omega_z = \dot{\phi}s\theta + \dot{\psi} . \tag{A.9}$$

Hence, $\mathbf{\Omega}$ is a linear function of the time-derivatives of the three Cardan angles that orient $S$ in the global system $B$. Differentiating the local $R$ components of $\mathbf{\Omega}$ in Equation A.9 with respect to time yields

$$\dot{\Omega}_x = \ddot{\phi}c\theta c\psi - \dot{\phi}\dot{\theta}s\theta c\psi - \dot{\phi}\dot{\psi}c\theta s\psi + \ddot{\theta}s\psi + \dot{\theta}\dot{\psi}c\psi ,$$

$$\dot{\Omega}_y = -\ddot{\phi}c\theta s\psi + \dot{\phi}\dot{\theta}s\theta s\psi - \dot{\phi}\dot{\psi}c\theta c\psi + \ddot{\theta}c\psi - \dot{\theta}\dot{\psi}s\psi , \tag{A.10}$$

$$\dot{\Omega}_z = \ddot{\phi}s\theta + \dot{\phi}\dot{\theta}c\theta + \ddot{\psi} .$$

# APPENDIX B

## Virtual Work Concepts

The virtual work done by the external forces and moments acting on a rigid segment $S$ is obtained by determining the work done by these kinetic quantities, treated like constant vectors, during kinematically admissible virtual displacements of the points of force application and kinematically admissible virtual rotations of the segment (Greenwood, 1988).

## Virtual Displacements and Virtual Rotations

A kinematically admissible virtual displacement of Point $A$ fixed in $S$ is an infinitesimal (first-order differential) displacement of $A$ that takes place with time held constant and is consistent with the constraints (e.g., Equation 8.7), but is otherwise arbitrary. A kinematically admissible virtual rotation of $S$ is an infinitesimal rotation of $S$ that takes place with time held constant and is consistent with the constraints, but is otherwise arbitrary.

A virtual displacement of Point $A$ fixed in $S$ is found directly from $\mathbf{v}^A$, the $3 \times 1$ vector that denotes the velocity of $A$. Recall that $\mathbf{v}^A$ can always be expressed, in terms of $\mathbf{v}^G$ and $\mathbf{\Omega}$ (Greenwood, 1988):

$$\mathbf{v}^A = \mathbf{v}^G + (\mathbf{\Omega} \times \mathbf{r}^{A/G});$$

where $\mathbf{r}^{A/G}$ locates $A$ relative to $G$. Recall also that Equations 8.8, 8.9, and 8.10 indicate that $\mathbf{v}^G$ and $\mathbf{\Omega}$ are both linear functions of the six components of $\dot{\mathbf{q}}$. Hence, it is always possible to express $\mathbf{v}^A$ in the form

$$\mathbf{v}^A = \mathbf{D}^A\dot{\mathbf{q}} + \mathbf{e}^A, \tag{B.1}$$

where $\mathbf{D}^A$ is the appropriate point-dependent $3 \times 6$ matrix function of $\mathbf{q}$, and $\mathbf{e}^A$, if present, is the corresponding $3 \times 1$ vector function of $\mathbf{q}$. Rewriting Equation B.1 in differential form yields

$$d\mathbf{r}^{A/O} = \mathbf{D}^A d\mathbf{q} + \mathbf{e}^A dt. \tag{B.2}$$

The virtual displacement of $A$ fixed in $S$, denoted by $\delta\mathbf{r}^A$, is obtained by replacing each $d$ in Equation B.2 with $\delta$, and then setting $\delta t = 0$. Hence

$$\delta\mathbf{r}^A = \mathbf{D}^A\delta\mathbf{q}, \tag{B.3}$$

where the $6 \times 1$ vector $\delta\mathbf{q} = [\delta X^G, \delta Y^G, \delta Z^G, \delta\phi, \delta\theta, \delta\psi]^T$ denotes the virtual change in the segment's generalized coordinate vector $\mathbf{q}$.

The virtual rotation of $S$ is found directly from $\mathbf{\Omega}$, the $3 \times 1$ vector that denotes the angular velocity of $S$. Because Equations 8.9 and 8.10 indicate that $\mathbf{\Omega}$ is a linear function of the components of $\dot{\mathbf{q}}$, $\mathbf{\Omega}$ can always be expressed

$$\Omega = J\dot{q} + h, \tag{B.4}$$

where $J$ is the appropriate segment-dependent $3 \times 6$ matrix function of $q$, and $h$, if present, is the corresponding $3 \times 1$ vector function of $q$. Using the convenient (but unjustified) assumption that $\Omega$ can be expressed as the derivative of a $3 \times 1$ vector, $\pi$, Equation B.4 can be written in differential form

$$d\pi = Jdq + hdt. \tag{B.5}$$

The virtual rotation of $S$, denoted by $\delta\pi$, is obtained by replacing each $d$ in Equation B.5 with $\delta$, and then setting $\delta t = 0$. Hence

$$\delta\pi = J\delta q. \tag{B.6}$$

If the segment's generalized coordinate vector $q$ is holonomically constrained by Equation 8.7, then the time-derivative of Equation 8.7 must also be satisfied during system motion. This derivative of Equation 8.7 is a linear function of the components of $\dot{q}$ and can always be written in the form

$$\dot{c} = K\dot{q} + 1 = 0, \tag{B.7}$$

where $K$ is the appropriate segment-dependent $m \times 6$ matrix function of $q$, and $1$, if present, is the corresponding $m \times 1$ vector function of $q$. Rewriting equation B.7 in differential form yields

$$Kdq + 1\, dt = 0. \tag{B.8}$$

The vector constraint equation that must be satisfied by all kinematically admissible virtual changes in the segment's generalized coordinate vector, denoted by $\delta q$, is obtained by replacing each $d$ in Equation B.8 with $\delta$, and then setting $\delta t = 0$. Hence, kinematically admissible $\delta q$ must satisfy

$$K\delta q = 0. \tag{B.9}$$

## Virtual Work and Generalized Force

The virtual work done by Force $F^A$ applied at Point $A$ fixed in $S$, denoted by $\delta W^F$, is the work done by $F^A$ during a kinematically admissible virtual displacement of $A$. Thus

$$\delta W^F = (F^A)^T \delta r^A, \tag{B.10}$$

where $F^A$ is treated as a constant vector during $\delta r^A$, and the superscript $T$ denotes the transpose. Similarly, the virtual work done by a couple with resultant moment $M^A$ applied to $S$ at Point $A$, denoted by $\delta W^M$, is the work done by $M^A$ during a kinematically admissible virtual rotation of $S$. Thus

$$\delta W^M = (\mathbf{M}^A)^T \, \delta\boldsymbol{\pi}, \tag{B.11}$$

where $\mathbf{M}^A$ is treated as a constant vector during $\delta\boldsymbol{\pi}$.

The total virtual work done on $S$ by the external force-couple system, denoted by $\delta W$, is simply the algebraic sum of the virtual work done by each external force and each external moment that acts on $S$. Thus

$$\delta W = \Sigma(\delta W^F) + \Sigma(\delta W^M) = \Sigma\{(\mathbf{F}^A)^T \delta\mathbf{r}^A\} + \Sigma\{(\mathbf{M}^A)^T \delta\boldsymbol{\pi}\}, \tag{B.12}$$

where the superscript $A$ represents any point at which an external force $\mathbf{F}^A$ or external moment $\mathbf{M}^A$ is applied to $S$. Substituting Equations B.3 and B.6 into B.12 yields

$$\delta W = (\Sigma\{(\mathbf{F}^A)^T\mathbf{D}^A\} + \Sigma\{(\mathbf{M}^A)^T\mathbf{J}\})\delta\mathbf{q} = \mathbf{Q}^T\delta\mathbf{q} = \delta\mathbf{q}^T\mathbf{Q}, \tag{B.13}$$

where

$$\mathbf{Q} = \Sigma\{(\mathbf{D}^A)^T\mathbf{F}^A\} + \Sigma(\mathbf{J}^T\mathbf{M}^A) \tag{B.14}$$

is the $6 \times 1$ generalized force vector corresponding to the segment's $6 \times 1$ generalized coordinate vector $\mathbf{q}$.

Equation B.14 is used to determine $\mathbf{Q}$ once the external forces ($\mathbf{F}^A$) and external moments ($\mathbf{M}^A$) acting on $S$ have been identified, and once the relevant kinematic quantities (i.e., $\mathbf{D}^A$ for each force $\mathbf{F}^A$, and $\mathbf{J}$ for the segment) have been determined. In the absence of constraint equations (Equation 8.7), the components of $\mathbf{q}$ are independent and the process of determining $\mathbf{Q}$ from Equation B.14 is straightforward, although sometimes challenging. However, if the components of $\mathbf{q}$ are constrained by Equation 8.7, the components of $\delta\mathbf{q}$ are also dependent and must satisfy Equation B.9 in order to be kinematically admissible. In such circumstances (holonomic constraints), the derivation of Lagrange's equations is modified by invoking the Lagrange mutliplier theorem (Haug, 1992). This theorem assures the existence of a unique but unknown $m \times 1$ Lagrange multiplier vector $\boldsymbol{\mu}$ that is directly related to the generalized force vector $\mathbf{Q}^c$ associated with the constraints specified by Equation 8.7. The virtual work done by these holonomic constraints, denoted by $\delta W^c$, can be expressed

$$\delta W^c = (\mathbf{Q}^c)^T\delta\mathbf{q} = \delta\mathbf{q}^T\mathbf{Q}^c, \tag{B.15}$$

where

$$\mathbf{Q}^c = -\mathbf{K}^T\boldsymbol{\mu} \tag{B.16}$$

represents the $6 \times 1$ generalized force vector corresponding to the holonomic constraints expressed by Equation 8.7, the $m \times 6$ constraint matrix $\mathbf{K}$ is obtained from Equation B.7, and the negative sign in Equation B.16 is conventional. When $\delta W^c$ from Equation B.15 is included in the total virtual work expression given by Equation B.13, the total generalized force vector $\mathbf{Q}$ in Equation B.14 becomes

$$Q = \Sigma\{(D^A)^T F^A\} + \Sigma(J^T M^A) - K^T \mu. \qquad (B.17)$$

Hence, if the generalized coordinate vector **q** is constrained by Equation 8.7, the corresponding generalized force vector **Q** is determined from Equation B.17 in terms of the unknown $m \times 1$ Lagrange multiplier vector **μ**.

To avoid including the effect of the gravity force twice in Lagrange's equations (i.e., in the gravitational potential energy $V$ and again in the generalized force **Q**), the gravity force is deleted from the first sum on the right-hand side of Equation B.17 for each segment, leaving what will be referred to here as the active generalized force vector and denoted by **Q′**. Note that for a mechanical system composed of two or more rigid segments interconnected by smooth spherical joints, the joint resultant forces that act on adjacent segments do zero net virtual work on the system during a kinematically admissible δ**q**. Hence, they will not contribute to **Q′** for the system and can therefore be ignored when **Q′** is constructed using Equation B.17. In contrast, however, the joint resultant moments that act on adjacent segments generally do some net virtual work on the system during a kinematically admissible δ**q**. Hence, joint resultant moments should not be ignored when **Q′** is constructed using Equation B.17.

# APPENDIX C

## Lagrange's Equations for a Single Rigid Segment $S$

Recall that

$$\mathbf{q} = [X^G, Y^G, Z^G, \phi, \theta, \psi]^T = [q_1, \ldots, q_6]^T \qquad (C.1)$$

is the generalized coordinate vector that specifies the configuration of $S$ in the global system $B$, and the axes of the local system $R: Gxyz$ are principal axes of inertia embedded in $S$ at $G$, with the $z$-axis a longitudinal axis of geometric and mass symmetry. Hence, all three products of inertia vanish when the mass-center inertia matrix $\mathbf{I}^G$ is expressed in local $R$ components, and the corresponding principal moments of inertia can be expressed

$$I^G_{xx} = I^G_{yy} = I_t ; I^G_{zz} = I_l. \qquad (C.2)$$

The velocity of $G$, denoted by $\mathbf{v}^G$, is given in global $B$ components by Equation 8.8, and the angular velocity of $S$, denoted by $\Omega$, is given in local $R$ components by Equation 8.9, where these $R$ components of $\Omega$ are expressed in terms of the components of **q** and their derivatives in Equations 8.10. Using Equations 8.21 through 8.27, the Lagrangian $L$ for the motion of $S$ can now be expressed, in terms of the components of **q** and their derivatives,

$$L = (1/2)m[(\dot{X}^G)^2 + (\dot{Y}^G)^2 + (\dot{Z}^G)^2] + (1/2)I_t(\dot{\phi}^2 c^2\theta + \dot{\theta}^2)$$

$$+ (1/2)I_l(\dot{\phi}^2 s^2\theta + 2\dot{\phi}\dot{\psi}s\theta + \dot{\psi}^2) - mgZ^G. \qquad (C.3)$$

Using Equation C.3, the lefthand-side terms of Lagrange's equations (Equation 8.29) can now be obtained:

$$d(\partial L/\partial \dot{q}_1)/dt = d(\partial L/\partial \dot{X}^G)/dt = d(m\dot{X}^G)/dt = m\ddot{X}^G; \ \partial L/\partial q_1 = \partial L/\partial X^G = 0 \ .$$

$$d(\partial L/\partial \dot{q}_2)/dt = d(\partial L/\partial \dot{Y}^G)/dt = d(m\dot{Y}^G)/dt = m\ddot{Y}^G; \ \partial L/\partial q_2 = \partial L/\partial Y^G = 0 \ .$$

$$d(\partial L/\partial \dot{q}_3)/dt = d(\partial L/\partial \dot{Z}^G)/dt = d(m\dot{Z}^G)/dt = m\ddot{Z}^G; \ \partial L/\partial q_3 = \partial L/\partial Z^G = -mg \ .$$

$$d(\partial L/\partial \dot{q}_4)/dt = d(\partial L/\partial \dot{\phi})/dt = d[I_t\dot{\phi}c^2\theta + I_t(\dot{\phi}s^2\theta + \dot{\psi}s\theta)]/dt$$

$$= I_t(\ddot{\phi}c^2\theta - 2\dot{\phi}\dot{\theta}s\theta c\theta) +$$

$$I_t(\ddot{\phi}s^2\theta + 2\dot{\phi}\dot{\theta}s\theta c\theta + \ddot{\psi}s\theta + \dot{\theta}\dot{\psi}c\theta); \ \partial L/\partial q_4 = \partial L/\partial \phi = 0 \ .$$

(C.4)

$$d(\partial L/\partial \dot{q}_5)/dt = d(\partial L/\partial \dot{\theta})/dt = d(I_t\dot{\theta})/dt = I_t\ddot{\theta} \ ;$$

$$\partial L/\partial q_5 = \partial L/\partial \theta = -I_t\dot{\phi}^2 s\theta c\theta + I_t(\dot{\phi}^2 s\theta c\theta + \dot{\phi}\dot{\psi}c\theta) \ .$$

$$d(\partial L/\partial \dot{q}_6)/dt = d(\partial L/\partial \dot{\psi})/dt = d[I_t(\dot{\phi}s\theta + \dot{\psi})]/dt = I_t(\ddot{\phi}s\theta + \dot{\phi}\dot{\theta}c\theta + \ddot{\psi} \ ) \ ;$$

$$\partial L/\partial q_6 = \partial L/\partial \psi = 0 \ .$$

To obtain the righthand-side terms of Lagrange's equations (Equation 8.29), or the components of the active generalized force vector $\mathbf{Q}'$, the virtual work expression must first be constructed. Using Equation B.12 and deleting the gravity force $(-mg\mathbf{K})$ from the segment's external force-couple system (to avoid including it twice in Lagrange's equations), the segment's active virtual work $\delta W'$ can always be expressed

$$\delta W' = (\mathbf{F}')^T \delta \mathbf{r}^G + (\mathbf{M}^G)^T \delta \boldsymbol{\pi} \ , \tag{C.5}$$

where the segment's active resultant force $\mathbf{F}'$ is given, in terms of the segment's resultant force $\mathbf{F}$ (Equation 8.1), by

$$\mathbf{F} = (\mathbf{F}^P + \mathbf{F}^D + \mathbf{F}^A) - mg\mathbf{K} = \mathbf{F}' - mg\mathbf{K} \ . \tag{C.6}$$

The expressions for $\delta \mathbf{r}^G$ and $\delta \boldsymbol{\pi}$ are obtained from $\mathbf{v}^G$ and $\boldsymbol{\Omega}$, respectively, as described in Appendix B. Using $\mathbf{v}^G$ from Equation 8.8 and $\boldsymbol{\Omega}$ from Equation A.7, $\delta \mathbf{r}^G$ and $\delta \boldsymbol{\pi}$ can be expressed, in terms of the unit vectors of the global $B$, local $R$, and intermediate Cartesian reference frames $R_i$: $Gx_iy_iz_i$ ($i = 1, 2, 3$),

$$\delta \mathbf{r}^G = \delta X^G \mathbf{I} + \delta Y^G \mathbf{J} + \delta Z^G \mathbf{K} \ ; \tag{C.7}$$

$$\delta \boldsymbol{\pi} = \delta \phi \mathbf{i}_2 + \delta \theta \mathbf{j}_3 + \delta \psi \mathbf{k} \ . \tag{C.8}$$

The segment's active virtual work $\delta W'$, given by Equation C.5, can now be expressed in terms of the components of $\delta \mathbf{q}$:

$$\delta W' = \{(\mathbf{F}')^T \mathbf{I}\} \delta X^G + \{(\mathbf{F}')^T \mathbf{J}\} \delta Y^G + \{(\mathbf{F}')^T \mathbf{K}\} \delta Z^G$$

$$+ \{(\mathbf{M}^G)^T \mathbf{i}_2\} \delta \phi + \{(\mathbf{M}^G)^T \mathbf{j}_3\} \delta \theta + \{(\mathbf{M}^G)^T \mathbf{k}\} \delta \psi \ . \tag{C.9}$$

The equivalent form of $\delta W'$, given by Equation B.13, is used to identify the components of the active generalized force vector $\mathbf{Q}'$ corresponding to the generalized coordinate vector $\mathbf{q}$, where

$$\mathbf{Q}' = [Q'_X, Q'_Y, Q'_Z, Q'_\phi, Q'_\theta, Q'_\psi]^T = [Q'_1, \ldots, Q'_6]^T . \qquad (C.10)$$

Letting $\mathbf{F}$ and $\mathbf{F}'$ be expressed in global $B$ components such that

$$\mathbf{F} = F_X\mathbf{I} + F_Y\mathbf{J} + F_Z\mathbf{K} = \mathbf{F}' - mg\mathbf{K} \qquad (C.11)$$

and

$$\mathbf{F}' = F'_X\mathbf{I} + F'_Y\mathbf{J} + F'_Z\mathbf{K} = F'_X\mathbf{I} + F'_Y\mathbf{J} + (F_Z + mg)\mathbf{K} , \qquad (C.12)$$

and letting $\mathbf{M}^G$ be expressed in local $R$ components such that

$$\mathbf{M}^G = M^G_x\mathbf{j} + M^G_y\mathbf{i} + M^G_z\mathbf{k} , \qquad (C.13)$$

use of Equation B.13 and Equations C.9 through C.13 leads to the following results for the righthand-side terms of Lagrange's equations (Equation 8.29):

$$Q'_X = (\mathbf{F}')^T\mathbf{I} = F'_X = F_X ;$$

$$Q'_Y = (\mathbf{F}')^T\mathbf{J} = F'_Y = F_Y ;$$

$$Q'_Z = (\mathbf{F}')^T\mathbf{K} = F'_Z = F_Z + mg ; \qquad (C.14)$$

$$Q'_\phi = (\mathbf{M}^G)^T\mathbf{i}_2 = M^G_x c\theta c\psi - M^G_y c\theta s\psi + M^G_z s\theta ;$$

$$Q'_\theta = (\mathbf{M}^G)^T\mathbf{j}_3 = M^G_x s\psi + M^G_y c\psi ;$$

$$Q'_\psi = (\mathbf{M}^G)^T\mathbf{k} = M^G_z .$$

Using Equations C.4 and C.14, Lagrange's six scalar equations (Equation 8.29) for the motion of $S$ in $B$ can now be expressed:

$$m\ddot{X}^G = F_X ;$$

$$m\ddot{Y}^G = F_Y ;$$

$$m\ddot{Z}^G = F_Z ;$$

$$I_t(\ddot{\phi}c^2\theta - 2\dot{\phi}\dot{\theta}s\theta c\theta) + I_t(\ddot{\phi}s^2\theta + 2\dot{\phi}\dot{\theta}s\theta c\theta + \ddot{\psi}s\theta + \dot{\theta}\dot{\psi}c\theta) \qquad (C.15)$$

$$= M^G_x c\theta c\psi - M^G_y c\theta s\psi + M^G_z s\theta ;$$

$$I_t(\ddot{\theta} + \dot{\phi}^2 s\theta c\theta) - I_t(\dot{\phi}^2 s\theta c\theta + \dot{\phi}\dot{\psi}c\theta) = M^G_x s\psi + M^G_y c\psi ;$$

$$I_t(\ddot{\phi}s\theta + \ddot{\psi} + \dot{\phi}\dot{\theta}c\theta) = M^G_z .$$

# REFERENCES

Crowninshield, R.D., & Brand, R.A. (1981). The prediction of forces in joint structures: Distribution of intersegmental resultants. In D.I. Miller (Ed.), *Exercise and sports science reviews,* Vol. 9 (pp. 159-181). Philadelphia: Franklin Institute Press.

Greenwood, D.T. (1988). *Principles of dynamics* (2nd ed.). Englewood Cliffs, NJ: Prentice Hall.

Haug, E.J. (1992). *Intermediate dynamics.* Englewood Cliffs, NJ: Prentice Hall.

Kane, T.R., & Levinson, D.A. (1985). *Dynamics: Theory and applications.* New York: McGraw-Hill.

McGill, D.J., & King, W.W. (1989). *Engineering mechanics: An introduction to dynamics* (2nd ed.). Boston: PWS-Kent.

Wittenberg, J. (1977). *Dynamics of systems of rigid bodies.* Stuttgart: Teubner.

# Chapter 9

# Modeling Human Body Motions by the Techniques Known to Robotics

*Laurence Chèze*
*Joannes Dimnet*

Robotic science has developed to achieve the goal of replacing human tasks with working tools. The first robots looked like human arms. The large industrial use of mechanical systems as robots and manipulators has provoked various fundamental studies: to represent a robot by a mathematical form, to simulate the field of both three-dimensional displacements and velocities, and to calculate the set of instantaneous moments applied to each motor axis to obtain the desired motion at the level of the end segment. For industrial applications, all studies are mathematically sophisticated, reducing the time of data treatment so as to control the robot in real time.

There are several basic differences between robotics and biomechanical motion—at the level of the structure, the motorization, and the command mechanism equivalent to the human body. Only the structure is developed in this chapter, because it determines the equivalent robotic scheme of a body or a limb doing a specific task. The problem of robot's command is very particular in biomechanics: Doing a task in three-dimensional space, a subject has a redundant number of degrees of freedom. Several of them are tied together by coupling relationships. The command must take into account these coupling constraints, but it is necessary to make use of accurate geometric and kinematic data allowing the description of the temporal changes in each degree of freedom as well as precise identification of analog mechanism.

Human motion studies bring new constraints. Is it possible to adapt robotic techniques to analyze human motion? This chapter attempts to answer this question in part.

# BASES OF ROBOTICS RELEVANT TO BIOMECHANICAL MOTION

A robot may be considered an open chain consisting of elementary segments joined together. The robot's origin is the fixed body, and the end segment is free. Each segment is connected to the adjacent one by a joint with one degree of freedom (either rotation or translation). In robotics, the main problem concerns the location of objects in three-dimensional space. These objects are the constitutive links of the manipulator, the tools, and anything in the robot environment. The links and tools are described only by their positions and orientations through a representation that allows a mathematical manipulation. To describe both the position and the orientation of a robot link in space, a coordinate system or reference frame is rigidly attached to the corresponding segment.

## Geometric Models

Manipulators consist of rigid links connected by joints allowing relative motion between the neighboring segments. These joints have one degree of freedom; they are either revolute or prismatic joints, and they are instrumented with a position sensor and motorized. At the free end of the manipulator, the effector generally grips a tool. The instant position and orientation in space of the end effector must be computed from the joint sensor data.

Denavit and Hartenberg (1955) proposed a method allowing the partition of the chain into elementary segments and assigning a matrix to each segment. The coefficients of this matrix depend on four parameters, two of them describing the segment's geometry ($a_i$, $\alpha_i$) and the two others locating the segment with respect to the previous one ($d_i$, $\theta_i$) (Figure 9.1).

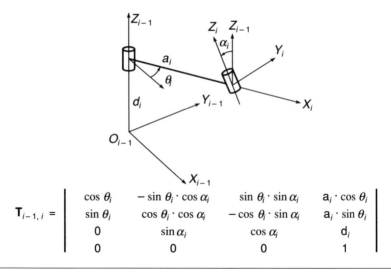

$$\mathbf{T}_{i-1,i} = \begin{vmatrix} \cos\theta_i & -\sin\theta_i \cdot \cos\alpha_i & \sin\theta_i \cdot \sin\alpha_i & a_i \cdot \cos\theta_i \\ \sin\theta_i & \cos\theta_i \cdot \cos\alpha_i & -\cos\theta_i \cdot \sin\alpha_i & a_i \cdot \sin\theta_i \\ 0 & \sin\alpha_i & \cos\alpha_i & d_i \\ 0 & 0 & 0 & 1 \end{vmatrix}$$

**Figure 9.1**    Elementary matrix with Denavit-Hartenberg notation.

The elementary matrix $\mathbf{T}_{i-1,i}$ defines the orientation and the position of the frame $F_i$ attached to the $i^{th}$ link versus the $i-1^{th}$ frame $F_{i-1}$. The first column of the homogeneous matrix gives the components of the unit vector $\overline{X}_i$ of the frame $F_i$ projected onto the frame $F_{i-1}$. The second and third columns give $\overline{Y}_i$ and $\overline{Z}_i$, respectively, so that the three first columns deal with the rotation matrix. The fourth column gives the vector $\overline{O_{i-1}O_i}$ projected onto $F_{i-1}$. These are the translation components of the displacement.

The $\mathbf{T}_{i-1,i}$ operator transforms the coordinates through the formula

$$\overline{O_{i-1}M}^{i-1} = \mathbf{T}_{i-1,i} \cdot \overline{O_iM}^i , \tag{9.1}$$

where $\overline{O_{i-1}M}^{i-1}$ represents the coordinates of any point $M$ in the frame $F_{i-1}$, and $\overline{O_iM}^i$ the coordinates of the same point $M$ in the frame $F_i$.

The location in space of the robot is obtained through the product of elementary matrices corresponding to each segment. Consider a robot consisting of $n$ links connected by $n$ joints, where the frame $F_0$ is attached to the base and the frame $F_n$ is attached to the tool.

The matrix $\mathbf{T}_{0,n}$ is obtained by the product $\mathbf{T}_{0,1} \cdot \mathbf{T}_{1,2}, \ldots , \mathbf{T}_{n-1,n}$. This matrix is called the *direct geometric model* of the robot: Each elementary matrix $\mathbf{T}_{i-1,i}$ incorporates the joint parameter $q_i$ (either rotation angle $\theta_i$ or translation $d_i$) so that both the location and orientation of the tool frame $F_n$ are given through the computation of $\mathbf{T}_{0,n}$.

$$\overline{OM}^0 = \mathbf{T}_{0,n} \cdot \overline{O_nM}^n. \tag{9.2}$$

$\overline{OM}^0$ represents the coordinates of a point $M$ in the base frame $F_0$, and $\overline{O_nM}^n$ the coordinates of the same point $M$ in the tool frame $F_n$.

Let an object be seized by the robot's grip. To move this object by a generalized displacement $\hat{X}$ (three finite translations along the fixed frame axes and three finite rotations around them), it is necessary to calculate, from the matrix $\mathbf{T}_{0,n}$, the set of articular variables $q_i$ of each robot joint. If it exists (i.e., when the matrix $\mathbf{T}_{0,n}$ is square and positive definite), this mathematical function is called the *inverse geometric model*.

The direct geometric model is always defined when the robot scheme is known, whereas the inverse geometric model is accessible in only a few cases (Paul, 1981). For the terminal segment to move with six degrees of freedom requires a robot with six axes. The determination of the inverse model function is easier if the three last rotation axes corresponding to the wrist in upper arm modeling pass through a single point.

# Kinematic and Dynamic Models

Let the generalized velocity of the tool be the vector $\hat{V}$. It consists in the linear velocity $\overline{V}$ and the angular velocity $\overline{\Omega}$ of the end effector. This generalized velocity vector $\hat{V}$ is related to the generalized joint velocity vector $\hat{q}$ by a matrix $\mathbf{J}$, in which each term is a vector, such that

$$\hat{V} = \mathbf{J} \cdot \hat{\dot{q}}. \tag{9.3}$$

This form is called the Jacobian matrix. The previous relation can be projected onto any frame.

For small displacements, the corresponding relation is

$$\widehat{\delta M^0} = \mathbf{J}^0 \cdot \widehat{\delta q}. \tag{9.4}$$

This *direct kinematic model* gives both the linear and angular small displacements $\widehat{\delta M}$ in the frame $F_0$ when the set of joint increments $\delta q$ is given.

The *inverse kinematic model* is expressed as

$$\widehat{\delta q} = (\mathbf{J}^0)^{-1} \cdot \widehat{\delta M^0}. \tag{9.5}$$

It may be computed if the matrix $\mathbf{J}^0$ can be inverted. Paul (1986) describes a method of resolution, valid when the number of the degrees of freedom of the task at the level of the terminal is different from the number of articular axes.

When forces and moments are applied on the end segment, it is possible to define a mathematical operator allowing the calculation of motor moment, which must be exerted on each joint contributing to the dynamic task.

Let $\hat{F}_t$ be the generalized force exerted by the environment at the level of the tool ($\hat{F}_t$ consists of a force $\overline{F}$ and a moment $\overline{M}_{(Ot)}$ reduced at the tool frame origin $Ot$). The virtual power developed at the level of the tool is

$$P = [\overline{V}\ \overline{\Omega}] \cdot \begin{bmatrix} \overline{F} \\ \overline{M}_{(Ot)} \end{bmatrix} \tag{9.6}$$

or

$$P = \hat{V}^t \cdot \hat{F}_t ,$$

where $\overline{V}^t$ is the transposed vector of $\overline{V}$.

The corresponding power is created at the level of each joint:

$$\dot{\theta}_1 C_1 + \dot{\theta}_2 C_2 + \ldots = \hat{\dot{q}}^t \cdot \hat{C} , \tag{9.7}$$

where $\dot{\theta}_i$ is the angular velocity and $C_i$ the motor torque at the $i^{th}$ joint, $\hat{\dot{q}}$ is the generalized joint angular velocity vector, and $\hat{C}$ the generalized joint torque vector.

Equaling the virtual power, we obtain

$$\hat{V}^t \cdot \hat{F}_t = \hat{\dot{q}}^t \cdot \hat{C} \cdot$$

But

$$\hat{V} = \mathbf{J} \cdot \hat{\dot{q}} ,$$

so

$$\hat{V}^t = \hat{q}^t \cdot \mathbf{J}^t \,.$$

Finally

$$\hat{C} = \mathbf{J}^t \cdot \hat{F}_t \,. \tag{9.8}$$

The generalized joint torque vector is obtained directly from the transposed Jacobian matrix and the generalized force at the level of the tool. Uicker (1984) has developed a technique that takes inertial forces into account.

Presently, research is oriented toward the effects of inertia and elastic properties of segment materials on the dynamic behavior of robots. Other investigators analyze the data input giving information about the external environment to improve the command of robots (Shimano, 1978; Whitney, 1972).

# THE SPECIFIC PROBLEMS
# IN BIOMECHANICAL MOTION

In robotics, the geometric model is completely defined; the rotation or translation axes are fixed relative to each motion segment, and the distances between two successive axes on the segment are precisely known. The notation of Denavit and Hartenberg (1955) can be easily applied. An elementary matrix form (defined in Figure 9.1) can be attached to each segment and the matrix corresponding to the entire robot computed. In biomechanics, the structure of joints is such that they work with a functional laxity that allows them to adapt to the task. This was observed on cadaveric specimens (Kinzel, Hall, & Hillberrg, 1972) as well as in living joints (Dimnet & Guingand, 1984).

## Identifying the Equivalent Mechanism of a Living Human Limb Movement

Numerous systems of motion analysis (Dimnet, Chèze, & Carret, 1990) are presently available. Our team tested both VICON and ELITE systems. Their general principles are quite similar: Reflecting markers are fixed on the skin. The trajectory of each marker is viewed from two separate cameras. A computer program allows identification of each marker's image centroid, then a numerical computation gives the marker's trajectory in three-dimensional space, onto a set of axes affixed to the room, this frame being previously calibrated. Under optimal conditions, the accuracy of identifying a point is about $L/3,000$, $L$ being the greatest size of the studied volume expressed in meters. This accuracy rapidly decreases when the studied subject is moving away from the calibrated area. These systems give the instant coordinates of a marker with respect to the laboratory frame. Accessing the joint's kinematics implies the numerical treatment of external data (the trajectories of the markers fixed on the skin) to obtain the relative displacement of internal bones. This deals with simplifying hypotheses and their corresponding models.

### Hypotheses

On each moving body segment, three markers are fixed on the skin as near as possible to anatomical points (Figure 9.2). The lines connecting the instant location

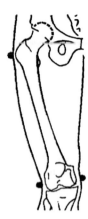

**Figure 9.2**   Positions of the external markers on the thigh.

of the three markers form a triangle, but the shape of each segment triangle varies in time because of motion of markers fixed on the skin with respect to the underlying soft tissues and bones. To be able to use the solid kinematic techniques, a mean solid triangle shape is defined for each segment, and this solid triangle is then fitted to the measured triangle according to a least-square criterion.

Consider the relative displacement between two successive segments. For the first segment, we know the instantaneous positions of the frame attached to the bone (we suppose that this segment has been considered as the second segment in a previous treatment: for example, the femur is the second segment if we are interested in the hip joint, but then it will be the first segment when we consider the knee joint). For the second segment, we only have the instantaneous positions of the solid triangle attached to the segment. The relative displacement of the second segment with respect to the first one results in two elementary motions: the articular motion of the second bone with respect to the first one, and the motion of the soft tissues with respect to the underlying bone.

Hypotheses must be proposed to decompose the only measurable motion (i.e., the motion of the solid triangle frame attached to the second segment with respect to the bone frame attached to the first segment) into two parts: the real articular motion between the two bones and the movement of soft tissues with respect to the internal structure. The articular motion of bones is assumed to be compatible with a mechanical joint: only rotations around either a point (hip, shoulder) or an axis (knee, elbow) are considered, while the disturbing motion of soft tissues is assumed to dissipate less energy during motion. In the articular motions, the location of rotation centers or axes can vary all along the movement.

### Preliminary Data Treatment

The motion analysis system used for three-dimensional computation of marker trajectories was the VICON system, but it was not the new one, which eliminates camera-image artifacts so that in a finite number of views, wrong reflective points can be substituted for real marker images, creating important disturbances on marker trajectories (Figure 9.3, a and b).

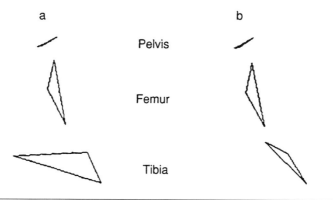

**Figure 9.3**   Sagittal view of the triangles formed by the markers during walking: (a) wrong image, (b) reconstructed image.

Because the distortion of the shape of the triangles representing the limb segments is often so important that the corresponding kinematics has no physical meaning, it is necessary to choose between two strategies: either to eliminate the view showing a wrong image of one marker, or to keep this view, replacing the wrong punctual image by a new point located in a position compatible with that occupied on the adjacent views. This last possibility has been retained.

So, we first filter the data, reducing the high frequency noises due to the sampling process (50 images a second) in the optic motion system. For this, the signal spectrum is calculated owing to the FFT (fast Fourier transform). This spectrum shows that the noise spreads on the whole frequency range when the useful signal remains in low frequencies: so we choose a low-pass filter. Then, in order to realize a good compromise between selectivity and complexity (involving long calculation times), we opt for a third-order Butterworth filter. The cut frequency (i.e., frequency at which both the spectrum density of the signal and that of the noise are equal) is automatically calculated from the frequency spectrum of the trajectory signal to adapt the filter to each kind of motion.

In a second step, we identify the "wrong" images (i.e., rough variations of the trajectory shape) on the trajectory of each marker, and we replace these points, using a local polynomial interpolation taking into account a finite number of images around the missing one. This preserves the maximum information, whereas a fitting algorithm at the beginning of the data treatment would alter the detection of local discontinuities in the movement (which are essential in pathologic motions).

To use solid kinematic techniques, we must make sure that the triangles representing each body segment do not show shape distortions during the movement. So, we calculate the imaginary solid triangle $A_S B_S C_S$ that best fits the real one defined by Markers $A$, $B$, and $C$, which are fixed on the body segment all along the motion, thus creating an intermediate frame, where the shape of triangles $ABC$ of each image can be compared by superposition. A statistic treatment, giving a special importance to the best triangle summit and side ("best" meaning the less varying ones) determines the solid triangle shape $A_S B_S C_S$ (Figure 9.4).

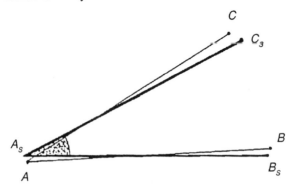

**Figure 9.4**   Optimization of the solid triangle position $A_s B_s C_s$ versus the real triangle $ABC$.

Then this shape is moved back to the instantaneous position of the triangle $ABC$ by minimizing both the linear distance between the best summit $A_S$ and the real position of the correspondent Marker $A$, and the angular distance between the best side $A_S B_S$ and the real correspondent side $AB$ (Chèze, 1990). After this last operation, a reference frame is affixed to the segment for each position.

## Accessing the Internal Articular Mechanism

The sampling frequency is determined by camera speed (50 Hz), so the time duration is measured using the number of images. If one considers the femur motion with respect to the pelvis, a matrix $^iT_{0,f}$ is available at each instant $i$. It locates the external femur frame $F_f$, obtained from the external solid triangle, versus the laboratory frame $F_0$. One also defines the matrix $^iT_{0,p*}$ giving the location of the internal frame $F_{p*}$, affixed to the pelvis at the time $i$ in the laboratory coordinate system. This location is obtained through a numerical process analogous to that described below.

It is then possible to compute the relative location of the external femur frame $F_f$ versus the internal pelvis frame $F_{p*}$ at the instant $i$ through the matricial relation:

$$^iT_{p*,f} = (^iT_{0,p*})^{-1} \cdot {}^iT_{0,f} . \tag{9.9}$$

The aim consists of finding the location of the internal femur frame $F_{f*}$ at the time $i$. For this, two anatomic situations are analyzed successively. They correspond to joints turning around one point and joints turning about an axis.

### Spherical Joints

These joints rotate around one point, which can migrate during the movement. They correspond to a simplified model of the hip or the shoulder. In this last case, the shoulder center is liable to move during the motion and the kinematic model should take this characteristic into account.

For the hip, the preliminary step consists of calculating the matrix ${}^{i}\mathbf{T}_{p*,f}$ defining, at the instant $i$, the relative location of the external femur frame versus the internal pelvis frame. The internal femur frame is assumed to move relative to the internal pelvis frame in a same way as a spherical joint. The angular rotation vector $\overline{\Omega}_i$ and the instant center location are computed successively.

From the matrix ${}^{i}\mathbf{T}_{p*,f}$, the angular magnitudes of displacements moving the femur relative to the pelvis are calculated at the time $i$. These angular displacements are assumed to be the result of three elementary BRYANT rotations. The first rotation of amplitude $\theta_1$ around the axis $\overline{X}_{p*}$ moves the frame $(\overline{X}_{p*}, \overline{Y}_{p*}, \overline{Z}_{p*})$ to the position $(\overline{X}^1\ \overline{Y}^1\ \overline{Z}^1)$. Then a second rotation of amplitude $\theta_2$ around the axis $\overline{Y}^1$ moves the frame $(\overline{X}^1\ \overline{Y}^1\ \overline{Z}^1)$ to the position $(\overline{X}^2\ \overline{Y}^2\ \overline{Z}^2)$. Finally, a third rotation of amplitude $\theta_3$ around the axis $\overline{Z}^2$ moves the frame $(\overline{X}^2\ \overline{Y}^2\ \overline{Z}^2)$ to the position $\overline{X}_f\ \overline{Y}_f\ \overline{Z}_f)$. The angular values are filtered and then differentiated to obtain the three components of the angular velocity $\overline{\Omega}_i$.

The hip center and the knee axis can be located roughly from frontal and sagittal X-ray images that show both the markers and the hip and knee. Spherical zones are experimentally evaluated (about 2 cm of radius) to delimit the migrating possibilities, either of the hip center or the knee axis, around this mean position.

The center location $S_i$ and the angular velocity $\overline{\Omega}_i$ are then improved through an optimizing procedure based on Simplex (Nelder & Mead, 1965). A simplex is an $n$-dimensional figure specified by giving its $n + 1$ vertices: a triangle in two dimensions, a tetrahedron in three dimensions, etc. This algorithm, at each step, calculates the values of the function to minimize at $n + 1$ points. Let the point $P_h$ be that at which the function value is highest and $P_l$ that at which it is lowest. Let $P_g$ be the center of mass of all points in the simplex except $P_h$, that is,

$$P_g = [\Sigma\ P_i - P_h]/n$$

for $i = 1, \ldots, n + 1$.

From the original simplex, a new one is formed by replacing the point $P_h$ by a better point, if possible. The first attempt to find a better point is made by reflecting $P_h$ with respect to $P_g$, producing

$$P^* = P_g + (P_g - P_h).$$

If $F(P^*) < F(P_l)$, a new point is tried at

$$P^{**} = P_g + 2(P_g - P_h).$$

If $F(P^*) > F(P_h)$, a new point is tried at

$$P^{**} = P_g - (P_g - P_h)/2\ .$$

The best of these new points then replaces $P_h$ in the simplex for the next step, unless none of them is better than $P_h$. In the latter case, a whole new simplex is formed around $P_l$, with dimensions reduced by a factor of 0.5

This algorithm searches in an ''intelligent'' direction, pointing from the highest value to the average of the lowest values. A convenient convergence criterion is based on the difference $F(P_h) - F(P_l)$. The iterations are stopped when this difference is less than a preset value.

In our case, we use the experimentally estimated center location and the calculated angular velocity as initial point in a 6-dimensional simplex. The function to minimize, depending on the six parameters (three center coordinates and three angular velocity components) corresponds to the difference between the observed motion and that compatible with rotation around a point.

Let $A_1B_1C_1$ and $A_2B_2C_2$ be two successive positions of the lower solidified frame versus the upper internal frame. The finite center of rotation $S_i$ and angular magnitude $\overline{\Omega}_i$ such as $\overline{\Omega}_i = 2 \cdot \overline{n}_i \cdot tg\ (\theta_i/2)$ moves $A_1B_1C_1$ to $A'_2B'_2C'_2$ through the RODRIGUES formula (Bishop, 1969; also see Appendix):

$$\overline{A_1A'_2} = 2t/(1 + t^2) \cdot (\overline{n}_i \times \overline{S_iA_1}) + 2t^2/(1 + t^2) \cdot [\overline{n}_i \times (\overline{n}_i \times \overline{S_iA_1})]$$

$$\overline{B_1B'_2} = 2t/(1 + t^2) \cdot (\overline{n}_i \times \overline{S_iB_1}) + 2t^2/(1 + t^2) \cdot [\overline{n}_i \times (\overline{n}_i \times \overline{S_iB_1})]$$

$$\overline{C_1C'_2} = 2t/(1 + t^2) \cdot (\overline{n}_i \times \overline{S_iC_1}) + 2t^2/(1 + t^2) \cdot [\overline{n}_i \times (\overline{n}_i \times \overline{S_iC_1})] \quad (9.10)$$

where $\overline{n}_i$ is the unit vector of the rotation axis, $t = tg(\theta_i/2)$, and $\theta_i$ is the finite angular displacement between the positions $i$ and $i + 1$.

The quantity $\delta = (\overline{A_2A'_2})^2 + (\overline{B_2B'_2})^2 + (\overline{C_2C'_2})^2$ is computed and depends on the location of the center $S_i$ and the angular velocity $\overline{\Omega}_i$. $S_i$ and $\overline{\Omega}_i$ are chosen by minimizing $\delta$, $S_i$ remaining inside the scatter zone but allowed to migrate throughout the movement. The trajectories of the points $A'$, $B'$, and $C'$ correspond to the internal articular movement. The difference between these trajectories and the observed ones corresponds to the soft tissues motion (see Figure 9.5).

## Axial Joint

In the case of an axial joint, the internal articular movement is the result of two components of displacement: one rotation around an axis, of which the direction

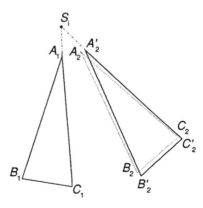

---

**Figure 9.5**   Decomposition of the joint motion in two elementary parts:
— The articular motion moves $A_1B_1C_1$ in $A'_2B'_2C'_2$
— The soft tissues motion moves $A'_2B'_2C'_2$ in $A_2B_2C_2$.

and position can vary, and a coupled translation. These joints may be treated as hinges, which can migrate like a spherical joint. The differences are that the direction of the rotation vector is less variable, and axis displacement is permitted (this translation is not obviously colinear to the rotation axis, unlike the helicoidal motion). This kinematic model is proposed to approximate the anatomical functioning of the knee: Guided by the menisci, the internal condyle motion can be compared to a spherical motion in which the center position varies during the movement.

The angular velocity $\overline{\Omega_i}$ is obtained in the same way as that in the spherical joint. The linear position of the center $H_i$ is situated inside a permitted zone previously defined. This center location $H_i$ and the angular velocity are improved when the finite displacement of the internal articular structure between the instants $i$ and $i + 1$ moves the points $A, B, C$ from the position $A_1B_1C_1$ to $A'_2B'_2C'_2$. This displacement involves a finite axial rotation around an axis (of unit vector $\overline{n_i}$, passing through $H_i$ and of magnitude $\theta_i$).

The corresponding vectors are

$$\overline{A_1A'_2} = 2t/(1 + t^2) \cdot (\overline{n_i} \times \overline{H_iA_1}) + 2t^2/(1 + t^2) \cdot [\overline{n_i} \times (\overline{n_i} \times \overline{H_iA_1})]$$

$$\overline{B_1B'_2} = 2t/(1 + t^2) \cdot (\overline{n_i} \times \overline{H_iB_1}) + 2t^2/(1 + t^2) \cdot [\overline{n_i} \times (\overline{n_i} \times \overline{H_iB_1})]$$

$$\overline{C_1C'_2} = 2t/(1 + t^2) \cdot (\overline{n_i} \times \overline{H_iC_1}) + 2t^2/(1 + t^2) \cdot [\overline{n_i} \times (\overline{n_i} \times \overline{H_iC_1})] \quad (9.11)$$

The quantity $\delta = (\overline{A_2A'_2})^2 + (\overline{B_2B'_2})^2 + (\overline{C_2C'_2})^2$ is then calculated for each interval of time $i, i + 1$. The center location $H_i$ and the angular velocity $\overline{\Omega_i}$ are chosen so that $\delta$ is minimized and $H_i$ stays inside the permitted zone. The trajectories of the points $A', B',$ and $C'$ represent the articular motion, which is the result of rotations $(n, \theta)$ around variable centers.

## Preliminary Results

The methods described above were tested on a normal subject.

### Step 1: Estimation of the Center Position

The joint centers are roughly located in the frame of external markers by experimental measurements on frontal and sagittal radiographs:

- The femoral head center in the pelvis frame
- The knee center in the femur frame (assuming that the knee center is situated along the tangent to the lower part of epicondyles, at the level of the tibial plateau crest)

### Step 2: Preliminary Data Treatment

- The filter effect on external marker trajectories (Figure 9.6, a and b)
- Changes in the shape of the triangle connecting the three markers that define a body segment (Figure 9.7) (these changes are due to the marker displacement rather than the movement of the underlying soft tissues)
- Comparison of marker trajectories before and after solidification (Figure 9.8, a and b)

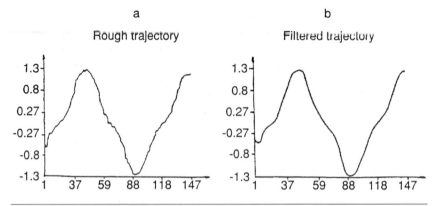

**Figure 9.6**   The filter effects on (a) rough and (b) filtered marker trajectory.

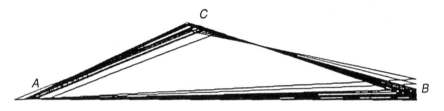

**Figure 9.7**   Distortion of the triangle shape due to marker slippage.

## Step 3: Determination of the Kinematic Model

The spherical joint method was used for the hip joint, whereas the axial joint method was used for the knee joint. We can display the following during walking:

- The displacement of solid triangles compatible with the kinematic model (the different triangles represent pelvis, femur, and tibia) and the center trajectories of both hip and knee joint (Figure 9.9)
- Angular changes in magnitude of flexion-extension, abduction-adduction, and internal-external rotation of the knee (Figure 9.10 a, b, and c)
- The influence of soft tissue movements displayed by the difference between an external solidified marker trajectory and the trajectory of the same marker affixed to the internal bone frame (Figure 9.11)

## Step 4: Motorization of the Model

If we wish to calculate articular moments and muscular efforts by using classical concepts of robotics, it is necessary to create a mechanical robot equivalent to the limb, and then to establish transfer functions between the kinematic pattern and the geometric robot and between the geometric robot and the anatomical components, such as bones and muscles (Figure 9.12).

Geometric center locations $H$ and $K$ derived from kinematic analysis and instant triangle positions ($ABC$) compatible with the internal motion are required

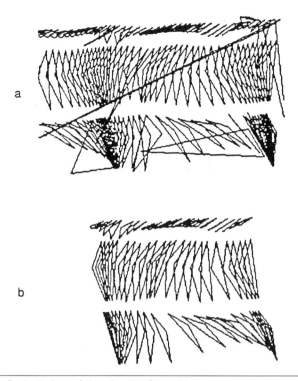

**Figure 9.8**    Sagittal view of the triangles formed by the markers during walking: (a) rough movement, (b) solidified movement.

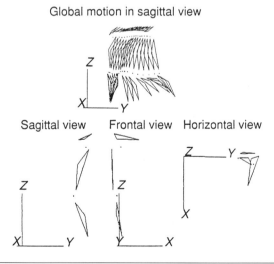

**Figure 9.9**    Views showing the triangles formed by the markers and the rotation centers.

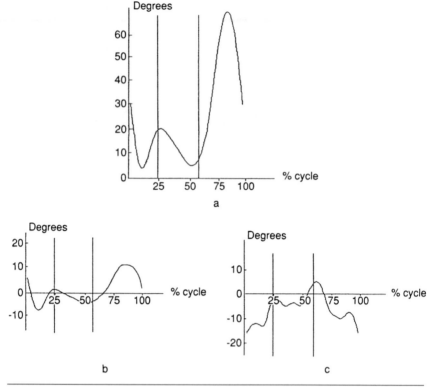

**Figure 9.10**   Angular displacements of the knee joint: (a) flexion-extension angle; (b) abduction-adduction angle; (c) internal-external rotation angle.

to define the global geometric robot pattern equivalent to the moving limb. It is necessary to introduce new anthropologic assumptions to improve the anatomical representation of a moving bone by a mechanical robot: the flexion axis of the hip joint is assumed to coincide with the normal axis $\overline{Z_P}$ to the external triangle, whereas the $\overline{X_P}$ axis coincides with the main direction $\overrightarrow{AB}$ of the femur. This is possible because markers are fixed next to anatomical points. We are then able to calculate, at each time, the geometrical parameters and the articular variables of the equivalent robot. An adimensional pattern of the different bones is presently studied. It locates geometrically muscular insertion areas and muscular force directions through the tendon directions. This pattern has to be fitted by using the kinematic model and corresponding points $H$ and $K$, identifying the equivalent mechanical robot, and looking for muscular force distribution that provokes a displacement analogous to that experimentally observed.

## Motorizing the Mechanism Equivalent to a Human Motion

The mechanism equivalent to the moving human limb consists of rotational joints, at least for real degrees of freedom corresponding to an active command by

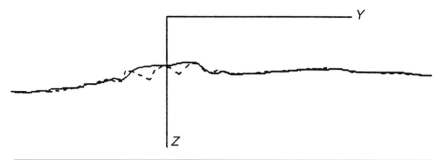

**Figure 9.11**  Trajectories of an external marker:
(- - - -) solidified trajectory,
(———) internal trajectory.

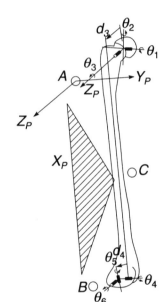

**Figure 9.12**  Superimposition of the kinematic, robotic, and anatomical models.

muscles. There may be some translation play, but they essentially correspond to small local displacements that allow the joint to adapt to the dynamic task.

In a robot, each axis is generally moved by a rotational activator that delivers a pure moment. In biomechanics, the motor effects are obtained through the action of agonist and antagonist muscles.

Once the equivalent mechanism is known for each task, it is necessary to identify the agonist and antagonist muscles that contribute actively to the task, and to define the application point and the directions of forces where the muscle tendons insert into the bones. Testing for each degree of freedom is not redundant because it is assumed that the only muscular activity is that of the agonist muscle.

This permits calculation of the magnitude of muscular force when the direction of force and the application point are known.

The antagonist muscle acts as a controller and stabilizer of the small displacements coupled with the principal motion. This occurs because of the nonlinearities created by the direction of the antagonist muscle force. Contrary to mechanical robotics in which the motor exerting a torque is placed between two successive segments, active biomechanical motorization is achieved by the forces exerted by muscles that are inserted either into two nonsuccessive segments (the biarticular muscles) or into bones adjacent to the joint (monoarticular muscles).

Taking into account the degrees of freedom and associated small displacements and magnitudes of the same number of muscular forces, it is possible to numerically solve a statically determined problem. However, in biomechanics, difficulties arise because in the motorization of a limb, the task changes constantly; thus, the task must be decomposed into different phases. For example, the muscular motorization during gait will be completely different in the stance phase than in the swing phase. So a specific model with corresponding numerical resolution must be adopted for each gait phase.

In any case, the results are the muscular force magnitude and the interarticular load. A complementary study takes into account the antagonist muscular forces, allowing the calculation of their influence on joint loads.

The biomechanics of motion may also be approached from a physiological point of view. Limb motion is produced by muscular actions. The exerted forces depend, in particular, on muscle lengths and stretching velocities. These parameters are directly controlled by the neuromuscular system, although it seems that angular displacement sensors are not present in human joints. The concept is that of a robot structure consisting of a series of segments joined in a linear chain (Figure 9.13). A new concept has been proposed (Merlet, 1990) consisting of two successive bodies, which represent the bony structure, connected by several actuators, which represent the muscles that act indirectly on the relative location of the bony segment (Figure 9.14).

In mechanical robots, the articular location $\alpha_i$ of each $i^{th}$ joint is directly measured, and the corresponding angular velocity $\dot{\alpha}_i$ is numerically computed.

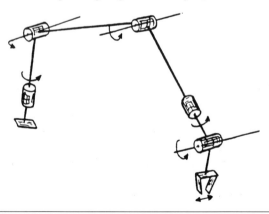

**Figure 9.13**   Series robotic structure.

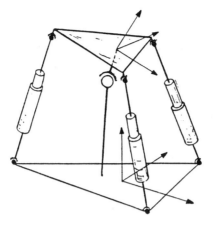

**Figure 9.14**   Parallel robotic structure.

The set of joint angles $\hat{\alpha}$ and angular velocities $\dot{\hat{\alpha}}$ are known at time $t$. The linear and angular velocities, $\overline{V}$ and $\overline{\Omega}$, of the actuator are calculated through the matricial direct relation

$$\hat{V} = \mathbf{J}_d \cdot \dot{\hat{\alpha}}, \qquad (9.12)$$

where $\hat{V}$ is the generalized velocity vector and $\mathbf{J}_d$ is the Jacobian direct matrix. The inverse procedure $\dot{\hat{\alpha}} = (\mathbf{J}_d)^{-1} \cdot \hat{V}$ allows calculation of the articular displacements.

In living motion studies, it is possible to compute the instantaneous configuration of the analog robot $\hat{\alpha}$ and the corresponding velocity vector $\dot{\hat{\alpha}}$ from the three-dimensional trajectories of external markers; the calculation schemes are shown in Figure 9.15, a-c.

The identification of the internal mechanism analogous to the human limb is not sufficient if computation of muscular forces is wanted. First, the robot equivalent of the limb must be defined. Then it is necessary to determine

- which muscles are active during the task, and
- what mechanical device best represents the active group of muscles.

## Identification of the Analog Muscular Parallel Robot

Figure 9.16 illustrates the stabilizing effect of muscles in a moving structure; only hip and knee joints are taken into account and are represented as rotational joints with three degrees of freedom. Six muscles with either mono- or biarticular action are necessary to move thigh and shank. The problem lies in determining the points at which muscles insert into bones, defining the direction of muscular forces, and then modeling the active muscles.

In this example, active muscles are modeled by straight segments. For each lower limb position defined by the generalized vector $\hat{\alpha}$ and moved with the

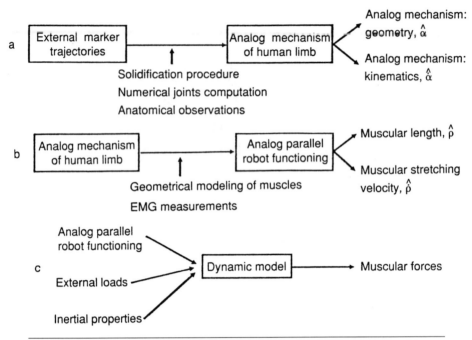

**Figure 9.15** Calculation schemes for (a) structural series robot identification, (b) muscular parallel robot identification, and (c) muscular force computation.

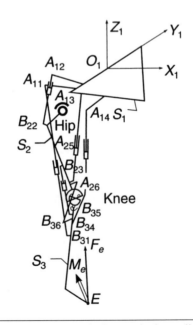

**Figure 9.16** Muscular parallel robot equivalent to the lower limb.

generalized velocity vector $\overset{\wedge}{\alpha}$, it is possible to compute step by step the generalized length $\overset{\wedge}{\rho}$ of muscles and the corresponding generalized stretching velocity $\overset{\wedge}{\dot{\rho}}$.

For groups of muscles modeled by linear actuators, let $\overline{A_iB_j}$ be the group of muscles that connects the bones $i$ and $j$ with $j$ equal to either $i + 1$ (if the considered muscle is monoarticular) or $i + 2$ (if the muscle is biarticular). Then, we can write

$$\overline{A_iB_j} = \rho_{ij} \cdot \overline{u}_{ij} = \overline{O_iA_i}^i + \overline{O_iO_j}^i + \mathbf{R}_{i,j} \cdot \overline{O_jB_j}^j , \qquad (9.13)$$

where $\overline{O_iA_i}^i$ and $\overline{O_jB_j}^j$ are the components of the insertion points $A_i$ and $B_j$ in the local frames $F_i$ and $F_j$, respectively, attached to the bones $i$ and $j$. $\overline{O_iO_j}^i$ and $\mathbf{R}_{i,j}$ define the relative position of the frame $F_j$ with respect to the frame $F_i$. They depend on the coefficients $a_{kl}$ of the homogeneous matrix $\mathbf{T}_{i,j}$ defined as

$$\mathbf{T}_{i,j} = \begin{array}{c} \\ X_i \\ Y_i \\ Z_i \\ \\ \end{array} \left| \begin{array}{cccc} X_j & Y_j & Z_j & \overline{O_iO_j}^i \\ a_{11} & a_{12} & a_{13} & a_{14} \\ a_{21} & a_{22} & a_{23} & a_{24} \\ a_{31} & a_{32} & a_{33} & a_{34} \\ 0 & 0 & 0 & 1 \end{array} \right|$$

With $\overline{u}_{ij}$ being a unit vector, $\rho_{ij}$ is obtained by

$$\rho_{ij}^2 = \overline{A_iB_j}^t \cdot \overline{A_iB_j} . \qquad (9.14)$$

$\overline{A_iB_j}^t$ being the transposed vector of $\overline{A_iB_j}$ ,

$$\rho_{ij}^2 = (\overline{A_iO_i}^i + \overline{O_iO_j}^i + \mathbf{R}_{i,j} \cdot \overline{O_jB_j}^j)^t \cdot (\overline{A_iO_i}^i + \overline{O_iO_j}^i + \mathbf{R}_{i,j} \cdot \overline{O_jB_j}^j) ;$$

$$\rho_{ij}^2 = \overline{A_iO_i}^{it} \cdot \overline{A_iO_i}^i + \overline{O_iO_j}^{it} \cdot \overline{O_iO_j}^i + \overline{O_jB_j}^{jt} \cdot \overline{O_jB_j}^j + $$
$$2 \, (\overline{A_iO_i}^{it} \cdot \overline{O_iO_j}^i) + 2 \, (\overline{A_iO_i}^i + \overline{O_iO_j}^i)^t \cdot \mathbf{R}_{i,j} \cdot \overline{O_jB_j}^j$$

$$\rho_{ij} \cdot \dot{\rho}_{ij} = \overline{O_iO_j}^{it} \cdot \dot{\overline{O_iO_j}}^i + \overline{A_iO_i}^{it} \cdot \dot{\overline{O_iO_j}}^i + $$
$$\overline{O_iO_j}^{it} \cdot \mathbf{R}_{i,j} \cdot \overline{O_jB_j}^j + (\overline{A_iO_i}^i + \overline{O_iO_j}^i)^t \cdot \dot{\mathbf{R}}_{i,j} \cdot \overline{O_jB_j}^j . \qquad (9.15)$$

Using $\dot{\overline{O_iO_j}}^i$ to design the derivative of the vector $\overline{O_iO_j}^i$ ,

$$\dot{\overline{O_iO_j}}^i = \dot{a}_{14}\overline{X}_i + \dot{a}_{24}\overline{Y}_i + \dot{a}_{34}\overline{Z}_i$$

$$\dot{\mathbf{R}}_{i,j} = \sum [\dot{\beta}_k \cdot d(\mathbf{R}_{i,j})/d(\beta_k)] ,$$

$\beta_k$ being the Bryant angles (with $k$ varying from 1 to 3).

So it is possible to express each $\dot{\rho}_{ij}$ as a function of $\overset{\wedge}{\alpha}$. The results are grouped in a matrix form

$$\overset{\wedge}{\dot{\rho}} = \mathbf{J}_i \cdot \overset{\wedge}{\dot{\alpha}} , \qquad (9.16)$$

$\dot{\alpha}_{ij}$ is the generalized joint velocity vector and $\mathbf{J}_i$ is the parallel-series Jacobian matrix.

## Computation of the Muscular Forces

Let $\hat{F}_e$ be the generalized external load reduced at the application point, $E$. $\hat{F}_e$ consists of a force $\overline{F}_e$ and moment $\overline{M}_{e(E)}$. Let $\hat{V}_e$ be the generalized velocity vector of the application point $E$ of the external load. The virtual power of the external force is

$$P_e = \hat{V}_e^t \cdot \hat{F}_e$$

with

$$\hat{V}_e^t = \hat{\alpha}^t \cdot \mathbf{J}^t , \tag{9.17}$$

where $\mathbf{J}^t$ is the transposed Jacobian matrix.

Let $f_{ij} = \overline{F}_{ij} \cdot \overline{u}_{ij}$ be the modulus of the load exerted by the muscles connecting $A_i$ and $B_j$. The corresponding virtual power is

$$P_{ij} = \dot{\rho}_{ij} \cdot f_{ij} .$$

For the set of active muscles, the virtual power is

$$P_m = \hat{\dot{\rho}}^t \cdot \hat{F}_m , \tag{9.18}$$

where $\hat{F}_m$ is the generalized muscular force vector and $\hat{\rho}$ the generalized stretching velocity vector with $\hat{\dot{\rho}}^t = \hat{\alpha}^t \cdot \mathbf{J}_i^t$.

In static conditions when the weight is neglected, it can be stated that

$$\hat{\alpha}^t \cdot \mathbf{J}^t \cdot \hat{F}_e + \hat{\alpha}^t \cdot \mathbf{J}_i^t \cdot \hat{F}_m = 0;$$

so

$$\mathbf{J}^t \cdot \hat{F}_e + \mathbf{J}_i^t \cdot \hat{F}_m = 0,$$

or

$$\hat{F}_m = -(\mathbf{J}_i^t)^{-1} \cdot \mathbf{J}^t \cdot \hat{F}_e . \tag{9.19}$$

If inertial forces are taken into account, the static conditions are modified. Let $\hat{F}_i$ be the generalized vector consisting of the linear inertial force $-M\overline{\gamma}(G)$ and the rotational inertial force $-\mathbf{I}\overline{\Omega}$ reduced in the segment center of mass $G$ ($\overline{\gamma}$ being the linear acceleration, $\mathbf{I}$ the inertial matrix, and $\overline{\Omega}$ the rotational acceleration).

The corresponding virtual power is

$$P_i = \hat{\alpha}^t \cdot \mathbf{J}_G^t \cdot \hat{F}_i , \tag{9.20}$$

where $\mathbf{J}_G$ is the Jacobian matrix calculated at the segment center of mass, $G$. The relation under dynamic conditions then becomes

$$\hat{\alpha}^t \cdot \mathbf{J}^t \cdot \hat{F}_e + \hat{\alpha}^t \cdot \mathbf{J}_i^t \cdot \hat{F}_m + \hat{\alpha}^t \cdot \mathbf{J}_G^t \cdot \hat{F}_i = 0,$$

or

$$\hat{F}_m = -(\mathbf{J}_i^t)^{-1} \cdot (\mathbf{J}^t \cdot \hat{F}_e + \mathbf{J}_G^t \cdot \hat{F}_i) . \qquad (9.21)$$

## DISCUSSION

There are two points to be discussed here: the validity of the transfer function from kinematic model to real anatomic model and the clinical application of information that comes from this new analysis.

## The Transfer Function

The results of the kinematic analysis can be easily validated. For this, we propose to use a double pendulum (Figure 9.17), in which the rotation axes are precisely known, and analyze its rotation from the computation of the displacements of three markers fixed on each segment. We have constructed such a pendulum, and the kinematic validation is in progress.

On the other hand, we cannot *directly* validate the changeover from the kinematic model to the anatomic model. The kinematic model gives the instantaneous positions of the hip center, $H$, and the corresponding positions of the knee center, $K$. But it is impossible, using only these two positions, to reconstruct the mean direction of the femur. We need either frontal and sagittal radiographs showing both the external markers and the shape of the bone, or we must have recourse to morphometric data.

The anatomic model is essential in establishing a dynamic model of the moving limb; muscular actions actually depend on the location of the insertions and the directions of the muscles. To determine exactly where the muscles are attached to bones and the directions in which the tendons move (which correspond to the muscular force directions), we must have a good idea of the anatomy of the limb.

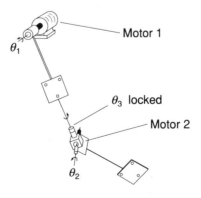

**Figure 9.17**   Double pendulum used for clinical validations.

## An Example of Clinical Application

Clinicians are interested in several measurements on the frontal radiograph of a subject in stance phase (Figure 9.18). They measure the angles between anatomical axes (e.g., the femoral axis and the tibial axis) and the mechanical axis defined by the segment that binds the joint centers (in this case, the ankle and the hip joint centers). They also calculate the orthogonal distance between the knee center and this mechanical axis (Figure 9.19). These measurements can be computed during the stance phase of a subject's walking.

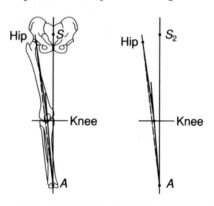

**Figure 9.18**   Clinical measurements on frontal radiograph.

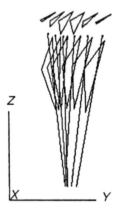

**Figure 9.19**   Mechanical axis in stance phase (sagittal view).

## CONCLUSION

The current techniques in robotics must be considerably modified before being applied to the study of living human movement. First, the mechanical robot equivalent of a moving limb must be identified; then it must be motorized.

The robot's mechanical scheme is known, while its equivalence to a living limb mechanism must be obtained from the kinematic measurements of marker displacements fixed externally on body segments. This is particularly difficult because of the ability of living subjects to adapt their internal structure to the task either through the elasticity of the bones or by changes in articular motion. Thus a very accurate motion analysis is necessary to recognize the degrees-of-freedom displacements and the small associated movements.

The motorization of the moving-limb equivalent mechanism requires complete knowledge of the muscles activated throughout the various phases of the studied task. This involves a multistep calculation procedure. First, the agonist muscles are introduced for large displacements, then the antagonist muscles for stabilizing effects. When the mechanical equivalent is properly constructed, robotics techniques may be used to calculate the muscular forces and to predict the interarticular loads.

## ACKNOWLEDGMENT

This work was supported by the Direction de la Recherche et des Etudes (DRET) of the French Ministry of Defense.

## APPENDIX A

Consider the finite rotation of axis $(O,u)$, of magnitude $\theta$, moving the point $M_0$ to $M_1$. From the RODRIGUES formula,

$$M_0M_1 = 2tg(\theta/2)u \times (OM_0 + OM_1)/2 . \tag{A.1}$$

Let $t = tg(\theta/2)$. Then

$$M_0M_1 = 2tu \times (OM_1 - M_0M_1/2)$$

$$= 2tu \times OM_1 - tu \times M_0M_1 ; \tag{A.2}$$

so

$$u \times M_0M_1 = 2tu \times (u \times OM_1) - tu \times (u \times M_0M_1)$$

$$= 2tu \times (u \times OM_1) - tu(u \cdot M_0M_1) + tM_0M_1 . \tag{A.3}$$

$(u \cdot M_0M_1)$ is equal to zero because the two vectors are perpendicular, so

$$u \times (M_0M_1) = 2tu \times (u \times OM_1) + t(M_0M_1) .$$

According to Equation A.2,

$$M_0M_1 = 2tu \times OM_1 - 2t^2\, u \times (u \times OM_1) - t^2M_0M_1$$

$$\tag{A.4}$$

$$M_0M_1(1 + t^2) = 2tu \times \mathrm{OM}_1 - 2t^2u \times (u \times OM_1) .$$

Then

$$M_0M_1 = [2t/(1 + t^2)] \, (u \times OM_1) - [2t^2/(1 + t^2)]u \times (u \times OM_1) . \qquad (A.5)$$

In the same way

$$M_0M_1 = [2t/(1 + t^2)] \, (u \times OM_0) + [2t^2/(1 + t^2)] \, u \times (u \times OM_0) . \qquad (A.6)$$

# REFERENCES

Bishop, K.E. (1969). Rodrigues' formula and the screw axis. *Journal of Engineering for Industrial Transactions of ASME,* **91**(B), 179-185.

Chèze, L. (1990). La robotique et l'étude des mouvements humains [Robotics and human motion study]. Unpublished DEA thesis, Universite Villeurbanne Cedex, Claude Bernard-Lyon I, France.

Denavit, J., & Hartenberg, R.S. (1955). A kinematic notation for lower pair mechanisms based on matrices. *ASME Journal of Applied Mechanics,* **22**, 215-221.

Dimnet, J., & Guingand, M. (1984). The finite displacement vector's method. *Journal of Biomechanics,* **17**, 387-394.

Dimnet, J., Chèze, L., & Carret, J.P. (1990). Determination of the three dimensional functioning of joints in living subjects from VICON system data treatment: Application to the hand-arm system. In J.C. Goh & A. Nather (Eds.), *Proceedings of the 6th International Conference on Biomechanical Engineering* (pp. 81-85). Singapore: National University of Singapore.

Kinzel, G.L., Hall, A.S., & Hillberrg, B.M. (1972). Measurement of the total motion between two body segments: Analytical development. *Journal of Biomechanics,* **5**, 93-105.

Merlet, J.P. (1990). *Les robots parallèles* [The parallel robots]. Paris: Hermes.

Nelder, J.A., & Mead, R. (1965). A simplex method for function minimization. *Computation Journal,* **7**, 308.

Paul, R.P. (1981). *Robot manipulators: Mathematics, programming and control.* Cambridge, MA: MIT Press.

Paul, R.P. (1981). Kinematic control equations for simple manipulators. *IEEE Transactions on Systems of Man and Cybernetics,* 303-309.

Shimano, B.E. (1978). *The kinematic design and force control of computer controlled manipulators* (Internal report: Artificial Intelligence Laboratory, Stanford University, AIM 313).

Uicker, J. (1984). Dynamic behavior of spatial linkages. *ASME Mechanics,* **5**(68), 1-15.

Whitney, D.E. (1972). The mathematics of coordinated control of prostheses and manipulators. *Transactions of ASME Journal of Dynamic Systems,* 303-309.

# Chapter 10

# Estimation of Muscle and Joint Forces

*Kai-Nan An*
*Kenton R. Kaufman*
*Edmund Y-S. Chao*

The actual loads carried by joints have significant implications. Joint cartilage degeneration and capsuloligamentous laxity are often related to the magnitude and pattern of load transmission at the joint. In total joint arthroplasty, the wear and deformation of the articulating surfaces, the stress distribution in an implant, the mechanical behavior of the bone-implant interface, and the load-carrying characteristics of the remaining bone are intimately related to the joint load. In fractures, the joint and muscle loads play an important role in the bone union. Knowledge of the magnitude and manner of joint loading encountered by the human body is important in determining the possible mechanism and prevention of injury during occupational and sports activities. Calculating internal muscle and joint forces provides additional useful information for the design of implants, surgery, and rehabilitation programs.

In the past two decades, numerous analytical and experimental techniques have been developed for estimating muscle and joint forces. Analytically, the determination of muscle and joint forces involves two steps (Figure 10.1):

1. The determination of intersegmental forces and moments at the joint based on given or measured kinematic and kinetic data (inverse dynamic problem)
2. The partitioning of intersegmental forces and moments into muscle and joint constraint forces and moments (force distribution problem)

This chapter discusses and illustrates the concepts of these two steps.

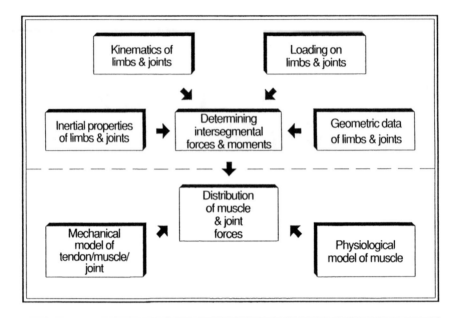

**Figure 10.1**   Analytic determination of the muscle and joint forces can be achieved in two steps: determination of intersegmental forces and moments and distribution of the forces.

## DETERMINATION OF INTERSEGMENTAL FORCES AND MOMENTS

In biomechanics, the unknown muscle and joint forces are commonly determined mathematically, because they cannot be easily measured directly. On the other hand, motion can be measured using experimental techniques. Determining the intersegmental forces and moments based on kinematic data requires solution of the inverse dynamic problem (Chao & Rim, 1973; Figure 10.2). Derivation of the equations of motion can be based on either Newtonian or Lagrangian formulas.

A simplified solution assumes that kinematic effects are negligible, allowing a "quasi-static" analysis. Static equilibrium is a condition in which a body is at rest (i.e., with no motion in relation to surrounding objects) and the external and internal forces and moments are balanced. Both translation and rotation equilibrium need to be maintained for each body segment. Thus, the equations for static equilibrium are

$$\Sigma \mathbf{F} = 0 \qquad\qquad (10.1)$$

and

$$\Sigma \mathbf{M} = 0 \qquad\qquad (10.2)$$

for each segment of the body.

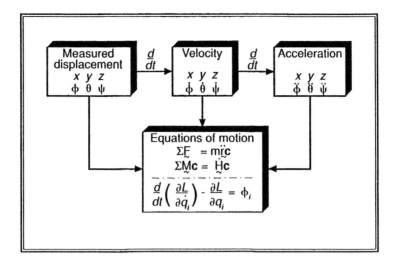

**Figure 10.2**   The inverse dynamic problem.

The summation of forces and moments includes the intersegmental force, $F_I$, and intersegmental moment, $M_I$, and the externally applied force and moment on the segment distal to the joint. To illustrate the procedure, a simplified two-dimensional problem is considered. For example, in Figure 10.3, when a 20-N weight is held in the hand a distance of 30 cm from the elbow joint center, and the 15-N forearm weight having its center of mass 15 cm from the elbow, the intersegmental force can be calculated from the force equilibrium equations,

$$F_{Ix} + L_x + W_x = 0 \tag{10.3}$$

and

$$F_{Iy} + L_y + W_y = 0, \tag{10.4}$$

where $W_x = W \sin 30°$, $W_y = -W \cos 30°$, $L_x = L \sin 30°$, $L_y = -L \cos 30°$, $W = 15$ N, and $L = 20$ N, which gives $F_{Ix}$ and $F_{Iy}$ to be $-17.5$ N and $30.31$ N, respectively. The negative sign indicates that the calculated force is in the negative direction of the coordinate system in Figure 10.3.

Similarly, the intersegmental moment $M_{Iz}$ can be calculated from the moment equilibrium equation,

$$M_{Iz} + L_y \times 30 + W_y \times 15 = 0, \tag{10.5}$$

giving

$$M_{Iz} = 714.5 \text{ N} \cdot \text{cm}.$$

For dynamic situations, the musculoskeletal body can be modeled by a number of rigid body segments interconnected at the joints in motion. If the displacement

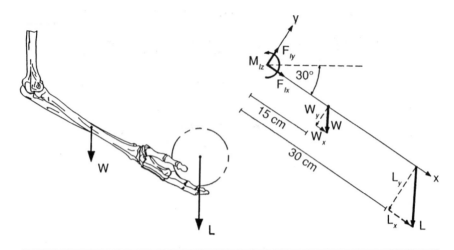

**Figure 10.3**    Static force problem with a weight in the hand when the forearm is 30° below the horizontal.

histories, mass properties of the segments, and the externally applied forces are known, then the intersegmental forces and moments acting on the limb can be determined by applying the equations of motion for the system. If a 3-D problem is considered, three additional equilibrium equations should be included for calculating $F_{Iz}$, $M_{Ix}$, and $M_{Iy}$.

An intersegmental resultant force and moment can be considered to act at the proximal and distal end of each body segment. These forces and moments are the net kinetic effect that the adjoining body segments have on each other. It is important to note that these intersegmental resultants are conceptual kinetic quantities that are not necessarily physically present in any single anatomic structure. Rather, the intersegmental resultant force and moment represent the vector sum of all the forces in the anatomic structure and the vector sum of all the moments produced by those forces.

The intersegmental resultant can be determined uniquely based on the known body segment mass and inertial and external loads. When the musculoskeletal system is engaged in either static or dynamic functions, three types of forces are active: muscle and tendon forces, joint contact forces, and capsuloligamentous forces. For convenience of modeling and computation, joint contact and ligamentous forces are frequently combined to form the resultant joint force and moment. The problem of calculating the forces within anatomic structures can be thought of as a distribution problem.

## DISTRIBUTION OF MUSCLE AND JOINT FORCES

With the intersegmental resultant force, $F_I$, and moment, $M_I$, data calculated, the muscle force, $F_m$, joint constraint force, $F_J$, and joint constraint moment, $M_J$, can then be calculated based on the concept of equilibrium,

$$F_I = \Sigma \, F_m + F_J \qquad\qquad (10.6)$$

and

$$M_I = \Sigma \, M_m + M_J \, . \qquad\qquad (10.7)$$

Partitioning of these intersegmental resultant forces and moments is generally called the *force distribution problem*. Unfortunately, the number of unknown variables of muscle forces and joint constraint forces and moments usually exceeds the number of available equations. This is primarily because of the redundant nature of anatomic structures: There are multiple muscles that can execute synergistic functions. Mathematically, this produces an indeterminate problem that has no unique solution. The difference between the number of unknown variables and the number of equations represents the degree of redundancy. In order to resolve this indeterminate problem, the degree of redundancy must be reduced by either introducing constraint equations or by decreasing the number of unknown variables.

Therefore, an important consideration in muscle and joint force determination is accounting for the number of unknown variables versus constraints in the equations of motion. In general, this decoupling procedure is achieved based on the concept of degrees of freedom (DOF) of the joint. Human joints, which have both capsuloligamentous and joint articular constraints, can move freely in several directions of translation or rotation. The possible modes of joint movement represent the rotation and translational DOF of the joint. For example, the shoulder joint and elbow joint have been considered to have three and one rotational DOF, respectively. Therefore, the associated moment equilibrium equations consist of only the unknown muscle force variables. With the muscle force determined, the joint constraint forces and moments can then be determined by using the remaining equations of motion.

Decoupling of the constraint equations makes the procedure for solution easier and more comprehensible. However, under some conditions, it is inappropriate to use the decoupling procedure to solve muscle and joint forces independently. For example, it has recently been demonstrated (An, Himeno, Tsumura, Kawai, & Chao, 1990) that the magnitude and direction of the resultant joint force determined in this way may result in an unstable joint, thereby requiring additional antagonistic muscle force to achieve stability.

# REDUCTION METHOD

The goal of the reduction method is to reduce the degree of redundancy by reducing the number of unknown forces until the number of unknown forces is equal to the number of equations. Muscles with similar functions or common anatomic insertions and orientations can be grouped together, whereas qualitative electromyographic data can be used to eliminate inactive muscles. Although this method gives the joint force, the detailed behavior of individual muscles is lost from the solution.

In the example of lifting a weight in the hand, let us assume that all of the elbow flexors are grouped as one muscle (Figure 10.4). The moment equilibrium equation for flexion-extension consists of only two variables, the intersegmental

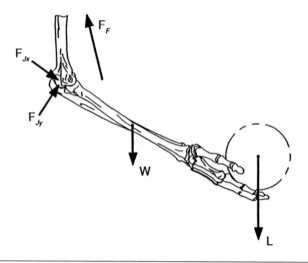

**Figure 10.4**   Determination of muscle forces about the elbow joint with the flexors grouped as a single muscle. The dimensions are the same as shown in Figure 10.3.

moment, which has been calculated from the inverse dynamic problem, and the unknown muscle force. In this simplified consideration, the flexor force can be uniquely determined from

$$M_{Iz} = 714.5 \text{ N} \cdot \text{cm} = F_F \times \sin 45° \times 3 \text{ cm},$$

hence

$$F_F = 337 \text{ N}.$$

With the muscle force determined, the joint constraint forces, $F_{Jx}$ and $F_{Jy}$, can then be determined based on the two force equilibrium equations,

$$F_{Ix} = -17.5 \text{ N} = F_{Jx} - F_F \times \cos 45°,$$

hence

$$F_{Jx} = 220.5 \text{ N}$$

to the right; and

$$F_{Iy} = 30.31 \text{ N} = F_{Jy} + F_F \times \sin 45°,$$

hence

$$F_{Jy} = -208.0 \text{ N}$$

that is, 208.0 N downward.

An alternative method of reducing the degree of redundancy is to increase the number of constraint equations. This is usually accomplished by assuming a force distribution between muscles based on anatomic consideration of the muscles' physiological cross-sectional areas or based on physiological observation such as quantitative EMG measurements.

Let us consider the same example of lifting a weight in the hand. However, this time in the model, three muscles—the biceps, brachialis, and brachioradialis—are considered instead (Figure 10.5). With the forearm in the position shown, the lever arms of the biceps (BIC), brachialis (BRA), and brachioradialis (BRD) are 4.6 cm, 3.4 cm, and 7.5 cm, respectively. What muscle forces are needed to maintain the forearm in the position shown?

By equilibrium of moments about the elbow,

$$M_{Iz} = 714.5 \text{ N} \cdot \text{cm} = 4.6 \times F_{BIC} + 3.4 \times F_{BRA} + 7.5 \times F_{BRD} .$$

Note that there is only one equation, but there are three unknowns. It is not possible to solve this problem, which is termed an *indeterminate* problem. That is to say, the equations contain more unknown values of forces in the anatomic structures than there are equations to describe the joint behavior. Physiologic constraints allow us to eliminate any solutions in which muscle forces are negative or unrealistically high. Still, there is an infinite number of solutions that can satisfy the equation, but it may be difficult or impossible to know which solution is the correct one. In order to distribute the forces in the muscles and joint to obtain a unique solution, we may have to introduce additional constraint equations. For example, based on physiological considerations, the muscle force might be assumed to be proportional to the physiological cross-sectional areas (PCSA) of the muscles (An, Hui, Morrey, Linscheid, & Chao, 1981):

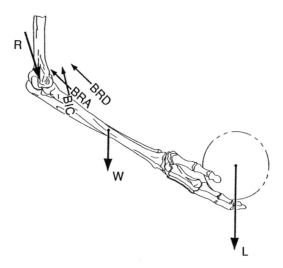

**Figure 10.5**   Determination of muscle forces about the elbow joint with three major flexors, the biceps, brachialis, and brachioradialis, considered in the solution.

$$F_{BIC} / F_{BRD} = 4.6 / 1.5$$

and

$$F_{BIC} / F_{BRA} = 4.6 / 7.0 .$$

Now there are three equations, which will make the problem uniquely solvable. The muscle forces, $F_{BIC}$, $F_{BRD}$, and $F_{BRA}$, are found to be 58.5 N, 19.1 N, and 89.0 N, respectively.

It should also be remembered that the number of constraint equations can be increased by considering the equilibrium conditions of the adjacent joints as well. For cases in which a muscle crosses two joints, it is important to solve the equilibrium equations of both joints with the same value for such muscle forces.

# OPTIMIZATION METHODS

The distribution problem at a joint is, typically, an indeterminate problem, because the number of muscles, ligaments, and articular contact regions available to transmit force across a joint, in many cases, exceeds the minimum number of equations required to generate a determinate solution. An infinite number of possible solutions exists for the indeterminate equations. Determinate solutions are obtainable only with significant simplification of the functional anatomy. One method of solution without such simplification is that of seeking an optimum solution, that is, a solution that maximizes or minimizes some process or action.

In this method, a solution to the redundant equation system is obtained by formulating an objective function and utilizing a mathematical optimization technique. The objective function provides the basis for comparison of candidate solutions; the "best" solution is sought by the optimization algorithm. The optimization approach assumes that the load sharing between the muscles follows certain designated rules during learned motor activities and that the strategy of muscle recruitment is governed by certain physiologic criteria that achieve functional efficiency. The objective function should correspond to these physiologic criteria. For example, it seems reasonable that the controlled process of human locomotion involves the optimization of body function. This general notion was first mentioned by the Weber brothers (1836), who stated that man walks in "the way that affords us the slightest energy expenditure for the longest time and with the best results." Given a particular physical activity to be performed with a large number of potential force-carrying structures at a joint, an individual may exercise considerable discretion in generating force in the actively controlled muscles. The criteria that the individual chooses, either consciously or unconsciously, to determine the control of muscle action may vary considerably with the nature of the physical activity to be performed and the physical capabilities of the individual. Muscle control in sprint running may serve to maximize velocity, whereas in walking, the control process may serve to maximize endurance. In a painful pathologic situation, such as degenerative joint disease, muscular control may serve to minimize pain. If this pain is due to joint surface pressure, the appropriate optimization criterion may be to minimize the articular surface contact force. Muscular control may also serve to minimize the forces transmitted by passive joint structures, such as ligaments. The possible optimization criteria are many, and the choice of a criterion to solve a particular distribution problem may not be obvious.

In order to solve an optimization problem, its format must be specified. This is done by:

(a) defining the cost function,
(b) identifying the constraint functions,
(c) specifying the design variables, and
(d) setting the appropriate bounds for the design variables.

The formulation of this method can be summarized as follows:

Minimize

$$J = f(x_1, x_2, \ldots, x_n)$$

subject to

$$g_j(x_1, x_2, \ldots, x_n) = 0 \quad (j = 1, 2, \ldots, m)$$

and

$$0 \leq x_i \leq U_i \quad (i = 1, 2, \ldots, n)$$

where $J$ is the optimal criterion (cost function), which can be a linear or nonlinear function of the variables $x_1, \ldots, x_n$. The function $g$ represents the equations of motion and certain equality constraint relationships. The $x_i$ stands for the independent variables, which are the unknown joint and muscle forces. These variables may also be subject to inequality constraints. With either linear or nonlinear functions, the optimization problem can be solved mathematically, and a number of numerical schemes and programs have been developed for this purpose.

It is important to understand both the mathematical behavior and the physiologic significance of this general method of solving the indeterminate distribution problem. The way in which an optimal solution is found and the general characteristics of that solution are best introduced by a simple example.

Consider the problem:

Minimize

$$J = 2x + y$$

subject to

$$2x - y = 2$$
$$x + 3y \leq 9$$
$$x \geq 0$$
$$y \geq 0.$$

A graphical representation of this problem and its solution are shown in Figure 10.6. The penalty function described by the line $J = 2x + y =$ decreases as it moves leftward into the constraint area. It continues to move until it intersects the inequality constraint line, $y = 0$. Any further movement of this line would violate the $y \geq 0$ constraint; therefore, the minimum value of the penalty function has been found at $x = 1$ and $y = 0$, where $J = 2$.

Selection and justification of optimal criteria have been major problems. A wide variety of cost functions have been used with different degrees of success, and these are discussed briefly in the next two sections. The criteria are grouped according to the nature of the optimization method—linear or nonlinear programming. Further, these criteria may include single or multiple objective functions.

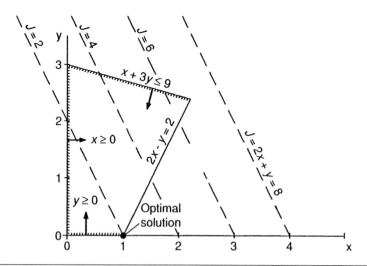

**Figure 10.6**   Graphic solution of the optimization problem: min $J = 2x + y$; subject to $2x - y = 2$; $x + 3y \leq 9$; $x \geq 0$; $y \geq 0$. The shaded area indicates the region of feasible solution of the corresponding constraints.

## Linear Criteria

Mathematical formulations of physiologic criteria for load sharing began appearing in earnest about 25 years ago. MacConaill (1967) defined a principle of minimal total muscular force, which postulated that no more total muscular force than is both necessary and sufficient to maintain a posture or perform a motion would be used. Accordingly, this would minimize the sum of the muscles forces. This objective function has been used extensively (e.g. Pedotti, Krishnan, & Stark, 1978; Seireg & Arvikar, 1975).

For instance, in the example of weight lifting using three elbow flexors, the muscle force distribution can be determined by

Minimize

$$J = F_{BIC} + F_{BRD} + F_{BRA}$$

subject to

$$M_{Iz} = 714.5 \text{ N} \cdot \text{cm} = 4.6 \text{ } F_{BIC} + 3.4 \text{ } F_{BRA} + 7.5 \text{ } F_{BRD}$$

Thus,

$$F_{BIC} = 0, \text{ } F_{BRA} = 0, \text{ and } F_{BRD} = 95.27 \text{ N}.$$

It is obvious in this example based on the linear objective function that two of the unknown variables have a zero value, and only one muscle, the brachioradialis, develops any force. This characteristic of linear programming, that in the solution only a few active muscles with nonzero value are predicted simultaneously, is not always physiologically consistent. In the above example, when a planar model for a single joint is used, minimization of a linear objective function will result in the prediction that only one muscle (the muscle with the largest lever arm) is active. However, in most activities based on electromyographic observation, it is known that several muscles crossing the joint are active simultaneously.

One way to improve the predictions of muscle forces with linear criteria is to formulate additional constraints. An inequality constraint that limits the force per unit area (stress) in each muscle was used by Crowninshield (1978), Crowninshield and Brand (1981), Pedotti et al. (1978), and Patriarco, Mann, Simon, and Mansour (1981). This constraint results in the prediction that one muscle starts to be active when another muscle reaches its maximum stress. Initially, only the muscle with the largest arithmetic product of moment arm and physiologic cross-sectional area is active until it reaches its limit, at which moment the muscle with the second largest product is activated.

Recently, three new and somewhat similar approaches that consider muscle intensity have been suggested. Schultz, Haderspeck, Warwick, and Portillo (1983) described lumbar trunk models that minimized the compression on the spine while minimizing the maximum muscle contraction intensity. They solved this problem by using a sequence of linear programs, each with a different upper bound of muscle intensity selected by a step-wise procedure: They began with low muscle intensities and gradually increased the muscle intensity until the first feasible solution was found. The difficulties of this approach were the substantial computational requirements and the instability of solutions as intensity values were changed. In spite of these shortcomings, this model provided results that were in better agreement with experimental measurements than those from less sophisticated models.

An, Kwak, Chao, and Morrey (1984) suggested an improved technique based on minimizing the upper bound of muscle stress. In this approach, additional inequality conditions were introduced, namely,

$$F_m / PCSA_m \leq \sigma.$$

The left-hand side represented the stress developed by the $m$th muscle, where $F_m$ is the force and $PCSA_m$ is its physiologic cross-sectional area. The single variable, $\sigma$, on the right-hand side represented the upper bound for all of the muscle stresses. The unique solution for distributing muscle forces was then obtained by minimizing $\sigma$.

These approaches make use of the linear optimization method and hence guarantee that the solution will converge on a minimum which satisfied the objective criterion for all the muscles.

For more realistic modeling, the muscle physiology based on length-tension or velocity-tension relationships can also be considered and implemented in the inequality constraints.

$$0 \leq F_m \leq (\alpha \, \hat{F}_m^a + \hat{F}_m^p) \cdot PCSA \cdot \sigma$$

in which $\sigma$ represents maximum muscle stress which must be specified; $\hat{F}_m^a$ is the normalized muscle active force based on the length-velocity-tension relationship; $\hat{F}_m^p$ is the normalized muscle passive force characteristics; and $\alpha$ represents the upper bound of the activation level of the muscle (An, Kaufman, & Chao, 1989; Kaufman, An, Litchy, & Chao, 1991a, 1991b). The unique solution for muscle forces will be obtained by minimizing $\alpha$.

## Nonlinear Criteria

Nonlinear objective functions can predict synergism, even without the formulation of additional constraints. Gracovetsky, Farfan, and Lamy (1977) defined an objective function of the sum of squared shear stresses in the vertebral column and predicted forces during weight lifting. For the analysis of walking, Pedotti et al. (1978) used the sum of squared muscle forces,

$$\Sigma \, (F_m)^2,$$

which is a second-order (power) criterion.

In terms of proper selection of the power, several investigators have experimentally sought to characterize the relationship between muscle force and endurance. These studies agree, in general, that the endurance time is approximately inversely proportional to the value of muscle force raised to some power, but with a large spread in the reported values of this power. Crowninshield and Brand (1981) proposed a cost function expressed as the sum of muscle stresses raised to the third power, which appeared to correspond to maximum endurance of muscles.

## Multiple Performance Criteria

Recently, Bean, Chaffin, and Schultz (1988) presented a novel scheme for using linear programming to calculate muscle contraction forces in models of musculoskeletal system biomechanics. They suggested a dual-objective problem with two sequential linear programs each having a unique solution. First, like An et al. (1984), they minimized the upper bound of muscle stress to define the optimal muscle intensity value from this solution. Second, they solved a linear program to minimize the sum of the muscle forces using the optimal muscle intensity predicted in the first stage. Thus, the first linear program determined the lowest muscle intensity value, which allowed feasible solutions. The second linear program chose among these solutions to minimize muscle force. Hence, this scheme addressed both the stated objectives.

## GENERAL REMARKS

Two important points with respect to the analytical determination of muscle and joint forces must be emphasized. First, in general, the muscle and joint forces can be determined independently based on the decoupled groups of equilibrium equations and constraint conditions providing one assumes that the resultant joint constraint force and moment have no limitations. Unfortunately, this assumption may not always be true. For example, if either articular surface or capsuloligamentous structures are damaged, the solution constraints may have to be modified. In this situation, the muscle force would not be independent of the resultant joint force determination. The magnitude and direction of the resultant joint force has tremendous implications for joint stability (An et al., 1990). The resultant joint force is dependent on the distribution of the muscle forces. Therefore, solving for the muscle and joint forces independently may not be appropriate.

Second, the method of solution can be important. When using the optimization method, either linear or nonlinear procedures can be used. If the entire system of objective function and constraint conditions consists of linear terms of unknown variables, then the well-established linear programming algorithm can be used to obtain the solution efficiently. In contrast, nonlinear optimization is usually more involved and less efficient than linear programming. Also, convergence of the solution to a unique minimum is not always guaranteed.

Based on our experience, the solution for the distribution of muscle forces may be very sensitive to the specific optimization method or objective function used. On the other hand, the magnitude of the resultant joint forces may not differ as much (Brand, Pedersen, & Friederich, 1986). Therefore, depending on the purpose of the study, selection of the optimal criteria may not be critical. Nevertheless, improvement of the methodology for solution of the indeterminate problem and experimental verification of the solution, by direct (Schuind, Garcia-Elias, Cooney, & An, 1992) or indirect (Funk, An, Morrey, & Daube, 1987) measurements, should still be actively pursued in the future. Every proposed solution should be reviewed critically.

## REFERENCES

An, K.-N., Himeno, S., Tsumura, H., Kawai, T., & Chao, E.Y. (1990). Pressure distribution on articular surfaces: Application to joint stability evaluation. *Journal of Biomechanics, 23*, 1013-1020.

An, K.-N., Hui, F.C., Morrey, B.F., Linscheid, R.L., & Chao, E.Y. (1981). Muscles across the elbow joint: A biomechanical analysis. *Journal of Biomechanics, 14*, 659-669.

An, K-N., Kaufman, K.R., & Chao, E.Y. (1989). Physiological considerations of muscle force through the elbow joint. *Journal of Biomechanics, 22*, 1249-1256.

An, K-N., Kwak, B.M., Chao, E.Y., & Morrey, B.F. (1984). Determination of muscle and joint forces: A new technique to solve the indeterminate problem. *Transactions of the American Society of Mechanical Engineering, 106*, 364-367.

Bean, J.C., Chaffin, D.B., & Schultz, A.B. (1988). Biomechanical model calculations of muscle contraction forces: A double linear programming method. *Journal of Biomechanics, 21*(1), 59-66.

Brand, R.A., Pedersen, D.R., & Friederich, J.A. (1986). The sensitivity of muscle force predictions to changes in physiological cross-sectional area. *Journal of Biomechanics, 19*(8), 589-596.

Chao, E.Y., & Rim, K. (1973). Application of optimization principles in determining the applied moments in human leg joints during gait. *Journal of Biomechanics, 6*, 497-510.

Crowninshield, R.D. (1978). Use of optimization techniques to predict muscle forces. *Journal of Biomechanical Engineering, 100*, 88-92.

Crowninshield, R.D., & Brand, R.A. (1981). A physiologically based criterion of muscle force prediction in locomotion. *Journal of Biomechanics, 14*, 793-801.

Funk, D.A., An, K.N., Morrey, B.F., & Daube, J.R. (1987). Electromyographic analysis of muscles across the elbow joint. *Journal of Orthopaedic Research, 5*, 529-538.

Gracovetsky, S., Farfan, H.F., & Lamy, C. (1977). A mathematical model of the lumbar spine using an optimization system to control muscles and ligaments. *Orthopaedic Clinics of North America, 8*, 135-153.

Kaufman, K.R., An, K.N., Litchy, W.J., & Chao, E.Y. (1991a). Physiological prediction of muscle forces—I. Theoretical prediction. *Neuroscience, 40*(3), 781-792.

Kaufman, K.R., An, K.N., Litchy, W.J., & Chao, E.Y. (1991b). Physiological prediction of muscle forces—II. Application to isokinetic exercise. *Neuroscience, 40*(3), 793-804.

MacConaill, M.A. (1967). The ergonomic aspects of articular mechanics. In F. G. Evans (Ed.), *Studies on the anatomy and function of bones and joints* (pp. 69-80). Berlin: Springer.

Patriarco, A.G., Mann, R.W., Simon, S.R., & Mansour, J.M. (1981). An evaluation of the approaches of optimization models in the prediction of muscle forces during human gait. *Journal of Biomechanics, 14*(8), 513-525.

Pedotti, A., Krishnan, V.V., & Stark, L. (1978). Optimization of muscle-force sequencing in human locomotion. *Mathematical Biosciences, 38*, 57-76.

Schuind, F., Garcia-Elias, M., Cooney, W.P., & An, K.N. (1992). Flexor tendon forces: In vivo measurements. *Journal of Hand Surgery, 17A*, 291-298.

Schultz, A., Haderspeck, K., Warwick, D., & Portillo, D. (1983). Use of lumbar trunk muscles in isometric performance of mechanically complex standing tasks. *Journal of Orthopaedic Research, 1*(1), 77-91.

Seireg, A., & Arvikar, R.J. (1975). The prediction of muscular load-bearing and joint forces in the lower extremities during walking. *Journal of Biomechanics, 8*, 89-102.

Weber, W., & Weber, E. (1836). *Mechanik der menschlichen Gehwerkzeuge* [Mechanics of human locomotion apparatus]. Gottingen: Fisher-Verlag.

# Chapter 11

# Analytic Representation of Muscle Line of Action and Geometry

*Michael Raymond Pierrynowski*

Commenting on the present state of biomechanical modeling, Panjabi, in a letter to the editor of the *Journal of Biomechanics* (1979), stated that "generally much effort is spent in formulating the governing equations of the model. In contrast, the physical properties data is screened from the available literature, which is often meager, inaccurate, incomplete and outdated." Fifteen years later, one finds that little has changed since 1979. This chapter consolidates a variety of published quantitative anatomical information pertaining to the human that may prove useful in biomechanical modeling and analyses and perhaps may form the nucleus of some future, complete data set. (Anatomical values deemed most useful include the geometries of the muscles and tendons and the lines of action of these structures in relation to the skeletal framework.)

An appendix entitled "A Survey of Human Musculotendon Actuator Parameters" by Yamaguchi, Sawa, Moran, Fessler, and Winters (1990) represents a recent effort to tabulate quantitative human muscle data. These authors present 48 pages of quantitative human muscle data from the lower and upper extremities, trunk, hand, head, and neck obtained from 24 " . . . leading researchers." These data are presented as nearly as possible in the original form found in the published sources (i.e. the authors did not manipulate the data to define an average individual). However, this chapter uses methods to scale the published anatomical data to a standard person and discusses these techniques.

The human body contains over 200 bones and 600 muscles. Although many of these structures are of limited interest to most biomechanical investigators (e.g., cranial bones and inner ear stapes), a useful total-body musculoskeletal model of the human body is still a formidable undertaking. Considering that several reported musculoskeletal models include bone, muscle, and joint surfaces (Buford & Thompson, 1987; Delp et al., 1990; Huiskes et al., 1985; Little, Wevers, Sui, & Cooke, 1986; Shiba et al., 1988; Wismans, Veldpaus, & Janssen,

1980) and enfleshed musculature for graphic and quantitative purposes (Wood, Meek, & Jacobsen, 1989a), a complete biomechanical model of the human is still beyond present day capabilities. Therefore, this chapter is limited to a description of the lower extremity, shoulder, and elbow. Data on other muscle groups are available (see Yamaguchi et al., 1990), but these will not be reported here.

Although efforts were taken to ensure the accuracy of the data contained in this chapter, errors may have occurred. Users must be cautious when employing these data in their work. It must also be remembered that life assumes a wide variety of form and function not only across species but also within a given lifeform. There is substantial variability of human musculoskeletal morphology, even before considering variations such as muscular anomalies (Mecalister, 1875). This must temper our attempt to compile a representative data base of muscle location and geometry.

## SKELETAL AND MUSCLE LINE-OF-ACTION MODELING

To represent the human musculoskeletal system as a set of rigid skeletal members located in 3-D space, embedded skeletal reference systems and attached musculature with lines of action and defined geometrical characteristics are first scaled to a given subject using superficial bony landmarks, then defined as the subject moves through 3-D space.

This section defines the skeletal reference systems, superficial bony landmarks, and muscle line-of-action points and suggests several methods that can be used to map these data to a specific individual. All the provided information, which has been gleaned from many published data sources, has been normalized to a 1.70-m, 70-kg reference individual.

## System and Axes Definition— Skeletal Reference Systems

Biomechanical modeling of the human body generally assumes that we are composed of rigid links (Dempster, 1955; Panjabi, White, & Brand, 1974). Rigid body analyses are performed because they are computationally easy and give acceptable results (Kinzel & Gutkowski, 1983). However, several human body segments (shoulder, forearm, foot) do not behave as rigid bodies and should be analyzed as deformable elements. Salathe, Arangio, and Salathé (1986) present such a model for the human foot.

All rigid segments moving in 3-D space have six degrees of freedom (Kinzel, Hall, & Hillberry, 1972). Researchers typically link these rigid segments with joints that ignore longitudinal rotations (Dapena, 1978) or joint translations (Winter, 1984) to (a) simplify the model, (b) mask errors inherent to the kinematic data acquisition system, or (c) facilitate the interpretation of their results. This simplified link-segment model is then analyzed using engineering dynamics (Paul, 1981).

Labeling these remaining degrees of freedom is noteworthy in its lack of standards or convention. Space- and body-fixed Cartesian reference systems (Kane, Likins, & Levinson, 1983) and helical reference systems (Shiavi et al., 1987) have been used, with no clearly defined advantage of one system over the other. However, the use of unit quaternions (Haruz, 1990; Horn, 1987) to represent rotations can simplify the manipulation of multiple rotations and rotating frames of reference. Most investigators use a laboratory-fixed orthogonal $XYZ$ Cartesian axes system, but there is no accepted standard regarding which of the many variations one should use (Kane et al., 1983). Some biomechanists employ a system in which the subject moves forward in the $XY$ plane, but others use the $XZ$ plane, and most other global reference systems (GRS) variations have been employed.

This chapter uses a modified coordinate system recommended by the International Workshop on Human Subjects for Biomechanical Research on anatomical frames of reference for biomechanics (see Reynolds & Hubbard, 1980). This GRS reference system is orthogonal, follows the right-hand rule, and is defined such that when an individual is in the anatomical position, $X$ is posterior to anterior, $Y$ is right to left, and $Z$ inferior to superior. The recently formed (1990) International Society of Biomechanics Committee for Standardization and Terminology has yet to endorse an alternative system, but they have proposed a Cartesian skeletal reference system in which $+X$ is forward, $+Y$ vertical, and $+Z$ to the right, emanating from the mass centroid of each segment. This chapter modifies the ISB standard to ignore the arbitrary $XYZ$ designation in favor of $FOU$, where $F$ = forward ($+X$), $O$ = outward ($+Y$), and $U$ = upward ($+Z$). This terminology conforms to a standardization proposal suggested by the CAMARC group (see Paul, 1990). This chapter also modifies this standard such that each of the subject's segments are mathematically rotated to the vertical position about a body-fixed $O$, then $F$, axis (no $U$ rotation) originating at the proximal joint, until each segment's distal joint falls directly below ($-U$) the proximal joint (anatomically, this position is not possible for many segments; Tupling & Pierrynowski, 1987). Exceptions are the feet, clavicles, and scapula, because rotating these segments to a vertical position does not make sense. Figure 11.1 illustrates the $FOU$ reference system for the right humerus. Note that the defined skeletal reference systems use a left-handed coordinate system for the right segments and a right-handed coordinate system for the left segments; midline segments are arbitrarily selected.

## Scaling to a Subject—Superficial Bony Landmarks

Scaling techniques must be used to transform skeletal anthropometric data from one individual to another (Brand et al., 1982; Lew & Lewis, 1977; Lewis, Lew, & Zimmerman, 1980; Sommer, Miller, & Pijanowski, 1982; White, Yack, & Winter, 1989). Theoretically, nonhomogeneous scaling is preferred to homogeneous scaling, but in reality it is difficult to locate accurately enough superficial bony landmarks to perform this type of transformation. Therefore, the preferred scaling method is to locate several points of prominence on a skeletal segment and the

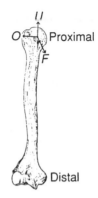

**Figure 11.1**   The skeletal reference system for the right humerus. The outward (*O*) and upward (*U*) axes are shown. The forward (*F*) axis is directed toward the reader from the proximal joint and is orthogonal to the *O* and *U* axes. Note that the humerus is rotated such that the distal joint is directed downward (−*U*) from the proximal joint—a vertical anatomical position.

corresponding points, measured in vivo, on a subject, and then to calculate a transformation matrix that manipulates the data from the disarticulated reference skeletal segment into that of the subject. When applied to a group of at least four points on a body segment, a rigid-body translation and rotation are performed with a homogeneous deformation of the initial segment (skeletal bone) into the final segment (subject). These transformations are then used to locate the skeletal reference system, superficial bony landmarks, and muscle line-of-action points within the subject. The theory underlying this transformation is found in Pierrynowski and Morrison (1985b), Rogers and Adams (1990) and Veldpaus, Woltring, and Dortmans (1988).

Although this transformation technique is accurate (Lew & Lewis, 1977), it can be time consuming for the subject, unless automated body landmark identification procedures are used. In a clinical setting, the data set for each skeletal segment can be transformed (scaled, translated, and rotated) to a subject if the locations of and distances between joint centers and superficial bony landmarks are known (Brand et al., 1982; White et al., 1989) or by using a set of measured distances and calculating the transformation from the standard set to the subject of interest (Dul & Johnson, 1985).

## Muscle Lines of Action

The lines of action of the locomotor muscles have been modeled as straight lines joining origin to insertion (Brand et al., 1982; Crowninshield, Johnston, Andrews, & Brand, 1978; Dostal & Andrews, 1981; Hardt, 1978; Hogfors, Sigholm, & Herberts, 1987; Koolstra, van Eijden, Weijs, & Naeije, 1988). However, many muscles are deflected over the skeletal framework or other muscles and pass through tendon sheaths (Delp et al., 1990; Wood, Meek, & Jacobsen, 1989b). Seireg and Arvikar (1973) and White et al. (1989) represented muscles that crossed the ankle in two parts to account for their large change in direction. Frigo and Pedotti (1978) improved the one or two straightline representations of the muscles by using lines and arcs of circles to model the course of 11 major muscles. Pierrynowski and Morrison (1983) used up to five straight and curvilinear sections to model 47 human locomotor muscles. Other investigators use the centroid line of the muscle (Jensen & Davy, 1975; Jensen & Metcalf, 1975; Wood et al., 1989a, 1989b). Recently, Koolstra, van Eijden, & Weijs (1989) presented a simple, accurate technique in which serial-parallel computed tomography (CT) or magnetic resonance imaging (MRI) sections were used to estimate the in vivo line of action of muscle in 3-D space.

Balancing simplicity and accuracy, in this chapter each muscle is represented by a 3-D line or curve that connects the centroid of its area of origin to the centroid of its area of insertion. Between these two endpoints additional anchor points are defined based on anatomical considerations. The coordinates of these points are presented relative to the appropriate skeletal segment and its embedded reference system. These points are joined either by straight lines (whenever a muscle runs freely from point to point) or by curves (when a muscle is forced to alter its course either over bony prominences or through tendon sheaths). For those nonlinear sections, a parametric cubic polynomial is fitted, the theory of which is given in Appendix A. For each muscle, summing the length of all sections, whether straight or curved, gives its length, which is required in a later section.

## Moment Arm Calculations

The calculation of a muscle's moment arm is not difficult if accurate axes of rotation and muscle line-of-action data are available. However, if anatomical data are precise to the typical 2 cm, then moment arm data may contain large errors (An, Takahashi, Harrigan, & Chao, 1984). In addition, because muscles are generally modeled by straight lines running from origin to insertion, the influence of anatomical structures near the joints (condyles, bursa, other muscles, and tendons) tends to alter the location of the muscle. As an alternative, muscle moment arms may be measured directly using X-ray (Gregor, Komi, Browning, & Jarvinen, 1991) or CT techniques (Nemeth & Ohlsen, 1989) or using the equivalency of muscle and joint work (Spoor, van Leeuwen, Meskers, Titulaer, & Huson, 1990). However, these techniques are time-intensive and require much specialized equipment. The best automated method would be a computer model that considers the muscles, bones, and joint structures to be 3-D deformable entities that cannot move through each other (see Delp et al., 1990). This type of computer-rendered anatomical human model will provide more accurate estimates of moment arm data and should be an area of future research.

## Skeletal and Muscle Line-of-Action Data Bases

Data for human musculoskeletal modeling are presented in this section. Data sources, methods, and the tabulated values (skeletal reference system, superficial bony landmarks, muscle lines of action) are given for the lower and upper extremities. The data for each region of the body were combined to yield a representative data base, using a weighted average, based on the number of specimens in the original data sets. Because very few identical landmarks are provided to scale one published source to another, corresponding skeletal reference systems, superficial bony landmarks, and muscle line-of-action points were used to define segment-specific transformations (see a later section) from each data source to a 1.70-m reference individual. In this way a partial data set could be formulated for the skeletal reference system, superficial bony landmarks, and muscle line-of-action points.

## Lower Extremity

An abundance of data exists for biomechanical modeling of the human lower extremity. Unfortunately, many of these sources are internally incomplete, use simple musculoskeletal models, and lack appropriate scaling information to transform the data to another individual. However, 10 papers were used to compile skeletal reference systems, superficial bony landmarks, and muscle line-of-action data.

**Data Sources.**   Three theses provided information on the lower extremity. Arvikar's 1971 master's thesis furnished anatomical information derived from a skeleton diagram, for skeletal reference systems and muscle line-of-action data of the lower extremity (pelvis, thigh, shank, foot) and 29 muscles. This work is noteworthy in that it modeled many ankle muscles by four points to account for their lines of action across this joint. Arvikar continued his work by modeling the total human body, but his doctoral thesis (Arvikar, 1974) does not contain skeletal reference systems, superficial bony landmarks, or muscle line-of-action information. Hardt (1978) used several anatomy atlases and observations on cadavers to present two-point muscle line-of-action data on 31 muscles. Proctor (1980), modeling the ankle joint, dissected five cadavers to obtain 10 superficial bony landmark and muscle line-of-action points close to the ankle for 10 muscles.

Six published papers delivered a variety of information. A hip model, derived from one male skeleton, presented skeletal reference systems, scaling, and two-point muscle line-of-action data for the pelvis and thigh segments (Dostal & Andrews, 1981). Burdett (1982) published information collected from five cadavers on a simplified line-of-action model of 11 ankle muscles. The location of the lines of action of 47 lower extremity muscles were collected from three cadavers by Brand et al. (1982). Computed tomography was used to obtain the two-point line-of-action points of the hip extensor muscles from 20 patients (Nemeth & Ohlsen, 1985). The tibial insertion points for 13 knee muscles were obtained by Mikosz, Andriacchi, & Andersson (1988) from an atlas of anatomy. White et al. (1989) gave skeletal reference systems and line-of-action data on 40 lower extremity muscles acquired from several dry skeletons (6 pelvis, 9 thigh-&-shank, 1 foot), 1 cadaver foot, and 6 subjects and provided appropriate scaling information.

The final data source is Pierrynowski (1982). This work contains the skeletal reference systems (pelvis, thigh, shank, talus, foot, digits), scaling information

(23 superficial bony landmarks), and muscle line-of-action data from 47 lower extremity muscles collected from a disarticulated, dried male skeleton. Each muscle's LOA was modeled using straight and curved sections anchored by up to 6 points.

**Results.** The coordinates defining the proximal and distal joint centers (when available) for the pelvis, thigh, shank, talus, foot, and digit segments are given in Table 11.1. Note that the patella is not modeled. The origin of the segment ensemblage is the midpoint of the hips and the pelvis; thigh and shank segments were rotated such that their distal joint falls below their proximal joint (−U). Their other orientations approximate the anatomical position. The talus, foot, and digit segments, considered as one unit, have their origin at the ankle (tibiotarsal) joint and are rotated such that these segments are in the anatomical position with the third metatarsal phalanx oriented forward (+F) of the tibiotarsal joint. The axes about which the lower extremity segments rotate about each other (hip, knee, tibiotarsal, subtalar, and MTP joints) are also defined in Table 11.1. These data are given as 50-mm vectors directed in the positive F, O, and U directions. In addition, the anatomical label typically assigned to rotations about these axes is provided.

The names and coordinates of the superficial bony landmarks for the lower extremity segments are listed in Table 11.2. Obtaining the 3-D location of at least four of these points for each segment on the subject in any position allows one to scale, translate, and rotate skeletal reference systems and muscular lines of action to any subject of interest.

The muscular system of the human lower limb was divided into six groups, according to the joints acted upon, and several muscles were partitioned into two or more distinct structures based on functional considerations (e.g., Gottschalk, Kourosh, & Leveau, 1989). In all, 48 "muscles" were deemed significant. Anchor points defining the line of action of these structures are presented in Table 11.3. Each coordinate is defined within the same reference systems found in Table 11.1.

## Upper Extremity

Little modeling of the shoulder complex has been done until very recent times. Most investigators have only considered the movements of the glenohumeral joint (Braune & Fisher, 1888; Fick, 1911), but recently the thoracic-clavicular-scapular movements have been measured (Doddy, Waterland, & Freedman, 1970; Engin & Williams, 1968; Shoup, 1976) using either an open- or closed-chain mechanism (Dvir & Berme, 1978). Whichever model was considered, all of these studies are hampered by poor kinematics because the motion of the underlying shoulder skeleton is quite difficult to measure due to lack of superficial bony landmarks which are not influenced by skin and muscle movement. Fortunately, the elbow is not as complex even though it has separated, nonorthogonal, flexion-extension and supination-pronation axes.

Only four papers were found that provided good quantitative information on the upper extremity. Yet even these, compared to those describing the lower extremity, were incomplete. This made it difficult to compile a representative model of the upper extremity.

**Table 11.1   Skeletal Reference Systems for the Lower Extremity**

|  |  | F (Forward) | O (Outward) | U (Upward) |
|---|---|---|---|---|
| **PELVIS** | | | | |
| Proximal point | Pubic crest | 55 | 0 | −3 |
| 50 mm axes | F | 122 | 0 | −3 |
| | O | 55 | 0 | 48 |
| | U | 55 | 51 | −3 |
| Distal points | Right hip | 0 | 71 | 0 |
| | Left hip | 1 | −70 | 1 |
| **THIGH** | | | | |
| Proximal point | Hip | 0 | 71 | 0 |
| 50 mm axes | F | 51 | 71 | −1 |
| | O | 0 | 71 | 51 |
| | U | 0 | 122 | 0 |
| Distal point | Knee | 0 | 71 | −406 |
| **SHANK** | | | | |
| Proximal point | Knee | 0 | 71 | −406 |
| 50 mm axes | F | 50 | 71 | −406 |
| | O | 0 | 71 | −356 |
| | U | 0 | 122 | −406 |
| Distal point | Tibiotarsal | 0 | 71 | −808 |
| **TALUS** | | | | |
| Proximal point | Tibiotarsal | 0 | 71 | −808 |
| 50 mm axes | F | 50 | 69 | −808 |
| | O | 0 | 71 | −757 |
| | U | 2 | 121 | −808 |
| Distal point | Subtalar | −6 | 78 | −822 |
| **FOOT** | | | | |
| Proximal point | Subtalar | −7 | 77 | −822 |
| 50 mm axes | F | 23 | 46 | −782 |
| | O | −34 | 85 | −770 |
| | U | 21 | 119 | −821 |
| Distal point | Metatarsophalangeal | 114 | 71 | −843 |
| **DIGITS** | | | | |
| Proximal point | Metatarsophalangeal | 114 | 71 | −843 |
| 50 mm axes | F | 163 | 85 | −840 |
| | O | 108 | 71 | −780 |
| | U | 125 | 124 | −853 |

*Note.* Origin at the midpoint of the hips joint centers: all values in mm.

**Table 11.2  Superficial Bony Landmarks for the Lower Extremity**

|  | F (Forward) | O (Outward) | U (Upward) |
|---|---|---|---|
| **PELVIS** | | | |
| Right pubic tubercle | 60 | 22 | -13 |
| Left pubic tubercle | 58 | -20 | -14 |
| Left anterior superior iliac spine | 62 | -111 | 67 |
| Right anterior superior iliac spine | 64 | 109 | 69 |
| Left posterior superior iliac spine | 11 | -123 | 94 |
| Right posterior superior iliac spine | 13 | 119 | 95 |
| Coccyx | -99 | -3 | -22 |
| **THIGH** | | | |
| Greater trochanter | -4 | 125 | -3 |
| Lateral condyle—anterior | 34 | 96 | -406 |
| Lateral condyle—distal | 19 | 94 | -424 |
| Lateral condyle—posterior | -6 | 90 | -395 |
| Medial condyle—medial | 1 | 37 | -390 |
| Medial condyle—distal | 16 | 45 | -424 |
| Medial condyle—posterior | -15 | 51 | -392 |
| **SHANK** | | | |
| Medial condyle—medial | -2 | 36 | -390 |
| Lateral condyle—distal | 19 | 93 | -425 |
| Tibial tuberosity | 39 | 81 | -468 |
| Fibular head | -1 | 112 | -445 |
| Lateral malleolus—tip | -5 | 95 | -789 |
| Medial malleolus—tip | -2 | 47 | -779 |
| **TALUS** | | | |
| Medial anterior upper talus | 21 | 51 | -812 |
| **FOOT** | | | |
| Medial Achilles tendon | -48 | 81 | -834 |
| Lateral Achilles tendon | -40 | 103 | -835 |
| Proximal fifth metatarsal | 42 | 108 | -866 |
| Proximal first metatarsal | 112 | 36 | -831 |
| Distal third metatarsal | 114 | 71 | -843 |
| Distal fifth metatarsal | 101 | 92 | -856 |
| Lateral malleolus—tip | -6 | 95 | -791 |
| Medial malleolus—tip | 0 | 45 | -781 |
| Heel | -49 | 90 | -839 |
| Navicular | 17 | 49 | -833 |

*Note.* Origin at the midpoint of the hips joint centers; all values in mm.

**Table 11.3  Lower Extremity Muscle Line-of-Action Points and the Segments to Which They are Attached**

| Muscle | | On segment | Point 1 | | | Point 2 | | | Point 3 | | | Point 4 | | | Point 5 | | | Point 6 | | |
|---|---|---|---|---|---|---|---|---|---|---|---|---|---|---|---|---|---|---|---|---|
| | | | F | O | U | F | O | U | F | O | U | F | O | U | F | O | U | F | O | U |
| Psoas major | PM | 1 1 2 | 25 | 8 | 210 | 29 | 74 | 23 | −14 | 83 | −60 | | | | | | | | | |
| Iliacus | IC | 1 1 2 | −6 | 83 | 122 | 39 | 81 | 38 | −8 | 89 | −77 | | | | | | | | | |
| Gemellus superior | GMS | 1 2 | −55 | 48 | 8 | −8 | 110 | −4 | | | | | | | | | | | | |
| Gemellus inferior | GMI | 1 2 | −62 | 55 | −10 | −8 | 110 | −4 | | | | | | | | | | | | |
| Obturator externus | OBE | 1 2 | 0 | 31 | −32 | −22 | 112 | −16 | | | | | | | | | | | | |
| Obturator internus | OBI | 1 2 | −29 | 44 | −29 | −13 | 113 | −3 | | | | | | | | | | | | |
| Piriformis | PIR | 1 2 | −73 | 31 | 58 | −7 | 124 | −1 | | | | | | | | | | | | |
| Quadratus femoris | QF | 1 2 | −42 | 63 | −30 | −28 | 110 | −31 | | | | | | | | | | | | |
| Pectineus | PEC | 1 2 | 38 | 42 | 6 | −8 | 98 | −88 | | | | | | | | | | | | |
| Adductor longus | AL | 1 2 | 42 | 13 | −20 | 8 | 86 | −225 | | | | | | | | | | | | |
| Adductor magnus (anterior) | AMa | 1 2 | −17 | 36 | −41 | −6 | 106 | −110 | | | | | | | | | | | | |
| Adductor magnus (middle) | AMm | 1 2 | −35 | 45 | −47 | 4 | 96 | −232 | | | | | | | | | | | | |
| Adductor magnus (posterior) | AMp | 1 2 | −54 | 45 | −50 | 7 | 65 | −367 | | | | | | | | | | | | |
| Adductor brevis (superior) | ABs | 1 2 | 18 | 23 | −28 | −1 | 98 | −115 | | | | | | | | | | | | |
| Adductor brevis (inferior) | ABi | 1 2 | 19 | 23 | −27 | −4 | 97 | −144 | | | | | | | | | | | | |
| Gluteus minimus (anterior) | GMINa | 1 2 | 34 | 106 | 71 | −1 | 126 | −20 | | | | | | | | | | | | |
| Gluteus minimus (middle) | GMINm | 1 2 | −5 | 89 | 85 | 0 | 127 | −20 | | | | | | | | | | | | |

| Muscle | Abbr | | | | | | | | | | | | | | | |
|---|---|---|---|---|---|---|---|---|---|---|---|---|---|---|---|---|
| Gluteus minimus (posterior) | GMINp | 1 | 2 | | -34 | 68 | 62 | 0 | 127 | -20 | | | | | | |
| Gluteus medius (anterior) | GMEDa | 1 | 2 | | 30 | 114 | 89 | -23 | 127 | -6 | | | | | | |
| Gluteus medius (middle) | GMEDm | 1 | 2 | | -22 | 78 | 121 | -23 | 126 | -6 | | | | | | |
| Gluteus medius (posterior) | GMEDp | 1 | 2 | | -67 | 53 | 72 | -23 | 126 | -6 | | | | | | |
| Gluteus maximus (deep) | GMAXd | 1 | 2 | | -96 | 35 | 79 | -7 | 113 | -98 | | | | | | |
| Gluteus maximus (superficial) | GMAXs | 1 | 2 | 3 | -97 | 36 | 81 | -6 | 133 | -24 | 20 | 102 | -432 | | | |
| Tensor fascia latae | TFL | 1 | 2 | 3 | 50 | 110 | 66 | -6 | 133 | -24 | 21 | 102 | -430 | | | |
| Semimembranosus | SM | 1 | 3 | 3 | -55 | 65 | -29 | -17 | 63 | -448 | | | | | | |
| Semitendinosus | ST | 1 | 3 | 3 | -63 | 58 | -40 | -5 | 59 | -468 | 16 | 59 | -516 | | | |
| Gracilis | GR | 1 | 3 | | 18 | 11 | -31 | 16 | 59 | -498 | | | | | | |
| Sartorius | SAR | 1 | 3 | 3 | 55 | 103 | 66 | 20 | 53 | -438 | | | | | | |
| Rectus femoris | RF | 1 | 2 | 3 | 42 | 87 | 38 | 42 | 72 | -396 | 21 | 60 | -467 | | | |
| Biceps femoris (long) | BFl | 1 | 3 | | -67 | 52 | -30 | 4 | 114 | -452 | (38 | 67 | -446) | 37 | 67 | -448 |
| Biceps femoris (short) | BFs | 2 | 3 | | 6 | 97 | -205 | 4 | 114 | -452 | | | | | | |
| Vastus lateralis | VL | 2 | 2 | 3 | -1 | 119 | -45 | 42 | 72 | -396 | (37 | 66 | -447) | 36 | 66 | -449 |
| Vastus intermedius | VI | 2 | 2 | 3 | 5 | 106 | -46 | 42 | 72 | -396 | (37 | 67 | -446) | 36 | 67 | -448 |
| Vastus medialis | VM | 2 | 2 | 3 | 7 | 89 | -44 | 42 | 72 | -396 | (37 | 66 | -447) | 36 | 66 | -449 |
| Popliteus | POP | 2 | 3 | | -1 | 106 | -414 | -5 | 102 | -442 | 0 | 63 | -514 | | | |
| Gastrocnemius (medial) | GM | 2 | 5 | | -10 | 97 | -391 | -47 | 93 | -840 | | | | | | |

(continued)

**Table 11.3**  (continued)

| Muscle | | On segment | Point 1 | | | Point 2 | | | Point 3 | | | Point 4 | | | Point 5 | | | Point 6 | | |
|---|---|---|---|---|---|---|---|---|---|---|---|---|---|---|---|---|---|---|---|---|
| | | | F | O | U | F | O | U | F | O | U | F | O | U | F | O | U | F | O | U |
| Gastrocnemius (lateral) | GL | 2  5 | -8 | 61 | -388 | -47 | 93 | -840 | | | | | | | | | | | | |
| Plantaris | PLT | 2  5 | -8 | 93 | -393 | -47 | 93 | -840 | | | | | | | | | | | | |
| Soleus | SOL | 3  5 | 1 | 80 | -488 | -47 | 93 | -840 | | | | | | | | | | | | |
| Tibialis anterior | TA | 3  3  5 | 22 | 81 | -515 | 22 | 68 | -751C | 30 | 42 | -811 | 51 | 37 | -830 | | | | | | |
| Tibialis posterior | TP | 3  3  5 | 7 | 93 | -466 | -9 | 53 | -787C | 17 | 47 | -839 | 44 | 47 | -841 | | | | | | |
| Peroneus longus | PL | 3  3  5  5 | 11 | 108 | -468 | -16 | 88 | -794C | 7 | 96 | -839 | 22 | 94 | -854C | 35 | 79 | -857 | 65 | 55 | -846 |
| Peroneus brevis | PB | 3  3  5 | 0 | 101 | -622 | -14 | 88 | -794C | 18 | 104 | -834 | 43 | 106 | -859 | | | | | | |
| Peroneus tertius | PT | 3  3  5 | 0 | 95 | -673 | 11 | 87 | -754C | 36 | 86 | -825 | 67 | 86 | -846 | | | | | | |
| Extensor digitorum longus | EDL | 3  3  5  6 | 11 | 104 | -466 | 17 | 83 | -754C | 61 | 74 | -816 | 114 | 70 | -840C | 128 | 69 | -844 | 151 | 68 | -856 |
| Extensor hallucis longus | EHL | 3  3  5  6 | 5 | 96 | -572 | 19 | 79 | -753C | 58 | 45 | -804 | 107 | 38 | -827C | 122 | 34 | -831 | 151 | 28 | -836 |
| Flexor digitorum longus | FDL | 3  3  4  6 | 2 | 73 | -544 | -13 | 58 | -783C | 3 | 58 | -838 | 112 | 69 | -857C | 124 | 70 | -859 | 149 | 68 | -865 |
| Flexor hallucis longus | FHL | 3  4  6 | -5 | 104 | -545 | -25 | 72 | -818C | -2 | 71 | -845 | 105 | 42 | -857C | 121 | 38 | -856 | 150 | 32 | -852 |

*Note.* Segments: 1 = pelvis, 2 = thigh, 3 = shank, 4 = talus, 5 = foot, 6 = digits. Origin at the midpoint of the hip joint centers; all values in mm. The C represents two points joined by a curvilinear section (see Appendix A) as opposed to a straightline section. Additional information must be used to define the line of action of the quadriceps muscles near the knee because the patella was not included in the model. For this reason, the third point defining the location of the quadriceps muscles is braced "()" indicating that these values only pertain to when the standard individual is in the anatomical position. Several suggestions are contained in Appendix B to define this point during knee movement.

**Data Sources.** Two papers by Wood et al. (1989a, 1989b) provided the most complete set of information on the shoulder region. Using one male cadaver (1.8 m, 78 kg), they modeled the thorax, clavicle, scapula, upper-arm, and forearm segments and gave skeletal reference systems data for the sternoclavicular, acromioclavicular, shoulder, and elbow joints. Additionally, the line of action of 30 shoulder muscles were provided both as straightline representations from origin to insertion (Wood et al., 1989b, p. 312) and as centroid trajectories for each muscle (Wood et al., 1989a, p. 289). Although their segment, muscle, and skeleton surface modeling approach is quite representative and captures the nonlinear line-of-action of the muscles, these data are not included in this discussion. Those constructing shoulder models that require detailed line-of-action information should refer directly to the work by these authors. Finally, although several of the skeletal reference systems and muscle line-of-action points could be used for superficial bony-landmark purposes, they did not publish any additional superficial bony-landmark coordinates, although they noted their existence (Wood et al., 1989b, p. 310).

Hogfors et al. (1987) modeled the human shoulder complex as the thorax, clavicle, scapula, and humerus. Using information from three cadavers and two skeletons, they detailed the linear muscle line-of-action data for 21 muscles, several of which were partitioned into functional units. Some skeletal reference systems and superficial bony-landmark points were provided along with several bone dimensions useful for scaling purposes.

Park (1977) performed a mathematical analysis of the shoulder joint during lifting movements and reported the origins of 15 muscles and parts of muscles, obtained from one cadaver (1.90 m, 79.2 kg). Four anthropometric measurements were also provided from which several skeletal reference systems and superficial bony-landmark points were derived.

**Results.** The coordinates defining the endpoints (when appropriate) of the thorax, clavicle, scapula, upper arm, and forearm are presented in Table 11.4. The origin of the whole segment ensemblage is the midpoint of the sternoclavicular joints with the shoulder joint positioned to the right ($+O$). The upper arm was positioned such that the shoulder's skeletal reference system aligned with the $FOU$ axes, and the forearm was extended about the elbow's flexion-extension axis ($O$). The axes about which these segments rotate are given as 50-mm vectors directed from the applicable joint centers and are labeled using the convention employed by Wood et al. (1989b).

The line-of-action points for 34 muscles that act on the shoulder mechanism or elbow are presented in Table 11.5. Each muscle was represented by a line running between two points located on the indicated segment, and each coordinate is defined in the reference system found in Table 11.4.

No direct superficial bony-landmark data are provided, because few were available from the sources used. Segment lengths, using the skeletal reference systems points, should be used for scaling purposes.

# Dangers, Pitfalls, and Limitations

Scaling anatomical data to a living subject is fraught with inaccuracies. We all incorporate unique anatomical differences, which cannot be modeled using simple

**Table 11.4  Skeletal Reference Systems for the Upper Extremity**

| **THORAX** | | | | |
|---|---|---|---|---|
| Proximal point | Midsternoclavicular | 0 | 0 | 0 |
| 50 mm axes | F | 50 | 0 | 0 |
| | O | 0 | −50 | 0 |
| | U | 0 | 0 | 50 |
| Distal point | Sternoclavicular | 0 | 25 | 0 |
| **CLAVICLE** | | | | |
| Proximal point | Sternoclavicular | 0 | 25 | 0 |
| 50 mm axes | F | −40 | 52 | −12 |
| | O | 30 | 64 | −12 |
| | U | −3 | 42 | 47 |
| Distal point | Acromioclavicular | 1 | 181 | 0 |
| **SCAPULA** | | | | |
| Proximal point | Acromioclavicular | 1 | 181 | 0 |
| 50 mm axes | F | −39 | 208 | −13 |
| | O | 31 | 219 | −12 |
| | U | −2 | 197 | 47 |
| Distal point | Shoulder | 4 | 181 | −34 |
| **UPPER ARM** | | | | |
| Proximal point | Shoulder | 4 | 181 | −34 |
| 50 mm axes | F | −43 | 182 | −34 |
| | O | 6 | 232 | −34 |
| | U | 6 | 182 | −84 |
| Distal point | Elbow | 11 | 201 | −322 |
| **FOREARM** | | | | |
| Proximal point | Elbow | 11 | 201 | −322 |
| 50 mm axis | O | 10 | 250 | −321 |
| | Proximal radioulnar | 11 | 199 | −310 |
| 50 mm axis | U | 14 | 196 | −359 |
| Distal point | Wrist | 11 | 139 | −531 |

*Note.* Origin at the midpoint of the sternoclavicular joints; all values in mm.

or homogeneous scaling methods (Isman & Inman, 1969). Also, it is difficult to locate specific body landmarks even when the subject is still. In addition, musculoskeletal models require accurate estimates of joint centers and muscle line-of-action coordinates, and until better data collection and techniques of analysis become available, the location and orientation of the anatomical reference systems are at best accurate to 1 cm to 2 cm and 4° to 8° for adults (Apkarian, Naumann, & Cairnes, 1989; White et al., 1989).

Although the muscle line-of-action data presented in this chapter provides adequate average estimates, substantial individual deviations are possible (Luchansky & Paz, 1986; Mecalister, 1875). Throckmorton (1989) reported that averaged experimental muscle line-of-action data are not sufficiently accurate to

**Table 11.5   Upper Extremity Muscle Line-of-Action Points and the Segments to Which They Are Attached**

| Muscle | | On segment | Point 1 | | | Point 2 | | |
|---|---|---|---|---|---|---|---|---|
| | | | F (mm) | O (mm) | U (mm) | F (mm) | O (mm) | U (mm) |
| Trapezius (lower) | TRPL | 1 | -135 | 105 | -135 | -91 | 153 | -37 |
| Trapezius (middle) | TRPM | 1 | -113 | 90 | 21 | -61 | 181 | -17 |
| Trapezius (upper) | TRPU | 1 | -73 | 89 | 64 | -17 | 162 | 2 |
| Rhomboid major | RMBJ | 1 | -116 | 101 | -4 | -108 | 138 | -89 |
| Rhomboid minor | RMBN | 1 | -109 | 84 | 33 | -106 | 133 | -27 |
| Levator scapulae | LS | 1 | -94 | 70 | 72 | -71 | 131 | -37 |
| Pectoralis minor | PMIN | 1 | 65 | 63 | -59 | 6 | 140 | -17 |
| Subclavius | SUBC | 2 | -5 | 46 | 4 | -6 | 113 | 7 |
| Serratus anterior (lower) | SANL | 1 | 48 | 135 | -114 | -101 | 129 | -141 |
| Serratus anterior (upper) | SANU | 1 | 0 | 121 | 0 | -95 | 126 | -50 |
| Sternocleidomastoid | SCM | 2 | -11 | 89 | 225 | -4 | 150 | -15 |
| Sternohyoid | SHY | 2 | -35 | 17 | 142 | 8 | 163 | -2 |
| Omohyoid | OMH | 3 | -37 | 63 | 63 | -43 | 143 | -31 |
| Pectoralis major (clavicular) | PMJC | 2 | 10 | 59 | 11 | 16 | 192 | -118 |
| Pectoralis major (sternocostal) | PMJS | 1 | 19 | -9 | -40 | 19 | 193 | -98 |
| Pectoralis major (abdominal) | PMJA | 1 | 75 | -3 | -114 | 23 | 192 | -93 |
| Deltoid anterior (acromial) | DELA | 2 | -3 | 203 | -19 | 18 | 192 | -175 |
| Deltoid middle (clavicular) | DELC | 3 | 12 | 165 | 7 | 18 | 191 | -155 |
| Deltoid posterior (scapular) | DELS | 3 | -64 | 166 | -28 | 13 | 194 | -160 |

(continued)

**Table 11.5**   *(continued)*

| Muscle | | On segment | Point 1 F (mm) | Point 1 O (mm) | Point 1 U (mm) | Point 2 F (mm) | Point 2 O (mm) | Point 2 U (mm) |
|---|---|---|---|---|---|---|---|---|
| Latissimus dorsi | LATD | 1 | -54 | 175 | -205 | 16 | 170 | -86 |
| Teres major | TMAJ | 3 | -59 | 69 | 25 | 124 | 89 | -16 |
| Supraspinatus | SPSP | 3 | -80 | 134 | -1 | 15 | 179 | -12 |
| Infraspinatus | IFSP | 3 | -87 | 146 | -70 | 1 | 195 | -33 |
| Teres minor | TMIN | 3 | -63 | 162 | -101 | -3 | 188 | -42 |
| Subscapularis | SUBS | 3 | -94 | 137 | -65 | 7 | 163 | -27 |
| Coracobrachialis | COBR | 3 | 15 | 145 | -16 | 9 | 177 | -178 |
| Biceps (long) | BILH | 3 | 5 | 164 | -13 | 46 | 185 | -320 |
| Biceps (short) | BISH | 3 | 17 | 146 | -12 | 48 | 185 | -318 |
| Brachialis (lower) | BRCL | 4 | 4 | 187 | -250 | 1 | 171 | -356 |
| Brachialis (upper) | BRCU | 4 | 8 | 183 | -185 | 3 | 171 | -354 |
| Brachioradialis | BRAD | 4 | -1 | 189 | -248 | 36 | 153 | -530 |
| Triceps brachii (lateral) | TRIA | 4 | -10 | 189 | -162 | -12 | 184 | -312 |
| Triceps brachii (medial) | TRIM | 4 | -7 | 165 | -224 | -11 | 174 | -309 |
| Triceps brachii (long) | TRIO | 3 | -21 | 167 | -60 | -6 | 174 | -305 |

*Note.* Segments: 1 = thorax, 2 = clavicle, 3 = scapula, 4 = upper arm, 5 = forearm. Origin at the middle of the sternoclavicular joints; all values in mm.

calculate the temporomandibular joint forces; but Apkarian et al. (1989), Kepple, Arnold, and Stanhope (1991), and White et al. (1989) demonstrate that muscle line-of-action estimates have an accuracy comparable to that of locating segments in 3-D space. This suggests that it is not only the geometrical values that are in error when locating muscle line-of-action. However, as segment kinematic estimates improve, this will pressure musculoskeletal modelers to obtain better muscle line-of-action coordinates.

# MUSCLE MODELING

Skeletal muscle generates force when activated by the nervous system. How a muscle responds to this stimulus is partially related to its geometrical design (Edgerton, Apor, & Roy, 1990; Elftman, 1966; Gans, 1982; Gans & Bock, 1965). This section briefly reviews activation and contraction dynamics, focusing on those muscle geometric features required to formulate and use muscle models. How these data can then be scaled to a specific individual is discussed, concluding with human lower and upper extremity data useful for muscle modeling purposes.

## Muscle Geometry and Specific Tension

From a muscle-modeling perspective, skeletal muscles can be classified into two fiber orientations, parallel and pennate, with their fibers inserting into the skeletal framework either directly or via tendons (Gans, 1982). Various characteristics of these geometries can be defined as in Figure 11.2. Note that although some controversy exists as to whether most parallel-fibered muscles are composed of many short, overlapping fibers (Gans, Loeb, & De Vree, 1989; Loeb, Pratt, Chanaud, & Richmond, 1987) or fibers that extend end to end (Schwarzacher, 1959), most models equate fiber length (i.e., muscle fiber length) with the belly length (muscle belly length), because this mechanical structure behaves identically to a single, long fiber. In addition, pennation angle, muscle fiber length, and muscle belly length are defined when the muscle's sarcomeres are at their optimal force-generating length of 2.8 µm (Hoy, Zajac, & Gordon, 1990).

The maximum force output of a muscle, under isometric conditions at optimal length, is linearly related to the physiological cross-sectional area (pCSA) of the muscle fibers acting in parallel. The physiological cross-sectional area of a muscle can be estimated from

$$pCSA = (m \cdot \cos \alpha) / (Lf \cdot \rho) \qquad (11.1)$$

where $m$ = mass of the muscle belly, $\alpha$ = angle of pennation, $Lf$ = fiber length, and $\rho = 1,054 \, kg \cdot m^{-3}$, the density of muscle (Lieber & Blevins, 1989; Wickiewicz, Roy, Powell, & Edgerton, 1983).

The slope of the linear physiological cross-sectional area-force relationship is the specific tension (N · mm$^{-2}$). The specific tension of a variety of vertebrate muscle has been reported to range between 0.20 and 0.35 N · mm$^{-2}$ (Alexander & Vernon, 1975) with the higher values reported from single-fiber experiments

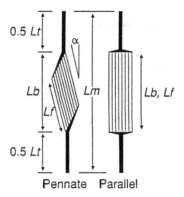

**Figure 11.2**    Geometry of pennate and parallel fibered muscles: Lb = muscle belly length, Lf = muscle fiber length, Lm = muscle length, Lt = tendon length, α = angle of pennation.

without the intervening effects of muscle pennation, incomplete muscle activation, or presence of elastic elements (Zajac, 1989). Although some controversy exists whether muscles of different fiber composition (fast, slow) may have different specific tensions (Alexander & Goldspink, 1977; Burke, Levine, Tsairis, & Zajac, 1973; Close, 1972; Powell, Roy, Kanim, Bello, & Edgerton, 1984; Wells, 1965), the peak, isometric, rest length and full-activation force output of a muscle can be estimated from its physiological cross-sectional area multiplied by $0.35 \text{ N} \cdot \text{mm}^{-2}$.

## Muscle Activation and Contraction Dynamics

Many muscle contraction models have been developed for use in musculoskeletal modeling. Winters and Stark (1987) suggested that these models can be divided into three categories: simple second-order systems, Hill-based lumped-parameter models, and Huxley-based distributed-parameter models. Upon examining the respective advantages and disadvantages of these models, they concluded that Hill-based models are clearly preferred for human movement analyses. Chapman (1985), Hatze (1981a), and Zajac (1989) provide excellent summaries of Hill-based models. Their reviews show that any biomechanically useful muscle model must incorporate the force-activation, force-length, and force-velocity relations of a contractile element modified by series, and possibly a lightly damped parallel, elastic component. The complexity of the models that link these mechanical elements is variable.

Mechanical models of muscle deal with combining the contractile, series elastic and parallel elastic elements, and the form of the equations used to describe their behavior. Audu and Davy (1985) performed a comparative study of four different muscle models that effectively summarized most mechanical muscle models. Their first model included a force generator (contractile element) with nonlinear force-activation relationship and force-length relationship modifiers in parallel with a spring and dashpot (parallel elastic element). The second configuration included a contractile element in series with a series elastic element, and these in parallel with a parallel elastic element. In this model, the contractile element incorporated a nonlinear force-activation relationship and linearized force-length and force-velocity equation. A third model was similar to the second except that the linear force-length and force-velocity relationships were replaced by nonlinear and hyperbolic equations. The last muscle model considered was that proposed by Hatze (1981a) and had the form of the second and third models but more complex force-activation relationship and force-length and force-velocity relationship equations.

Using these four models, Audu and Davy (1985) predicted the least time required for a weighted leg to kick a target. They discussed several advantages and disadvantages of the four models regarding complexity and practicality. Their results suggested that models of complexity similar to Model 3 or 4 *must* be used for biomechanical problems. They also noted the importance of obtaining appropriate, detailed force-activation, force-length, and force-velocity relationships, and that these curves can only be obtained from subject-specific experimental determinations.

Several investigators implicitly or explicitly agree that muscle modeling parameters must be collected from a specific individual (Hatze, 1981b; Hof & Van den Berg, 1981a, 1981b) though others remain undecided (Friederich & Brand, 1990; Hoy et al., 1990). However, Huijing (1985), Otten (1988), Pierrynowski and Morrison (1985a), Wickiewicz et al. (1983), and Zajac (1989)—among others—argue that appropriate scaling of muscle geometric features can provide the information necessary to model the contractile element (force-activation, force-length, force-velocity relationships), series-elastic and parallel-elastic element relationships without resorting to subject-specific experiments. A summary of these relationships follows.

## Force-Activation Relationship

Every skeletal muscle is composed of several neurologically independent units with distinct mechanical and biochemical properties. The force output of these motor units is controlled by recruiting more units or exciting already active units more frequently (Burke, 1981). A muscle, therefore, is a composite of many motor units, each with a neural excitation, which collectively develop force. The control signal is converted by a muscle's excitation-contraction dynamics into the signal that activates the muscle. This transformation is a complex, little understood process, which involves the propagation of the neural signal to release calcium ions from the sarcoplasmic reticulum into the interfilamentary space. Here they bind with a subunit of the troponin molecule. This binding allows the myosin head to interact with an active site on the actin filament, which results in generation of longitudinal force and movement (Esbashi & Endo, 1968). The number of active sites has been defined as the active state, or activation, of a muscle (Chapman, 1985; Ebashi & Endo, 1968; Hatze, 1981a). The potential of a muscle to generate force is linearly related to this active state.

234 Three-Dimensional Analysis of Human Movement

Many investigators have used a rectified, low-pass, filtered, time-delayed electromyographic (EMG) signal to estimate the activation of a muscle (Dowling, 1987; Hof, 1984; White & Winter, 1986). The form of the EMG-activation relationship has been reported to be linear (Lippold, 1952; Milner-Brown & Stein, 1975) or curvilinear (Bouisset, 1973; Le Brozec, Maton, & Cnockaert, 1980). If a linear relationship is assumed, the slope of the relationship varies with many factors in addition to muscle geometry. The type, size, orientation, and spacing of the electrodes, skin and fat thickness, the composition of the conducting jelly, and the amplifier specifications all affect the EMG-activation relationship (DeLuca & Van Dyke, 1975; Geddes, 1972; Lindstrom, 1970). For these reasons, the EMG-activation relationship is determined experimentally for each muscle of interest. This relationship can then be used as an index of muscle activation which, when modified by a dynamic muscle contraction model, estimates force output. The dynamic force output of a muscle can then be predicted during all contractile situations (Cnockaert, Lensel, & Pertuzon, 1975; Dowling, 1987; Galea & Norman, 1983; Hof, Pronk, & van Best, 1987; Inman, Ralston, Saunders, Feinstein, & Wright, 1952; Komi, 1973; McGill & Norman, 1986; Olney & Winter, 1985; Zahalak et al., 1976). (This modeling approach assumes that the muscle is not fatigued, because fatigue alters the form of the EMG-activation relationship; Krantz, Cassell, & Inbar, 1985.)

## Active and Passive Force-Length Relationships

The sliding filament theory of muscular contraction predicts that the area of overlap between the myosin and actin filaments should be linearly proportional to the force-producing capability of a fully activated muscle (Hanson & Huxley, 1953). This theoretical prediction was elegantly verified by Gordon, Huxley, and Julian in 1966. At reduced activation levels, the active force-length relationship is assumed to be a scaled version of the fully activated relationship (Hatze, 1981a), because it is assumed that the forces generated within a muscle (sarcomeres, fibers, motor units) sum linearly across the active parallel structures (Otten, 1988; Zajac, 1989). A passive muscle also resists external loads because of its parallel elastic component (Hatze, 1981a).

Several authors have proposed mathematical models to describe the active force-length relationship of muscle using linear, parabolic, Gaussian, and asymmetric equations (Audu & Davy, 1985; Bahler, 1968; Hatze, 1981a; Hof & Van den Berg, 1981a; Kaufman, An, & Chao, 1989; Otten, 1985; Pedotti, Krishnan, & Stark, 1978; Pierrynowski & Morrison, 1985a; Poliacu Proce & Otten, 1986; Winters & Stark, 1985; Woittiez, Huijing, Boom, & Rozendal, 1984; Woittiez, Huijing, & Rozendal, 1983; Zajac, 1989). However, few force-length relationship relations have been presented that only use muscle geometric features to define their parameters. Kaufman et al. (1989) present an equation that incorporates the physiological cross-sectional area, muscle belly length, and muscle fiber optimum length that gives good predictions for the active force-length relationship. Zajac (1989) and Hoy, Zajac, & Gordon (1990) develop and use an equation that requires physiological cross-sectional area, optimal muscle fiber length, and tendon length to predict the active and passive force-length relationships. Lastly, a set of muscle models was presented by Huijing and Woittiez (1984, 1985) and Woittiez et al. (1983, 1984) based on muscle belly length, muscle fiber length, angle of pennation, and physiological cross-sectional area. Their model was tested against a variety of active and passive force-length experimental data (Bobbert, Ettema, & Huijing, 1990; Heslinga & Huijing, 1990; Huijing & Woittiez, 1985) and was found to fit the measured data quite well *except* at short muscle lengths. All of these models imply that the active and passive force-length relationships can be predicted from the muscle geometrical values, muscle fiber length, tendon length, muscle belly length, angle of pennation, and physiological cross-sectional area.

## Force-Velocity Relationship

That an inverse relationship exists between muscle force and speed of contraction has been known for a long time (Hill, 1922). In 1938, Hill formulated his well-known equation that expressed this relationship in mathematical terms. After incorporating physiological constants proportional to physiological cross-sectional area and muscle fiber length for a muscle composed of a single fiber type (Ranatunga & Thomas, 1990; Wells, 1965), this equation has served as the model for most researchers interested in the muscle force-velocity relationship. Unfortunately, this equation does not model the eccentric (lengthening) side of the force-velocity relationship curve (FitzHugh, 1977; Hatze, 1981a), and the physiological constants vary with the force-activation and force-length relationships (Abbott & Wilkie, 1953; Bigland & Lippold, 1954).

Muscle models that examine the interdependence of geometry and the force-velocity relationship show the importance of muscle fiber length and fiber type. Sacks and Roy (1982) and Spector, Gardiner, Zernicke, Roy, and Edgerton (1980) present data showing a linear relationship between muscle fiber length and maximum muscle-shortening velocity when differences in fiber type are considered. Fiber composition results in an increase in the maximum muscle shortening velocity per unit fiber length 2.8 times higher when fast versus slow muscles are compared. Woittiez et al. (1984) published a 3-D geometric muscle model that predicts the force-length-velocity relationship from the muscle fiber length, tendon length, angle of pennation, physiological cross-sectional area, and maximum muscle-shortening velocity. A model proposed by Zajac (1989) requires similar geometric features. These published works suggest that the force-velocity relationship can be predicted from a muscle's geometry and its fiber composition. If one assumes that human muscles are heterogeneous with a constant ratio of slow-to-fast fibers, an average, maximum muscle-shortening velocity per unit muscle-fiber length can be used (Zajac, 1989).

# Required Muscle Geometries

Geometric features of a muscle determine its force-activation, force-length, and force-velocity relationships. These are the muscle, fiber, and tendon lengths, angle of pennation, and mass, and, derived from these, the muscle belly length, tendon slack length (i.e., tendon length divided by muscle length), and the physiological cross-sectional area. These features, in concert with the level of activation, specific tension, and the maximum shortening velocity per unit fiber length, allow one to predict the force output of a muscle under any condition of length and velocity. A method of scaling these data to a specific subject is the focus of the next section.

## Scaling to a Subject

Muscle geometrical data may be scaled to a specific individual based on total body height and mass, segment lengths and girths, somatotype, and skinfold thickness. Considering that transforming data from a published data base to a specific individual incorporates inaccuracy due to the large amount of individual variation, any scaling method can only be expected to provide approximate results. To this end a scaling method that is simple will likely give values as acceptable as more sophisticated techniques. The advantage will be that the simple method will require a reduced amount of subject involvement and computational effort.

To scale muscle geometric data from one individual to another one could use the equations of geometric similarity. This scaling method assumes that when two objects are geometrically similar they change proportions identically in all directions. For example, any object twice ($2^1$) the length, $L$, of another will have four times ($2^2$) the surface area, $A$, and eight times ($2^3$) the volume, $V$. Because the mass, $M$, of an object is directly proportional to its volume, $L \propto M^{1/3}$ and $A \propto M^{2/3}$ (Gunther, 1975).

McMahon (1984) summarized several published articles that refute the theory that one can use geometric similarity to scale animal structures. He developed a thesis in which size-dependent distortions in geometry must occur when scaling animal structures (allometric scaling). The mathematical development of this scaling method assumes two scaling factors: the first along the longitudinal axis, $l$, of a segment and the second along its transverse axis, $d$. If one defines $d \propto l^{3/2}$, this scaling theory is called elastic similarity. It can be derived if one assumes that bones are designed to resist buckling and bending under loads. In this scaling model, $l \propto d^{2/3}$, $l \propto M^{1/4}$, and $d \propto M^{3/8}$. A second allometric scaling theory assumes that bones are designed to experience constant stress and that $d \propto l^2$, $l \propto d^{1/2}$, $l \propto M^{1/5}$, and $d \propto M^{2/5}$ (McMahon, 1984).

The data cited by Gunther (1975) and McMahon (1984) support the theory of elastic similarity scaling for animal limb length, width, circumference, and weight. Therefore, to scale the muscle geometrical data presented in this chapter to any subject, elastic similarity scaling could be used with body height or mass. If the lengths of the subject's muscles were calculated using superficial bony-landmark and line-of-action points (discussed in a preceding section), better estimates of muscle geometries are available using muscle fiber length and tendon length as a percentage of the total muscle length (percent tendon length, percent muscle fiber length). Figure 11.3 illustrates the specific equations and data-base constants one could use to select the most appropriate transformations for the provided data.

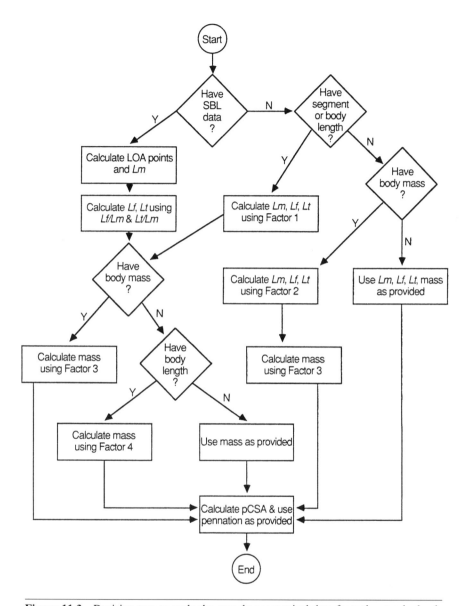

**Figure 11.3** Decision tree to scale the muscle geometrical data from the standard values (Tables 11.6 & 11.7) to a specific individual. There are four scaling factors:

Factor 1 = (individual height/reference height) · Table length values

Factor 2 = (individual mass/reference mass)$^{0.25}$ · Table length values

Factor 3 = (individual mass/reference mass) · Table mass

Factor 4 = (individual height/reference height)$^4$ · Table mass

*Note*: LOA = line of action, SBL = superficial bony landmarks, Lf = muscle fiber length, Lm = muscle length, Lt = tendon length, pCSA = physiological cross-sectional area.

# Muscle Geometry Data Bases

The geometry of the human musculature is available as qualitative information, but there is a dearth of reliable quantitative data. However, there are several notable exceptions that can form part of a useful total-body data base. Voss (1956) presented data originally compiled by von W. Theile on the masses of 172 muscles from 16 cadavers. However, no information was provided to scale these tabulated values to other individuals. Schumacher and Wolff (1966a, 1966b, 1966c) dissected 21 cadavers to obtain the mass and physiological cross-sectional area of 120 muscles from which fiber lengths can be calculated. Although little information is provided to transform these data to other individuals, scaling information and other muscle geometrical data is contained within a series of dissertations completed by Schumacher's students (Buß, 1963; Ladrick, 1963; Muller, 1966; Muller, 1967; Neumann, 1963; Rohmann, 1963).

An additional source of semiquantitative data on the human musculature is Eycleshymer and Schoemaker's 1970 atlas of cross-section anatomy. In this work, several cadavers were serially sectioned and meticulously rendered. Because the location of the "slices" are depicted with appropriate scaling information, the muscle belly and tendon lengths and volumes (mass) of most of the desired muscles and tendons can be determined by planimetry.

These three sources, together with others that provided detailed information on several or one muscle in some region of the body, were used to compile the muscle geometry data bases. (Papers that provided specific information on some region of the body are noted later.) These data were scaled to a 1.70-m, 70.0-kg reference individual using the most appropriate method outlined in Figure 11.3.

No effort was made to define muscle, muscle fiber, tendon, or muscle belly lengths "at rest." All of the reported data are for muscles situated within the body when the segments are in a vertically oriented anatomical position (see "System and Axes Definition").

## Lower Extremity

**Sources.** The muscle geometrical data reported by Pierrynowski and Morrison (1985a), transformed to a 1.78-m, 70.8-kg male (Pierrynowski, 1982) was the basis of this data base. Their data set was in turn compiled from the works of Alexander and Vernon (1975), Eycleshymer and Schoemaker's atlas (1970), Pedotti et al. (1978), Schumacher and Wolff and students (1966a, 1966b, 1966c), Voss (1956), and 19 additional sources. Here, four recent sources are also incorporated. Friederich and Brand (1990) provide data on the muscle belly length, muscle fiber length, physiological cross-sectional area, and angle of pennation for 47 muscles obtained from two cadavers of known age, height, and mass. The second source was that of Hoy et al. (1990) who present data, compiled from several sources, on the mass (derived from the peak muscle force), muscle fiber length, tendon length, and angle of pennation of 18 muscles and muscle groups. Their work, with that by Wickiewicz et al. (1983), who extensively dissected three cadavers and provides mass, muscle length, muscle fiber length, angle of pennation, and physiological cross-sectional area for 27 muscles, fail to provide cadaver identifying characteristics. Proctor's doctoral thesis (1980) provides data (mass, angle of pennation, muscle fiber length, physiological cross-sectional area) on 10 ankle-crossing muscles from five well-identified cadavers.

**Results.** The muscle, fiber, and tendon lengths, angle of pennation, and mass for 48 human, lower extremity muscles are presented in Table 11.6. In addition, the derived belly length, fiber and tendon lengths (a percentage of total muscle length), tendon slack length, and physiological cross-sectional area are given.

## Upper Extremity

**Sources.** In addition to the major sources already listed (Schumacher & Wolff, 1966a, 1966b, 1966c; Voss, 1956), four additional papers provided information on the geometry of the upper extremity muscles. An, Hui, Morrey, Linscheid, and Chao (1981) present muscle fiber length, mass, and physiological cross-sectional area (derived) for eight elbow muscles collected from four cadavers. Bassett, Browne, Morrey, and An (1990) provide muscle fiber length, muscle belly length, mass, and physiological cross-sectional area for 15 shoulder muscles. Edgerton et al. (1990) dissected four cadavers to obtain the muscle length, muscle fiber length, mass, and physiological cross-sectional area of seven elbow muscles. Unfortunately, all three of these sources fail to provide identifying information (e.g., height, mass, anthropometrics) of the cadavers. Wood et al. (1989a, 1989b) tabulated the muscle fiber length, physiological cross-sectional area, and mass of 30 shoulder muscles. However, none of these sources provides the angle of pennation of the muscles or the tendon length for the shoulder muscles.

**Results.** The muscle, fiber, and tendon lengths, mass, muscle belly length, fiber length-muscle length ratio, tendon length-muscle length ratio, and the physiological cross-sectional area for each of 34 shoulder and elbow muscles are contained in Table 11.7. Note the several missing values, representing items for which there was insufficient information in the literature.

# Dangers, Pitfalls, and Limitations

This section champions the thesis that the mechanical properties of a human striated muscle can be predicted from its geometry. Although a substantial body of published work shows relationships between a muscle's structure and its force-length–velocity-activation relationship, several factors have been ignored. Within the muscle, the mechanical effects of fiber type (Sacks & Roy, 1982; Spector et al., 1980), electromechanical delay (Norman & Komi, 1979), fatigue (Bigland-Ritchie, 1981), previous contractile history (Edman, Elzinga, & Nobel, 1978; Huijing & Ettema, 1988/89), and the inertia of the mass of the muscle (Hatze, 1981a) have not been considered. The parallel elastic components have been ignored because they are considered to be unimportant during the "normal" range of joint activity (Chapman, 1975; Engin, 1979; Herman, Schumerg, & Reiner, 1966). In contrast, the passive resistive properties of the joints (due to ligamentous constraints, skin-fascia influences, friction within the joint and bone,

**Table 11.6  Geometry of the Lower Extremity Muscles**

| Muscle | | Lm (mm) | Lf (mm) | Lt (mm) | α (deg) | mass (g) | Lb (mm) | Lf/Lm (%) | Lt/Lm (%) | pCSA (mm²) |
|---|---|---|---|---|---|---|---|---|---|---|
| Psoas major | PM | 292 | 190 | 54 | 5 | 278 | 238 | 65 | 18 | 1,383 |
| Iliacus | IC | 220 | 164 | 0 | 5 | 315 | 220 | 75 | 0 | 1,817 |
| Gemellus superior | GMS | 79 | 65 | 14 | 0 | 11 | 65 | 82 | 18 | 164 |
| Gemellus inferior | GMI | 77 | 61 | 16 | 0 | 15 | 61 | 79 | 21 | 236 |
| Obturator externus | OBE | 85 | 62 | 21 | 5 | 53 | 64 | 73 | 25 | 804 |
| Obturator internus | OBI | 76 | 56 | 20 | 17 | 69 | 56 | 74 | 26 | 1,126 |
| Piriformis | PIR | 128 | 73 | 23 | 7 | 88 | 105 | 57 | 18 | 1,132 |
| Quadratus femoris | QF | 49 | 49 | 0 | 0 | 35 | 49 | 100 | 0 | 684 |
| Pectineus | PEC | 119 | 199 | 0 | 0 | 67 | 119 | 100 | 0 | 535 |
| Adductor longus | AL | 220 | 146 | 0 | 4 | 235 | 220 | 66 | 0 | 1,518 |
| Adductor magnus (anterior) | AMa | 99 | 99 | 0 | 3 | 178 | 99 | 100 | 0 | 1,699 |
| Adductor magnus (middle) | AMm | 196 | 163 | 0 | 3 | 234 | 196 | 83 | 0 | 1,362 |
| Adductor magnus (posterior) | AMp | 323 | 194 | 81 | 3 | 342 | 242 | 60 | 25 | 1,674 |
| Adductor brevis (superior) | ABs | 116 | 116 | 0 | 0 | 70 | 116 | 100 | 0 | 571 |
| Adductor brevis (inferior) | ABi | 140 | 140 | 0 | 0 | 59 | 140 | 100 | 0 | 400 |
| Gluteus minimus (anterior) | GMINa | 100 | 79 | 19 | 7 | 80 | 81 | 79 | 19 | 960 |
| Gluteus minimus (middle) | GMINm | 112 | 84 | 28 | 0 | 94 | 84 | 75 | 25 | 1,065 |
| Gluteus minimus (posterior) | GMINp | 107 | 57 | 29 | 14 | 69 | 78 | 53 | 27 | 1,110 |
| Gluteus medius (anterior) | GMEDa | 110 | 83 | 23 | 5 | 190 | 87 | 76 | 21 | 2,159 |
| Gluteus medius (middle) | GMEDm | 136 | 116 | 20 | 0 | 192 | 116 | 85 | 15 | 1,572 |
| Gluteus medius (posterior) | GMEDp | 116 | 86 | 18 | 13 | 157 | 98 | 75 | 16 | 1,679 |
| Gluteus maximus (deep) | GMAXd | 213 | 154 | 39 | 3 | 324 | 174 | 72 | 18 | 1,986 |

| | | | | | | | | | |
|---|---|---|---|---|---|---|---|---|---|
| Gluteus maximus (superficial) | GMAXs | 580 | 171 | 409 | 0 | 393 | 171 | 29 | 71 | 2,185 |
| Tensor fascia latae | TFL | 517 | 139 | 204 | 2 | 75 | 313 | 27 | 39 | 516 |
| Semimembranosus | SM | 421 | 79 | 116 | 15 | 347 | 304 | 19 | 28 | 3,988 |
| Semitendinosus | ST | 484 | 175 | 196 | 4 | 173 | 288 | 36 | 40 | 938 |
| Gracilis | GR | 470 | 310 | 148 | 2 | 111 | 322 | 66 | 32 | 340 |
| Sartorius | SAR | 538 | 430 | 108 | 0 | 165 | 430 | 80 | 20 | 365 |
| Rectus femoris | RF | 488 | 88 | 186 | 10 | 317 | 302 | 18 | 38 | 3,357 |
| Biceps femoris (long) | BFl | 432 | 101 | 158 | 7 | 309 | 274 | 23 | 37 | 2,881 |
| Biceps femoris (short) | BFs | 248 | 146 | 96 | 15 | 163 | 152 | 59 | 39 | 1,024 |
| Vastus lateralis | VL | 411 | 110 | 138 | 11 | 814 | 273 | 27 | 34 | 6,880 |
| Vastus intermedius | VI | 407 | 106 | 87 | 6 | 606 | 320 | 26 | 21 | 5,368 |
| Vastus medialis | VM | 409 | 112 | 49 | 10 | 559 | 360 | 27 | 12 | 4,674 |
| Popliteus | POP | 111 | 109 | 2 | 0 | 132 | 109 | 99 | 2 | 1,150 |
| Gastrocnemius (lateral) | GL | 451 | 88 | 226 | 11 | 189 | 225 | 20 | 50 | 1,990 |
| Gastrocnemius (medial) | GM | 455 | 68 | 207 | 14 | 309 | 248 | 15 | 46 | 4,177 |
| Plantaris | PLT | 449 | 73 | 359 | 4 | 16 | 90 | 16 | 80 | 209 |
| Soleus | SOL | 356 | 49 | 227 | 26 | 679 | 129 | 14 | 64 | 11,868 |
| Tibialis anterior | TA | 334 | 99 | 217 | 9 | 215 | 117 | 30 | 65 | 2,040 |
| Tibialis posterior | TP | 414 | 43 | 252 | 17 | 169 | 162 | 10 | 61 | 3,622 |
| Peroneus longus | PL | 463 | 60 | 304 | 10 | 139 | 159 | 13 | 66 | 2,144 |
| Peroneus brevis | PB | 265 | 64 | 156 | 8 | 78 | 109 | 24 | 59 | 1,154 |
| Peroneus tertius | PT | 197 | 75 | 112 | 12 | 28 | 85 | 38 | 57 | 342 |
| Extensor digitorum longus | EDL | 468 | 101 | 344 | 11 | 114 | 124 | 22 | 74 | 1,050 |
| Extensor hallucis longus | EHL | 359 | 92 | 248 | 7 | 48 | 111 | 26 | 69 | 485 |
| Flexor digitorum longus | FDL | 451 | 48 | 311 | 11 | 51 | 140 | 11 | 69 | 991 |
| Flexor hallucis longus | FHL | 472 | 55 | 261 | 17 | 86 | 211 | 12 | 55 | 1,408 |

*Note.* Lm = muscle length, Lf = muscle fiber length, Lt = tendon length, Lb = muscle belly length, $\alpha$ = angle of pennation, pCSA = physiological cross-sectional area.

**Table 11.7   Geometry of the Upper Extremity Muscles**

| Muscle | | Lm (mm) | Lf (mm) | Lt (mm) | α (deg) | mass (g) | Lb (mm) | Lf/Lm (%) | Lt/Lm (%) | pCSA (mm²) |
|---|---|---|---|---|---|---|---|---|---|---|
| Trapezius (lower) | TRPL | 118 | | | | 139 | | | | |
| Trapezius (middle) | TRPM | 112 | | | | 62 | | | | |
| Trapezius (upper) | TRPU | 111 | | | | 55 | | | | |
| Rhomboid major | RMBJ | 93 | 93 | | | 39 | | 100 | | 402 |
| Rhomboid minor | RMBN | 78 | 78 | | | 7 | | 100 | | 80 |
| Levator scapulae | LS | 127 | 117 | | | 26 | | 92 | | 211 |
| Pectoralis minor | PMIN | 106 | 106 | | | 32 | | 100 | | 288 |
| Subclavius | SUBC | 67 | 57 | | | 3 | | 100 | | 42 |
| Serratus anterior (lower) | SANL | 152 | | | | 113 | | | | |
| Serratus anterior (upper) | SANU | 108 | | | | 21 | | | | |
| Sternocleidomastoid | SCM | 248 | 188 | | | 26 | | 76 | | 131 |
| Sternohyoid | SHY | 210 | 130 | | | 3 | | 62 | | 22 |
| Omohyoid | OMH | 124 | 114 | | | 3 | | 92 | | 25 |
| Pectoralis major (clavicular) | PMJC | 185 | 180 | | | 65 | | 97 | | 343 |
| Pectoralis major (sternocostal) | PMJS | 210 | 203 | | | 67 | | 97 | | 313 |
| Pectoralis major (abdominal) | PMJA | 203 | | | | 54 | | | | |

| Muscle | Abbr. | Lm | Lf | Lt | α | Lb | pCSA |
|---|---|---|---|---|---|---|---|
| Deltoid anterior (acromial) | DELA | 158 | 152 | 155 | | 96 | 967 |
| Deltoid middle (clavicular) | DELC | 164 | 139 | 61 | | 84 | 418 |
| Deltoid posterior (scapular) | DELS | 155 | 149 | 52 | | 96 | 331 |
| Latissimus dorsi | LATD | 138 | 135 | 169 | | 98 | 1,184 |
| Teres major | TMAJ | 189 | 152 | 64 | | 81 | 400 |
| Supraspinatus | SPSP | 106 | 106 | 36 | | 100 | 325 |
| Infraspinatus | IFSP | 107 | 106 | 79 | | 99 | 708 |
| Teres minor | TMIN | 88 | 78 | 17 | | 89 | 207 |
| Subscapularis | SUBS | 111 | 110 | 103 | | 99 | 887 |
| Coracobrachialis | COBR | 165 | 163 | 24 | | 99 | 140 |
| Biceps (long) | BILH | 310 | 191 | 38 | 14 | 62 | 191 |
| Biceps (short) | BISH | 310 | 181 | 30 | 20 | 58 | 156 |
| Brachialis (lower) | BRCL | 107 | | 24 | | | |
| Brachialis (upper) | BRCU | 170 | | 45 | | | |
| Brachioradialis | BRAD | 287 | 200 | 26 | 57 | 70 | 122 |
| Triceps brachii (lateral) | TRIA | 150 | 84 | 54 | 22 | 56 | 612 |
| Triceps brachii (medial) | TRIM | 120 | 63 | 41 | 12 | 49 | 611 |
| Triceps brachii (long) | TRIO | 246 | 102 | 73 | 73 | 42 | 675 |

*Note.* Lm = muscle length, Lf = muscle fiber length, Lt = tendon length, $\alpha$ = angle of pennation, Lb = muscle belly length, pCSA = physiological cross-sectional area. The missing values are not defined.

intersegment contact, and the parallel elastic components of the muscles) can be significant at the extremes of joint range of motion (Engin & Chen, 1986, 1987, 1988) and during normal activity (Mansour & Audu, 1986). Zchakaia (1926) showed that several lower extremity muscles shorten by about 30% of their fiber length after being sectioned, which suggests that muscles are always under tension. This is not always obvious, because agonist and antagonist parallel elasticities counterbalance each other. Another concern is that the standard geometrical data, largely derived from cadaver material, may not be representative because preservation, freezing, and embedding techniques all affect geometric features (Arnold & Worthman, 1974; Schumacher & Trommer, 1962). Lastly, it must be noted that because of the advanced age and unknown medical condition of most of the cadavers used to derive the data bases, one must be cautious when scaling these data to a younger population, because muscles in different regions of the cadaver bodies may have atrophied differently.

Finally, although scaled muscle geometrical values represent an average individual, they will never capture the nuances of any particular person (Brand, Pederson, & Friederich, 1986; Friederich & Brand, 1990). Therefore muscle geometrical data and derived mechanical properties cannot be predicted from an anatomical data base for any unique individual. This suggests that experimental methods must be used to estimate muscle mechanical properties in vivo (Hatze, 1981a, 1981b). The results of such experiments, performed on many individuals, may suggest regression equations derived from a variety of factors that will give better estimates of individual muscle properties. In the meantime, the data presented in this section can be used cautiously to provide estimates of individual muscle mechanical properties.

# FUTURE CONSIDERATIONS IN MUSCULOSKELETAL MODELING

This chapter presents quantitative musculoskeletal anatomy of the human lower and upper extremities. The lines of action and geometrical characteristics required for musculoskeletal biomechanical modeling were gleaned from the literature, then standardized and presented for scaling to any adult subject.

It is hoped that the work presented in this chapter will stimulate further work. Compiling information on the muscles' lines of action and geometries from the literature is only a starting point. Ideally, anatomical studies that examine the geometries, lines of action, and fiber compositions of the lower limb muscles should be performed on numerous cadavers, across a spectrum of subject ages (e.g., Spoor, van Leeuwen, de Windt, & Huson, 1989). Of great importance is the simultaneous collection of anthropometric data to scale these data to living individuals.

It is possible that medical imaging technology will provide the exciting possibility of whole-body, real-time, individualized, dynamic anthropometric measurement. Until that time however, the biomechanics community must develop a commonly accepted reference system (both laboratory fixed and segment fixed). In this way, a standardized musculoskeletal data base, in an easily interchangeable format, could be developed. If appropriate computer routines were also available, this data base could be logically extended to include new or refined information. All too often biomechanics laboratories develop software and information data bases specific to their location. Pooling of resources will likely occur in the next decade under the auspices of the International Society or World Congress of Biomechanics, or private concerns will make parts of this service available. Until standardization and pooling occurs, human musculoskeletal biomechanics research will remain unfocused and will continue to repeat much previous work.

## ACKNOWLEDGMENTS

This work was supported by the Children's Seashore House, the Children's Hospital of Philadelphia, and the University of Pennsylvania. I wish to acknowledge the helpful comments of A. Chapman, J. Morrison, and R. Wiens on a much earlier version of parts of this chapter. I thank C. Lee and C. Rooney for typing parts of the anatomical data base and for commenting on a draft of the manuscript. Last, I thank P. Allard for inviting me to contribute a chapter to this book and my friends, L. Carter and C. Garrity, for giving me the incentive to accept the invitation. I apologize to any group or individual who has published quantitative anatomical data that has not been incorporated within this chapter. Such omissions were not intentional.

## APPENDIX A

## Defining a Muscle Curvilinear Section

Define a pair of points through which the muscle must pass on either side of a joint as $P_1$ and $P_2$ and their corresponding tangent vectors as $T_1$ and $T_2$. These data are available from the muscle line-of-action data base, with appropriate scaling. A parametric cubic polynomial is then used to fit the curved section from $t = 0$ to $t = 1$ using a suitable step size (i.e., $\Delta t = 0.1$), such that

$$D(t) = A + Bt + Ct^2 + Dt^3, \tag{A.1}$$

where $D(t) = [x(t)\ y(t)\ z(t)]$ is the position vector of any point on the curve, and

$$A = P_1\ ;$$

$$B = T_1\ ;$$

$$C = 3(P_2 - P_1)/t^2 - (2T_1 + T_2)/(t + \Delta t)\ ; \tag{A.2}$$

$$D = 2(P_1 - P_2)/t^3 + (T_1 + T_2)/(t + \Delta t)^2\ .$$

# APPENDIX B

## Defining the Line of Action of the Quadriceps at the Knee

Additional constraints are necessary to define the location of the quadriceps muscles, because the patella was not included in the model. Each of these muscles can be assumed to arise from its respective origin (Point 1), converge and meet at a point 6 mm anterior to the most anterior and superior aspect of the femoral patellar surface (Point 2), curve to a third point (Point 3), and then be inserted into the tibial tuberosity (Point 4). Constraining the line through the third and fourth points to intersect the long axis of the tibia, in the sagittal plane, with an angle, $A$, in radians given by:

$$A = 0.2618 + 0.317\theta - 0.00084\theta^2 + 0.0000031\theta^3, \qquad (B.1)$$

where $\theta$ is the angle of the thigh relative to the shank (full extension is 0 radians). This equation was calculated from data collected from in vivo sagittal plane photographs of a male knee joint (Morrison, 1967). Letting the Point-3 to Point-4 distance be any small value, to define the direction of this section, the location of the third point can be calculated. Fitting a curve from Point 2 to 3 (see Appendix A) completely describes the location of each quadriceps muscle.

Although this method of describing the line of action of the quadricep muscle is acceptable for graphical and muscle length and velocity calculations, its accuracy for determining quadricep-moment arm values is suspect, because subject size and knee valgus-varus and torsions are not considered. To improve this estimate, the location of the second point 6 mm anterior to the femur could be scaled to the height of the subject (e.g., 6 mm is for a 1.80-m-tall male). Additionally, the data by Eijden, de Boer, and Weijs (1985) on the orientation of the distal part of the quadriceps through the patella as a function of the knee flexion-extension angle could be used. Last, for accurate quadriceps moment arm estimations, the published data can be employed directly (Kaufer, 1971).

# REFERENCES

Abbott, B.C., & Wilkie, D.R. (1953). The relation between velocity and shortening and the tension-length curve of skeletal muscle. *Journal of Physiology (Cambridge)*, **120**, 214-223.

Alexander, R.M., & Goldspink, G. (1977). *Mechanics and energetics of animal locomotion*. London: Chapman & Hall.

Alexander, R.M., & Vernon, A. (1975). The dimensions of knee and ankle muscles and the forces they exert. *Journal of Human Movement Studies*, **1**, 115-123.

An, K.N., Hui, F.C., Morrey, B.F., Linscheid, R.L., & Chao, E.Y. (1981). Muscles across the elbow joint: A biomechanical analysis. *Journal of Biomechanics*, **14**, 659-669.

An, K.N., Takahashi, K., Harrigan, T.P., & Chao, E.Y. (1984). Determination of muscle orientations and moment arms. *Journal of Biomechanical Engineering*, **106**, 280-282.

Apkarian, J., Naumann, S., & Cairnes, B. (1989). A three-dimensional kinematic and dynamic model of the lower limb. *Journal of Biomechanics*, **22**, 143-155.

Arnold, G., & Worthman, W. (1974). Zur querschnittsmessung von schnen unter biomechanischen gesichtspunkten. *Anatomischer Anzeiger*, **135**, 288-294.

Arvikar, R.J. (1971). A mathematical model for the musculoskeletal system for the lower extremities. Master's thesis, University of Wisconsin, Madison, WI, USA.

Arvikar, R.J. (1974). A comprehensive simulation of the human musculoskeletal system. Doctoral dissertation, University of Wisconsin, Madison, WI, USA.

Audu, M.L., & Davy, D.T. (1985). The influence of muscle model complexity in musculoskeletal motion modeling. *Journal of Biomechanical Engineering*, **107**, 147-156.

Bahler, A.S. (1968). Modeling of mammalian skeletal muscle. *IEEE Transactions on Biomedical Engineering*, **BME-15**, 249-257.

Bassett, R.W., Browne, A.O., Morrey, B.F., & An, K.N. (1990). Glenohumeral muscle force and moment mechanics in a position of shoulder instability. *Journal of Biomechanics*, **23**, 405-415.

Bigland, B., & Lippold, O.C.J. (1954). The relation between force, velocity, and integrated electrical activity in human muscles. *Journal of Physiology*, **123**, 214-224.

Bigland-Ritchie, B. (1981). EMG/force relations and fatigue of human voluntary contractions. *Exercise and Sport Sciences Reviews*, **9**, 75-117.

Bobbert, M.F., Ettema, G.C., & Huijing, P.A. (1990). The force-length relationship of a muscle-tendon complex: Experimental results and model calculations. *European Journal of Applied Physiology and Occupational Physiology*, **61**, 323-329.

Bouisset, S. (1973). EMG and muscular force in normal motor activities. In J.E. Desmedt (Ed.), *New developments in EMG and clinical neurophysiology* (pp. 547-583). Basel, Switzerland: Karger.

Brand, R.A., Crowninshield, R.D., Wittstock, C.E., Pederson, D.R., Clarke, C.R., & Krieken, F.M. van (1982). A model of lower extremity muscular anatomy. *Journal of Biomechanical Engineering*, **104**, 304-310.

Brand, R.A., Pederson, D.R. & Friederich, J.A. (1986). The sensitivity of muscle force predictions to changes in physiologic cross-sectional area. *Journal of Biomechanics*, **19**, 589-596.

Braune, W., & Fischer, O. (1888). Uber den Antheil, den die einzelne Gelenke des Schultergurtels am der Beweglichkeit des menschlichen Humerus haben. *Abh. math.-phys. Cl. d. k. Sachs Gesellsch. d. Wiss.*, **14**, 393-410.

Buford, W.L., & Thompson, D.E. (1987). A system for three-dimensional interactive simulation of hand biomechanics. *IEEE Transactions on Biomedical Engineering*, **BME-34**, 444-453.

Burdett, R.G. (1982). Forces predicted at the ankle during running. *Medicine and Science in Sports and Exercise*, **14**, 308-316.

Burke, R.E. (1981). Motor units: Anatomy, physiology, and functional organization. In American Physiological Society (Eds.), *Handbook of Physiology, The Nervous System*, pp. 611-619. Bethesda, MA: APA.

Burke, R.E., Levine, D.N., Tsairis, P., & Zajac, F.E. (1973). Physiological types and histochemical profiles in motor units of the cat gastrocnemius. *Journal of Physiology,* **234,** 723-748.

Buß, D. (1963). Morphologische und funktionelle Analyse des Musculus adductor magnus und Musculus gracilis. Doctoral dissertation, University of Rostock.

Chapman, A.E. (1975). The investigation of mechanical models of muscle based upon direct observations of voluntary dynamic human muscular contraction. Doctoral dissertation. University of London.

Chapman, A.E. (1985). The mechanical properties of human muscle. *Exercise and Sport Sciences Reviews,* **123,** 443-501.

Close, R.I. (1972). Dynamic properties of mammalian skeletal muscle. *Physiological Reviews,* **52,** 129-199.

Cnockaert, J.C., Lensel, G., & Pertuzon, E. (1975). Relative contribution of individual muscles to the isometric contraction of a muscular group. *Journal of Biomechanics,* **8,** 191-197.

Crowninshield, R.D., Johnston, R.C., Andrews, J.G. & Brand, R.A. (1978). A biomechanical investigation of the human hip. *Journal of Biomechanics,* **11,** 75-85.

Dapena, J. (1978). A method to determine the angular momentum of the human body about three orthogonal axes passing through its center of gravity. *Journal of Biomechanics,* **11,** 251-256.

Delp, S.L., Loan, J.P., Hoy, M., Zajac, F.E., Topp, E.L., & Rosen, J.M. (1990). An interactive graphics-based model of the lower extremity to study orthopaedic surgical procedures. *IEEE Transactions on Biomedical Engineering,* **37,** 757-767.

DeLuca, C.J., & Van Dyke, E.J. (1975). Derivation of some parameters of myoelectric signals recorded during sustained constant force isometric contractions. *Biophysical Journal,* **15,** 1167-1180.

Dempster, W.T. (1955). Space requirements of the seated operator: Geometrical, kinematic, and mechanical aspects of the body with special reference to the limbs. WADC Tech. Report 55-159, Wright Air Development Center, Air Research and Development Command. United States Air Force, Wright-Patterson Air Force Base, Dayton, OH.

Doddy, S.G., Waterland, J.C., & Freedman, L. (1970). Scapulohumeral goniometer. *Archives of Physical Medicine and Rehabilitation,* **51,** 711-713.

Dostal, W.F., & Andrews, J.G. (1981). A three-dimensional biomechanical model of hip musculature. *Journal of Biomechanics,* **14,** 803-812.

Dowling, J.J. (1987). The prediction of force in individual muscles crossing the human elbow joint. Doctoral dissertation. University of Waterloo, Waterloo, ON, Canada.

Dul, J., & Johnson, G.E. (1985). A kinematic model of the human ankle. *Journal of Biomechanical Engineering,* **7,** 137-143.

Dvir, Z., & Berme, N. (1978). The shoulder complex in elevation of the arm: A mechanism approach. *Journal of Biomechanics,* **11,** 219-225.

Edgerton, V.R., Apor, P., & Roy, R.R. (1990). Specific tension of the human elbow muscles. *Acta Physiologica Hungarica,* **73,** 205-216.

Edman, K.A.P., Elzinga, G., & Nobel, M.I.M. (1978). Enhancement of mechanical performance by stretch during tetanic contractions of vertebrate muscle fibres. *Journal of Physiology,* **281,** 139-155.

Eijden, T.M.G.J. van, de Boer, W., & Weijs, W.A. (1985). The orientation of the distal part of the quadriceps femoris muscle as a function of the knee flexion-extension angle. *Journal of Biomechanics, 18*, 803-809.

Elftman, H. (1966). Biomechanics of muscle with particular application to studies of gait. *Journal of Bone and Joint Surgery, 48-A*, 363-377.

Engin, A.E. (1979). Passive resistive torques about the long axes of major human joints. *Aviation, Space, and Environmental Medicine, 50*, 1052-1057.

Engin, A.E., & Chen, S-M. (1986). Statistical data base for the biomechanical properties of the human shoulder complex—II. Passive resistive properties beyond the shoulder complex sinus. *Journal of Biomedical Engineering, 108*, 222-227.

Engin, A.E., & Chen, S-M. (1987). Kinematic and passive resistive properties of human elbow complex. *Journal of Biomedical Engineering, 109*, 318-323.

Engin, A.E., & Chen, S-M. (1988). On the biomechanics of human hip complex in vivo—II. Passive resistive properties beyond the hip complex sinus. *Journal of Biomechanics, 21*, 797-805.

Engin, T.J., & Williams, A.S. (1968). Method of kinematic study of normal upper extremity movements. *Archives of Physical Medicine and Rehabilitation, 49*, 9-12.

Esbashi, S., & Endo, E. (1968). Calcium and muscle contraction. *Progress in Biophysics and Molecular Biology, 18*, 123-183.

Eycleshymer, A.C., & Schoemaker, D.M. (1970). *A cross-section anatomy*. New York: Appleton-Century-Crofts.

Fick, R. (1911). Handbuch der anatomie end mechanik der gelenke. Jena: Fischer.

FitzHugh, R. (1977). A model of optimal voluntary muscular control. *Journal of Mathematical Biology, 4*, 203-236.

Friederich, J.A., & Brand, R.A. (1990). Muscle fibre architecture in the human lower limb. *Journal of Biomechanics, 23*, 91-95.

Frigo, C. & Pedotti, A. (1978). Determination of muscle length during locomotion. In E. Asmussen & K. Jorgensen (Eds.), *Biomechanics VI* (pp. 355-360). Baltimore: University Park Press.

Galea, V., & Norman, R.W. (1983). Bone-on-bone forces at the ankle joint during rapid dynamic movement. In D.A. Winter, R.A. Noeman, R.P. Wells, K.C. Hayes, & A.E. Patla (Eds.), *Biomechanics IX-A* (pp. 71-76). Champaign, IL: Human Kinetics.

Gans, C. (1982). Fibre architecture and muscle function. *Exercise and Sport Sciences Reviews, 10*, 160-207.

Gans, C., & Bock, W.J. (1965). The functional significance of muscle architecture—A theoretical analysis. *Ergebrnsse der Anatomie und Entwicklungs Geschichte, 38*, 115-142.

Gans, C., Loeb, G.E. & De Vree, F. (1989). Architecture and physiological properties of the semitendinosus muscle in domestic goats. *Journal of Morphology, 199*, 287-297.

Geddes, L.A. (1972). *Electrodes and the measurement of bioelectric events*. New York: Wiley.

Gordon, A.M., Huxley, A.F., & Julian, F.J. (1966). The variation in isometric tension with sarcomere length in vertebrate muscle fibres. *Journal of Physiology, 184*, 170-192.

Gottschalk, F., Kourosh, S., & Leveau, B. (1989). The functional anatomy of tensor fascia latae and gluteus medius and minimus. *Journal of Anatomy,* **166,** 179-189.

Gregor, R.J., Komi, P.V., Browning, R.C., & Jarvinen, M. (1991). A comparison of the triceps surae and residual muscle moments at the ankle during cycling. *Journal of Biomechanics,* **24,** 287-298.

Gunther, B. (1975). Dimensional analysis and theory of biological similarity. *Physiological Reviews,* **55,** 659-699.

Hanson, J., & Huxley, H.E. (1953). Structural basis of the cross-striations in muscle. *Nature,* **172,** 530-532.

Hardt, D.E. (1978). A minimum energy solution for muscle force control during walking. Doctoral dissertation, Massachusetts Institute of Technology, Boston, USA.

Haruz, G. (1990). Representation of rotations by unit quaternions. *Ultramicroscopy,* **33,** 209-213.

Hatze, H. (1981a). *Myocybernetic control models of skeletal muscle: Characteristics and applications.* Pretoria: University of South Africa.

Hatze, H. (1981b). Estimation of myodynamic parameter values from observations on isometrically contracting muscle groups. *European Journal of Applied Physiology and Occupational Physiology,* **46,** 325-338.

Herman, R., Schumerg, H., & Reiner, S. (1966). A rotational joint apparatus: A device for study of tension-length relations of human muscle. *Medical Research Engineering,* **6,** 18-20.

Heslinga, J.W., & Huijing, P.A. (1990). Effects of growth on architecture and functional characteristics of adult rat gastrocnemius muscle. *Journal of Morphology,* **206,** 119-132.

Hill, A.V. (1922). The maximum work and mechanical efficiency of human muscles, and their most economical speed. *Journal of Physiology,* **56,** 19-41.

Hill, A.V. (1938). The heat of shortening and the dynamic constants of muscle. *Proceedings of the Royal Society of London,* **126,** 136-195.

Hof, A.L. (1984). EMG and muscle force: An introduction. *Human Movement Sciences,* **3,** 119-153.

Hof, A.L., Pronk, C.N.A., & Best, J.A. van (1987). Comparison between EMG to force processing and kinetic analysis for the calf muscle moment in walking and stepping. *Journal of Biomechanics,* **20,** 167-178.

Hof, A.L., & Van den Berg, Jw. (1981a). EMG to force processing II: Estimation of parameters of the Hill muscle model for the human triceps surae by means of a calfergometer. *Journal of Biomechanics,* **14,** 759-770.

Hof, A.L., & Van den Berg, Jw. (1981b). EMG to force processing III: Estimation of model parameters for the human triceps surae muscle and assessment of the accuracy by means of a torque plate. *Journal of Biomechanics,* **14,** 771-786.

Hogfors, C., Sigholm, G., & Herberts, P. (1987). Biomechanical model of the human shoulder—I. Elements. *Journal of Biomechanics,* **20,** 157-166.

Horn, B.K.P. (1987). Closed-form solution of absolute orientation using unit quaternions. *Journal of the Optical Society of America,* **4,** 629-642.

Hoy, M.G., Zajac, F.E. & Gordon, M.E. (1990). A musculoskeletal model of the human lower extremity: The effect of muscle, tendon, and moment arm on the moment-angle relationship of musculotendon actuators at the hip, knee, and ankle. *Journal of Biomechanics, 23*, 157-169.

Huijing, P.A. (1985). Architecture of the human gastrocnemius muscle and some functional consequences. *Acta Anatomica, 123*, 101-107.

Huijing, P.A., & Ettema, G.J.C. (1988/89). Length-force characteristics of aponeurosis in passive muscle and during isometric and slow dynamic contractions of rat gastrocnemius muscle. *Acta Morphologica Neerlando-Scandinavica, 26*, 51-62.

Huijing, P.A., & Woittiez, R.D. (1984). The effect of architecture on skeletal muscle performance: a simple planimetric model. *Netherlands Journal of Zoology, 34*, 21-34.

Huijing, P.A., & Woittiez, R.D. (1985). Notes on planimetric and three-dimensional muscle models. *Netherlands Journal of Zoology, 35*, 521-525.

Huiskes, R., Kremers, J., de Lange, A., Woltring, H.J., Selvik, G., & Rens, T.J.G. van. (1985). Analytically stereophotogrammetric determination of three-dimensional knee-joint geometry. *Journal of Biomechanics, 18*, 559-570.

Inman, V.T., Ralston, H.J., Saunders, J.B. de C.M., Feinstein, B. & Wright, E.W. (1952). Relation of human electromyogram to muscular tension. *EEG Clinical Neurophysiology, 4*, 187-194.

Isman, R.E., & Inman, V.T. (1969). Anthropometric studies of the human foot and ankle. *Bull. Prosthetic Res., 10-11*, 97-129.

Jensen, R.H., & Davy, D.T. (1975). An investigation of muscle lines of action about the hip: A centroid line approach vs. the straight line approach. *Journal of Biomechanics, 8*, 103-110.

Jensen, R.H., & Metcalf, W.K. (1975). A systematic approach to the quantitative description of musculo-skeletal geometry. *Journal of Anatomy, 119*, 209-221.

Kane, T.R., Likins, P.W., & Levinson, D.A. (1983). *Spacecraft dynamics.* New York: McGraw-Hill.

Kaufer, H. (1971). Mechanical function of the patella. *Journal of Bone and Joint Surgery, 53-A*, 1551-1560.

Kaufman, K.R., An, K-N., & Chao, E.Y.S. (1989). Incorporation of muscle architecture into the muscle length-tension relationship. *Journal of Biomechanics, 22*, 943-948.

Kepple, T.M., Arnold, A.S., & Stanhope, S.J. (1991, July). The estimation of muscle origins and insertions in a 3-D motion analysis and graphics program. International Symposium on 3-D Analysis of Human Movement, Montreal, Quebec, Canada.

Kinzel, G.L., & Gutkowski, L.J. (1983). Joint models, degrees of freedom, and anatomical motion measurement. *Journal of Biomechanical Engineering, 105*, 55-62.

Kinzel, G.L., Hall, A.S., & Hillberry, B.M. (1972). Measurement of the total motion between two body segments—I. Analytical development. *Journal of Biomechanics, 5*, 93-105.

Komi, P.V. (1973). Relationship between muscle tension, EMG, and velocity of contraction under concentric and eccentric work. In J.E. Desmedt (Ed.), *New Developments in Electromyography and Clinical Neurophysiology, Vol. 1* (pp. 596-606). Basel: Karger.

Koolstra, J.H., Eijden, T.M.G.J. van, & Weijs, W.A. (1989). An iterative procedure to estimate muscle lines of action in vivo. *Journal of Biomechanics,* **22,** 911-920.

Koolstra, J.H., Eijden, T.M.G.J. van, Weijs, W.A., & Naeije, M. (1988). A three-dimensional mathematical model of the human masticatory system predicting maximum possible bite forces. *Journal of Biomechanics,* **21,** 563-576.

Krantz, H., Cassell, J.F., & Inbar, G.F. (1985). Relation between electromyogram and force in fatigue. *Journal of Applied Physiology,* **59,** 821-825.

Ladrick, I. (1963). Morphologische und funktionelle Analyse des M. adductor longus et brevis und des M. pectineus. Doctoral dissertation, University of Rostock, Germany.

Le Brozec, S., Maton, B. & Cnockaert, J.C. (1980). The synergy of elbow extensor muscles during static work in man. *European Journal of Applied Physiology and Occupational Physiology,* **46,** 57-68.

Lew, W.D., & Lewis, J.L. (1977). An anthropometric scaling method with application to the knee joint. *Journal of Biomechanics,* **10,** 171-181.

Lewis, J.L., Lew, W.D., & Zimmerman, J.R. (1980). A nonhomogeneous anthropometric scaling method based on finite element principles. *Journal of Biomechanics,* **13,** 815-824.

Lieber, R.L., & Blevins, F.L. (1989). Skeletal muscle architecture of the rabbit hindlimb: Functional implications of muscle design. *Journal of Morphology,* **199,** 93-101.

Lindstrom, L. (1970). *On the frequency spectrum of EMG signals.* Res. Lab. Med. Electronics. Chalmers University, Goteborg, Sweden.

Lippold, O.J.C. (1952). The relation between integrated action potentials in a human muscle and its isometric tension. *Journal of Physiology,* **117,** 492-499.

Little, R.B., Wevers, H.W., Sui, D., & Cooke, T.D.V. (1986). A three-dimensional finite element analysis of the upper tibia. *Journal of Biomedical Engineering,* **108,** 111-119.

Loeb, G.E., Pratt, C.A., Chanaud, C.M., & Richmond, F.J.R. (1987). Distribution and innervation of short, interdigitated muscle fibres in parallel-fibred muscle of cat hindlimb. *Journal of Morphology,* **191,** 1-15.

Luchansky, E., & Paz, Z. (1986). Variations in the insertion of tibialis anterior muscle. *Anatomischer Anzeiger,* **162,** 129-136.

Mansour, J.M., & Audu, M.L. (1986). The passive elastic moment at the knee and its influence on human gait. *Journal of Biomechanics,* **19,** 369-373.

McGill, S., & Norman, R.W. (1986). Partitioning of the L4-L5 dynamic moment into disc, ligamentous, and muscular components during lifting. *Spine,* **11,** 666-676.

McMahon, T.A. (1984). *Muscle, reflexes, and locomotion.* Princeton, NJ: Princeton University Press.

Mecalister, A. (1875). Additional observations on muscular anomalies in human anatomy (3rd series), with a catalogue of the principle muscular variations hitherto published. *Proceedings of the Royal Irish Academy,* **5,** 1-134.

Mikosz, R.P., Andriacchi, T.P., & Andersson, G.B.J. (1988). Model analysis of factors influencing the prediction of muscle forces at the knee. *Journal of Orthopaedic Research,* **6,** 205-214.

Milner-Brown, H.S., & Stein, R.B. (1975). The relation between the surface electromyogram and muscular force. *Journal of Physiology,* **246**, 549-569.

Morrison, J.B. (1967). The forces transmitted by the human knee joint during activity. Doctoral dissertation, University of Strathclyde.

Muller, G. (1966). Ueber Struktur und Funktion der Glutaealmuskeln sowie des M. tensor fascia latae. Doctoral dissertation, University of Rostock, Germany.

Muller, K. (1967). Ueber Struktur und Funktion der inneren Huftmuskeln. Doctoral dissertation, University of Rostock, Germany.

Nemeth, G., & Ohlsen, H. (1985). In vivo moment arm lengths for hip extensor muscles at different angles of hip flexion. *Journal of Biomechanics,* **18**, 129-140.

Nemeth, G., & Ohlsen, H. (1989). Moment arms of hip abductor and adductor muscles measured in vivo by computer tomography. *Clinical Biomechanics,* **4**, 133-136.

Neumann, K. (1963). Struktur- und Funktionanalyse des Musculus quadriceps femoris. Doctoral dissertation, University of Rostock, Germany.

Norman, R.W., & Komi, P.V. (1979). Electromechanical delay in skeletal muscle under normal movement conditions. *Acta Physiologica Scandinavica,* **106**, 241-248.

Olney, S.J., & Winter, D.A. (1985). Prediction of knee and ankle moments of force in walking from EMG and kinematic data. *Journal of Biomechanics,* **18**, 9-20.

Otten, E. (1985). Morphometrics and force-length relations of skeletal muscle. In D.A. Winter, R.W. Norman, R.P. Wells, K.C. Hayes, & A.E. Patla (Eds.), *International Series on Biomechanics (ISB): Biomechanics IX-A* (pp. 27-32). Champaign, IL: Human Kinetics.

Otten, E. (1988). Concepts and models of functional architecture in skeletal muscle. *Exercise and Sport Sciences Reviews,* **16**, 89-137.

Panjabi, M. (1979). Validation of mathematical models. *Journal of Biomechanics,* **12**, 238.

Panjabi, M.M., White, A.A., & Brand, R.A. (1974). A note on defining body part configurations. *Journal of Biomechanics,* **7**, 385-387.

Park, Y-P. (1977). A mathematical analysis of the musculoskeletal system of the human shoulder joint. Doctoral dissertation, Texas Technical University.

Paul, J.P. (1990). Definition of the neuromuscular skeletal system for locomotion analysis. In *Models: Connection with experimental apparatus and relevant DSP techniques for functional movement analysis* (CAMARC, Computer-Aided Movement in a Rehabilitation Context, deliverable F) (pp. 35-42). Universita di Ancona, Ancona, Italy.

Paul, R.P. (1981). *Robot manipulators: Mathematics, programming, and control.* Cambridge, London: MIT Press.

Pedotti, A., Krishnan, V.V., & Stark, L. (1978). Optimization of muscle-force sequencing in human locomotion. *Mathematical Biosciences,* **38**, 57-76.

Pierrynowski, M.R. (1982). A physiological model for the solution of individual muscle forces during normal human walking. Doctoral dissertation, Simon Fraser University, Burnaby, BC, Canada.

Pierrynowski, M.R., & Morrison, J.B. (1983). Length and velocity patterns of the human locomotor muscles. In D.A. Winter, (Eds.), *Biomechanics IX A* (pp. 33-38). Champaign, IL: Human Kinetics.

Pierrynowski, M.R., & Morrison, J.B. (1985a). Estimating the muscle forces generated in the human lower extremity when walking: Theoretical aspects. *Mathematical Biosciences, 75*, 69-101.

Pierrynowski, M.R., & Morrison, J.B. (1985b). Estimating the muscle forces generated in the human lower extremity when walking: A physiological solution. *Mathematical Biosciences, 75*, 43-68.

Poliacu Proce, L., & Otten, E. (1986). The architecture and the length-force relations of skeletal muscles. *Acta Morphologica Neerlando-Scandinavica, 24*, 307.

Powell, P.L., Roy, R.R., Kanim, P., Bello, M.A., & Edgerton, V.R. (1984). Predictability of skeletal muscle tension from architectural determinations in guinea pig hindlimbs. *Journal of Applied Physiology: Respiration, Environment, and Exercise Physiology, 57*, 1715-1721.

Proctor, P. (1980). *Ankle joint biomechanics*. Doctoral dissertation, University of Strathclyde, Glasgow, Scotland.

Ranatunga, K.W., & Thomas, P.E. (1990). Correlation between shortening velocity, force-velocity relation, and histochemical fibre-type composition in rat muscles. *Journal of Muscle Research and Cell Motility, 11*, 240-250.

Reynolds, H.M., & Hubbard, R.P. (1980). Anatomical frames of reference and biomechanics. *Human Factors, 22*, 171-176.

Rogers, D.F., & Adams, J.A. (1990). *Mathematical elements for computer graphics* (second edition). New York: McGraw-Hill.

Rohmann, E. (1963). *Struktur- und Funktionanalyse des oberflaechlichen Flexoren des Unterschenkels (M. gastrocnemius, M. plantaris longus, und M. soleus)*. Doctoral dissertation, University of Rostock.

Sacks, R.D., & Roy, R.R. (1982). Architecture of the hind limb muscles of cats: Functional significance. *Journal of Morphology, 173*, 185-195.

Salathé, E.P., Arangio, G.A., & Salathé, E.P. (1986). A biomechanical model of the foot. *Journal of Biomechanics, 19*, 989-1001.

Schumacher, G.H. von, & Trommer, R. (1962). Die gegenseitige Abhangigkeit des Feucht- und Trockengewichtes und die Bedeutung des Trockengewichtes fur die Bestimmung der Arbeit von Skelettmuskeln. *Anatomischer Anzeiger, 111*, 175-188.

Schumacher, G.H. von & Wolff, E. (1966a). Trockengewicht und physiologischer Querschnitt der menschlichen Skelettmuskulatur. I. Trockengewichte. *Anatomischer Anzeiger, 118*, 317-330.

Schumacher, G.H. von, & Wolff, E. (1966b). Trockengewicht und physiologischer Querschnitt der menschlichen Skelettmuskulatur. II. Physiologische Querschnitte. *Anatomischer Anzeiger Bd., 119*, 259-269.

Schumacher, G.H. von, & Wolff, E. (1966c). Trockengewicht und physiologischer Querschnitt der menschlichen Skelettmuskulatur. III. Beziehungen zwischen Trockengewicht und physiologischen Querschnitt (schluß). *Anatomischer Anzeiger Bd., 119*, 270-283.

Schwarzacher, H.G. (1959). Uber die Lange und Anordnung der Muskelfasern in menschlichen Skeletmuskeln. *Acta Anatomica, 37*, 217-231.

Seireg, A., & Arvikar, R.J. (1973). A mathematical model for evaluation of forces in lower extremities of the musculoskeletal system. *Journal of Biomechanics,* **6**, 313-326.

Shiavi, R., Limbird, T., Frazer, M., Stivers, K., Strauss, A., & Abramovitz, J. (1987). Helical motion analysis of the knee—I. Methodology for studying kinematics during locomotion. *Journal of Biomechanics,* **20**, 459-469.

Shiba, R., Sorbie, C., Siu, D.W., Bryant, T., Cooke, T.D.V., & Wevers, H.W. (1988). Geometry of the humeroulnar joint. *Journal of Orthopaedic Research,* **6**, 897-906.

Shoup, T.E. (1976). Optical measurements of the centre of rotation for human joints. *Journal of Biomechanics,* **9**, 241-242.

Sommer, H.J., Miller, N.R., & Pijanowski, G.J. (1982). Three-dimensional osteometric scaling and normative modelling of skeletal segments. *Journal of Biomechanics,* **15**, 171-180.

Spector, S.A., Gardiner, P.F., Zernicke, R.F., Roy, R.R., & Edgerton, V.R. (1980). Muscle architecture and force-velocity characteristics of cat soleus and medial gastrocnemius: Implications for motor control. *Journal of Neurophysiology,* **44**, 951-960.

Spoor, C.W., Leeuwen, J.L. van, de Windt, F.H.J., & Huson, A. (1989). A model study of muscle forces and joint-force direction in normal and dysplastic neonatal hips. *Journal of Biomechanics,* **22**, 873-884.

Spoor, C.W., Leeuwen, J.L. van, Meskers, C.G.M., Titulaer, A.F., & Huson, A. (1990). Estimation of instantaneous moment arms of lower leg muscles. *Journal of Biomechanics,* **23**, 1247-1259.

Throckmorton, G.S. (1989). Sensitivity of temporomandibular joint force calculations to errors in muscle force measurements. *Journal of Biomechanics,* **22**, 455-468.

Tupling, S.J., & Pierrynowski, M.R. (1987). Use of Cardan angles to locate rigid bodies in three-dimensional space. *Computer Methods & Programs in Biomedicine,* **25**, 527-532.

Veldpaus, F.E., Woltring, H.J., & Dortmans, L.J.M.G. (1988). A least-squares algorithm for the equiform transformation from spatial marker co-ordinates. *Journal of Biomechanics,* **21**, 45-54.

Voss, H. von. (1956). Tabelle der Muskelgewichte des Mannes, berechnet und zusammengestellt nach den Untersuchungen von W. Theile (1884). *Anatomischer Anzeiger,* **103**, 356-360.

Wells, J.B. (1965). Comparison of mechanical properties between slow and fast mammalian muscles. *Journal of Physiology,* **178**, 252-269.

White, S.C., & Winter, D.A. (1986). The prediction of muscle force using EMG and a muscle model. *Proceedings of the North American Congress on Biomechanics.* Montreal, Canada, 25-27 August, I, 67-68.

White, S.C., Yack, H.J., & Winter, D.A. (1989). A three-dimensional musculoskeletal model for gait analysis. Anatomical variability estimates. *Journal of Biomechanics,* **22**, 885-893.

Wickiewicz, T.L., Roy, R.R., Powell, P.L., & Edgerton, V.R. (1983). Muscle architecture of the human lower limb. *Clinical Orthopaedics and Related Research,* **179**, 275-283.

Winter, D.A. (1984). Biomechanics of human movement with application to the study of human locomotion. *Critical Reviews in Biomedical Engineering, 9*, 287-314.

Winters, J.M., & Stark, L. (1985). Analysis of fundamental human movement patterns through the use of in-depth antagonistic muscle models. *IEEE Transactions on Biomedical Engineering,* **BME-32**, 826-839.

Winters, J.M., & Stark, L. (1987). Muscle models: What is gained and what is lost by varying model complexity. *Biological Cybernetics, 55*, 403-420.

Wismans, J., Veldpaus, F., & Janssen, J. (1980). A three-dimensional mathematical model of the knee joint. *Journal of Biomechanics, 13*, 677-685.

Woittiez, R.D., Huijing, P.A., Boom, H.B.K., & Rozendal, R.H. (1984). A three-dimensional muscle model: A quantified relation between form and function of skeletal muscles. *Journal of Morphology, 182*, 95-113.

Woittiez, R.D., Huijing, P.A., & Rozendal, R.H. (1983). Influence of muscle architecture on the force-length diagram: A model and its verification. *Pflugers Archiv. European Journal of Physiology, 397*, 73-74.

Wood, J.E., Meek, S.G., & Jacobsen, S.C. (1989a). Quantitation of human shoulder anatomy for prosthetic arm control—I. Surface modelling. *Journal of Biomechanics, 22*, 273-292.

Wood, J.E., Meek, S.G., & Jacobsen, S.C. (1989b). Quantitation of human shoulder anatomy for prosthetic arm control—II. Anatomy matrices. *Journal of Biomechanics, 22*, 309-325.

Yamaguchi, G.T., Sawa, A.G.U., Moran, D.W., Fessler, M.J., & Winters, J.M. (1990). A survey of human musculotendon actuator parameters. In J.M. Winters & S.L-Y. Woo (Eds.), *Multiple muscle systems: Biomechanics and movement organization* (pp. 717-773). New York: Springer-Verlag.

Zahalak, G.I., Duffy, J., Stewart, P.A., Litchman, H.M., Hawley, R.H., & Pasley, P.R. (1976). Partially activated human skeletal muscle: An experimental investigation of force, velocity, and EMG. *Journal of Applied Mechanics, 43*, 81-86.

Zajac, F.E. (1989). Muscle and tendon: Properties, models, scaling, and application to biomechanics and motor control. *Critical Reviews in Biomedical Engineering, 17*, 359-411.

Zchakaia, M. (1926). Uber die elastische Verkurzung der Skelettmuskeln nach dem Durchschneiden der distalen Sehne. *Pflugers Archives European Journal of Physiology, 215*, 457-458.

# Chapter 12

---

# Concepts in Neuromuscular Modeling

*Jack M. Winters*

In the three preceding chapters, we have considered aspects of musculoskeletal modeling. Now we extend this framework to consideration of *neuro*muscular modeling. These modeling approaches give insight into the neuromotor strategies that underlie human movement, which is their primary use. Often such models are used to complement experimental studies of human movement in which internal behavior (e.g., muscle forces) cannot be measured; that is, if model predictions are similar to measured data, a particular behavior can be assumed. Neuromuscular models also can be used to test hypotheses regarding neuromotor strategies; in such cases, inadequate model predictions are equally illuminating and can lead to new models and concepts of movement organization. To maximize insights from modeling studies, our models must strike a balance between simplicity and complexity.

Modeling is an evolving process, and we have entered an era in which 3-D movements can now be investigated; yet these same 3-D problems are difficult to understand. There have been recent advances in muscle and musculoskeletal modeling techniques; neural networks and optimization methods; computational speed and power (allowing complex 3-D modeling studies to be undertaken); and our evolving understanding of the relationships between defined movement-task goals and movement behavior. How should neuromuscular modeling techniques evolve to take advantage of these insights and advances? This chapter addresses this question by discussing two aspects: (1) development and contrast of basic approaches for neuromuscular modeling, and (2) identification of groups of movement tasks for which certain approaches to neuromuscular modeling appear to be most appropriate.

# BASES OF NEUROMUSCULAR MODELING

Figure 12.1 provides a simplified view of the neuromusculoskeletal system. From the perspective of this diagram, the musculoskeletal subsystem, which is the primary emphasis of this book, constitutes a relatively small portion of this overall system. Human 3-D movements must be planned by higher structures, and such planning—and all aspects of neuromotor execution—depend on sensory information of essentially mechanical variables and intricate interaction between neural and mechanical structures.

Communication between the neural and mechanical aspects of the system is through two processes: (1) the *actuator* of the system, skeletal muscle, that takes the output of the central nervous system (CNS) and drives the mechanical system (see chapter 11); and (2) the *sensors* of the mechanical system, which feed information back to the CNS regarding internal behavior and external performance. This chapter addresses modeling approaches at these interfaces. We will see that there is certainly not a ''best'' muscle model, nor a ''best'' formulation for modeling neural dynamics. Rather, different models evolve for addressing

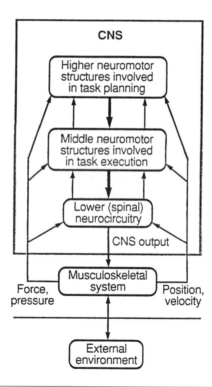

**Figure 12.1**   Simplified diagram emphasizing information processing throughout the neuromusculoskeletal motor system.

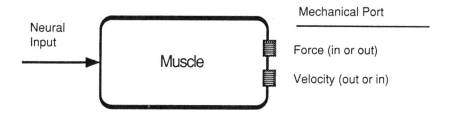

**Figure 12.2** Input-output view of an isolated muscle, showing unicausal (unidirectional) neural excitation and bicausal coupling at the mechanical port. For isolated muscle testing, either force or velocity is an input; the other is the output.

different strategic questions. Humans perform a rich variety of tasks—many goal directed—and have neural and mechanical systems that seem tuned to perform many tasks reasonably well, rather than a few perfectly. This argues against the concept of unifying principles that might govern all movement. To adequately address modeling options, this chapter considers certain classes of movement tasks and develops modeling approaches that help identify underlying movement strategies.

## STRUCTURAL POSSIBILITIES FOR ADVANCED NEUROMUSCULAR MODELS

There are several structural approaches to modeling the neuromuscular system. One is to model muscle (and usually skeletal) mechanics but not neural dynamics, usually by prescribing control inputs to a dynamic musculoskeletal model; this has been the most common approach (e.g., see Zajac & Winters, 1990). Another is to include *both* simplified neural dynamical elements and muscle elements as isolated entities within a larger model (e.g., Loeb & Levine, 1990). A third approach is to not explicitly distinguish between neural and muscle contributions, instead using neuromuscular entities (e.g., Feldman, Adamovich, Ostry, & Flanagan, 1990). Here we develop these approaches, starting first with a brief input-output review of muscle mechanics.

### Input-Output View of Muscle Modeling

Knowledge of muscle dynamics comes from careful input-output testing of muscles. From such a systems perspective, experiments with an isolated muscle require that two inputs be specified and then a single output be measured (Figure 12.2). One input represents a neural input, the other either muscle length or force; the one of the latter two that is not specified becomes the output. Most of the controlled experiments of muscle physiology can be considered "impulse" or "step" responses to one of the two inputs while the second remains constant (Winters, 1990). Though this is somewhat simplified (e.g., the neural input could be broken in "recruitment" and "firing rating" inputs; Hatze, 1977), it has proven useful (Winters, 1990).

Based on such systematic testing, a picture of muscle dynamic properties has emerged. To represent muscle properties in a model, two aspects must be considered: the assumed structure of the model, and properties of the elements *within* this structure. Described in chapter 11, the most popular class of structures for modeling muscle mechanics—evolving from that first proposed by A.V. Hill in 1938—involves a contractile element (CE) in series and in parallel with lightly damped elastic elements. The dynamic process of muscle excitation-activation must also be modeled, leading to the basic hybrid form of modeling structure shown in Figure 12.3. The schematics shown in Figure 12.4 a and b illustrate the basic nonlinear features of the CE force-length-velocity and series compliance relations commonly seen in experimental studies. Mathematical expressions are used to "fit" such data, and in the process assumptions are made, such as Hill's equation (for shortening velocity) or an exponential extension-force fit for series compliance (which assumes that stiffness increases linearly with force). The activation level scales the zero-velocity crossing level and, in some formulations, the unloaded peak CE shortening velocity, an approach that implicitly assumes an orderly recruitment of motoneurons (Winters & Stark, 1985).

In use for over 50 years, the popularity of this modeling framework relies, in part, on its conceptual simplicity, but also on its ability to predict muscle dynamic behavior reasonably well for many situations *without* requiring task-specific parameter variation. However, in its basic form, it does not predict some phenomena that can be seen in some refined experiments on isolated whole muscles or muscle fibers. These include yielding during transient lengthening (Joyce, Rack, & Westbury, 1969); force enhancement after transient lengthening (e.g., Edman, Elzinga, & Noble, 1978); subtle transient effects in response to sudden changes in length (McMahon, 1984); questions regarding the activation- versus force-dependence of series element compliance

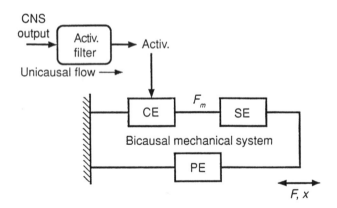

**Figure 12.3**  Schematic displaying the Hill model structure as it is now commonly used. The upper-left block represents excitation-activation dynamics, which are typically modeled as a first- or second-order unicausal filtering process with an input representing CNS output (neural excitation) and an output representing activation, which then becomes the input to the Hill muscle mechanical model. The series element (SE) and parallel element (PE) represent lightly damped passive springs; $F_m$ is contractile force; CE is contractile element; $F$ is output force; $x$ is length.

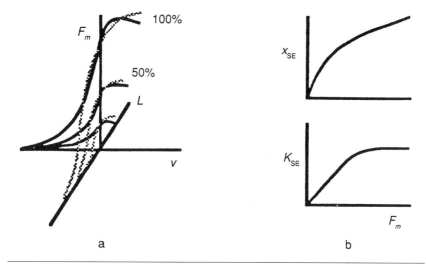

**Figure 12.4**   (a) The contractile element (CE) represents an instantaneous relation between muscle force ($F_m$), CE velocity ($v$) and length ($L$) that is modulated by the ongoing activation level (several levels shown here). (b) The SE extension ($x_{SE}$) and stiffness ($K_{SE}$) are shown as a nonlinear function of muscle force.

(Rack & Westbury, 1974); dependence of activation dynamics on muscle dynamics (Zahalak, 1990); and musculotendon series elastic properties that are considerably more complex than can be represented by a single spring (Ettema & Huijing, 1990). These effects can be dealt with in part within the Hill model framework but at the expense of added complexity (see Winters, 1990 for an in-depth assessment).

Use of the Hill model framework can be criticized from two opposing views. To some, it is considered incompatible with certain known physiology experiments and is not based on hypotheses for muscle contraction at the molecular level (summarized in Zahalak, 1990). From this perspective, a more suitable framework would be models using a biophysical framework as proposed by Huxley and colleagues (e.g., Huxley, 1957). To others—indeed, perhaps a majority of those who study human movement—the Hill model is already too complex for interpreting their data or developing unifying neuromotor theories. One's view of muscle depends largely on one's background and on the research questions that are being asked.

# One Possibility: Controlling Muscle Inputs (EMG Inputs)

Most forward dynamic models of human movement have employed musculoskeletal models that are controled by neural excitation signals that represent the CNS output (Zajac & Gordon, 1989). Thus, the nervous system is essentially isolated from a mechanical musculoskeletal model (see Figure 12.5). This is convenient for several reasons:

- The CNS output, via the motoneurons, is indeed the "final common way" of the nervous system that includes the contributions of higher centers, local spinal circuitry, and sensory feedback pathways.
- Information transfer at the neuronal level is unidirectional (opposed to bi-directional within the biomechanical system).
- This intermediate signal is loosely tied to electromyographic (EMG) activity, which can be experimentally measured.

Thus, a model is required for muscle dynamics but not for neural dynamics. This approach, documented elsewhere (e.g., Zajac & Gordon, 1989), possesses the inherent limitations common to deterministic, open-loop control systems. Using the Hill structure as a base, both quasi-static and dynamic models can be developed.

### Quasi-static Muscle Models

To be termed *quasi-static*, viscous influences (especially CE force-velocity) and mass accelerations must be negligible. Because the slope of the CE force-velocity relation is highest for low CE velocities (Figure 12.4, p. 261), forces may deviate as much as 20% above isometric for slow lengthening and by about 10% for slow shortening. The CE force-length relation, modulated by an activation-attachment-threshold input, is inherently nonlinear (Figure 12.6). However, for many applications, linear or bilinear springs provide useful representations over operating ranges of interest (see Figure 12.6). Several simpler approximations assume that the CE functions as a length-independent force generator (Figure 12.6) or as a force-independent position generator (e.g., the rack-and-pinion models described in Houk & Rymer, 1981, in which the force-length $CE_{LT}$ lines would be vertical). The CE-SE structure is shown connected to an inverted pendulum in Figure 12.6 to illustrate that because

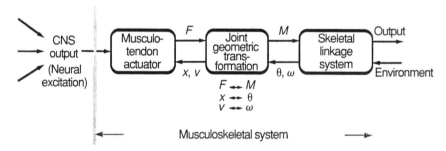

**Figure 12.5**   Schematic representing the common approach for studying human movement in which musculoskeletal dynamics are isolated from neural dynamics, with only musculoskeletal dynamics actually modeled. These inputs to the model include both those that represent the output of the central nervous system (CNS output) plus "inputs" from the environment (far right). If the coupled environmental system possesses its own dynamics, then the output-environment arrows at the mechanical interface become ill defined and the musculoskeletal subsystem is bicausally coupled to the external subsystem.

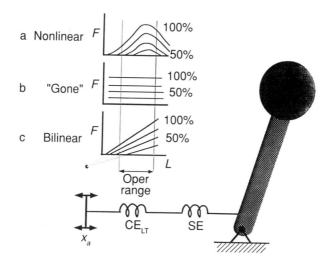

**Figure 12.6**   Static schematic of a Hill muscle model, interfaced to a pendulum with mass. Three possible CE force-length models are presented: (a) nonlinear, (b) linearized as a flat region (length-independent force generator), and (c) bilinearized.

the force across the CE and SE is the same, the strategy for holding equilibrium positions is very sensitive to the assumed $CE_{LT}$ relationship. For instance, if a length-independent force generator is assumed, this open-loop, spring-mass system is inherently unstable and prone to drifting from equilibrium.

In Figure 12.7a, spring-like contributions of muscle, aponeurosis, and tendon are considered separately. Extensions equal the tissue strain multiplied by the relevant length. Given the series arrangement, the forces in all elements are the same, and the element extensions add. Thus, any element with higher relative stiffness (i.e., lower compliance) could be neglected; in fact, compliances are of roughly the same order (Ettema & Huijing, 1990). Elastic properties of SE spring elements can be considered nonlinear, linear, or bilinear, depending on the goals of the analysis and on the force operating range for the task under investigation. Several interesting task-specific implications of SE element assumptions are illustrated in Figure 12.7b. For the idealized task of lifting a load, if the activation-dependent component of SE is represented by a bilinear spring, the steady-state muscle-SE relation becomes approximately vertical, that is, constant at about 2% strain and relatively independent of load. Because force in general depends on length and velocity as well as activation, this suggests a history-dependent SE contribution in which activated and deactivated muscles follow different SE force-extension paths, with paths during deactivation having higher extensions for a given force and being better represented by a more pronounced force-extension concavity than is normally assumed. For the idealized task of isometric contraction, the amount of CE shortening is equivalent to the amount of total SE extension, yet the relative contributions of activation-dependent and passive SE extensions is a function of activation.

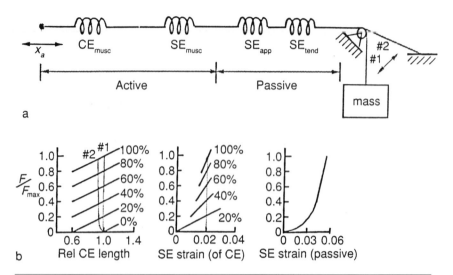

a

b

**Figure 12.7**   Quasi-static muscle "spring-like" behavior. (a) Series connections represent (from left to right) muscle CE force-length ("length-tension"); muscle SE, with stiffness a function of activation (i.e., number of attached bonds); aponeurosis; and tendon. Two idealized extremes for interaction with the environment are presented: In #1, muscle force is specified by the mass-pulley system and position is free to vary, whereas in #2, muscle length is specified ("isometric") and force is free to vary. (b) Representative assumptions for constitutive relations for these spring elements, with the two passive spring components lumped, due to our limited knowledge of aponeurosis properties. Here the CE force-length slope is assumed to be constant, with the conceptual spring offset $x_a$, or "threshold," sliding to the left with increased activation $[F = K_{LT} \cdot (x_{LT} - x_a)]$. The activation-dependent muscle SE assumes linear springs for those bonds that are attached, hence the slope increases directly with activation $[F = (F_a/0.02) \cdot x_{SE(musc)}]$; physiologically meaningful ranges are displayed. The passive spring includes a nonlinear, exponentially-shaped "toe region" followed by a linearized region.

Though there are notable exceptions (e.g., little tendon tissue in some spinal musculature), both muscle and tendon normally provide significant contributions to overall SE extension, with the muscle contribution being more complex and likely greater at lower activation and during deactivation than the tendon contribution, which is greater at higher activation and force. When a single lumped SE is desired, a nonlinear, exponential-shaped relation represents overall behavior considerably better than a linearized fit, especially in the low-activation region of primary importance in most activities of daily living (Winters, 1990). However, the relative contribution of muscle and tendon can vary, which seems consistent with the suspected functional roles of various musculotendonous units during life activities. Alexander and Ker (1990) have pointed out that the relative cross-section of tendons-versus-muscles varies and have suggested three classes of musculotendon units (see also Mungiole & Winters, 1990):

1. units with short tendons that are capable of producing a large amount of muscular work
2. units with long, thick tendons, which tend to stay within the nonlinear toe region (under 2% to 3% strain) and thus appear ideal for remote operation and impedance modulation (e.g., muscles in forearm)
3. units with relatively long, slender tendons that tend to operate in the more linear range during movement tasks and are better designed to store-release-transmit energy (e.g., many lower limb muscles)

## Dynamic Muscle Models: Bicausal Interaction

Can muscle be modeled simply by a smoothing (low-pass), unicausal (one-way) filter? Normally, it shouldn't be. As is evident from Figures 12.2 through 12.7 (see pp. 259-264), muscle force is very much a function of muscle length and velocity, with the mechanical port involving bicausal (two-way) force and position-velocity transmission. This conceptual foundation discourages the use of unicausal models. A possible exception are unloaded voluntary movements (e.g., certain types of eye movements).

Is the CE-SE structural arrangement necessary? Virtually always. Hannaford and Winters (1990) compared skeletal muscle to electrical, hydraulic, and pneumatic actuators and concluded that the primary distinctions are the presence of intrinsic (and nonlinear) series compliance and the interaction of CE-SE during dynamic tasks.

Can muscle behavior be linearized (e.g., Figure 12.8a) for certain tasks? Because of the simplicity and elegance of linear systems theory, there has been a tendency to linearize the Hill model over certain operating ranges. Winters and Stark (1987) illustrated cases in which a highly nonlinear, eighth-order, antagonistic muscle-joint model provided output trajectories for certain tasks that looked very much like the responses for a lower-order, linearized system. This can be explained by inspection of conceptual bilinear elements of Figure 12.8b. If the neural input is constant, the model becomes approximately linear. Similarly, for tasks that have no external loading and a small range of movement, the SE and CE force-length elements can often be linearized (or sometimes, even eliminated). Linearizing the CE force-velocity relationship to represent both shortening and lengthening muscle action is more dangerous, especially because antagonists are not effective at turning "off" because of their high viscosity (compare the CE force-velocity relations in Figure 12.8a to those in 12.8b for low activation). To some extent, this problem can be bypassed for single-direction movements by letting the antagonist be considerably less viscous than the agonist (e.g., Lehman & Stark, 1979). However, even with task-specific linearization, certain features cannot be captured.

This can be seen in the tracking task of Figure 12.9; despite *no* antagonist activation (EMG), the lengthening antagonist for the linearized model produces a resisting torque during the period between 50 and 100 ms. Additionally, the overall patterns for the linearized model are smoother because there is no capacity within the system for adaptive modulation of dynamic system properties. This helps explain the results obtained when optimally driving the two models. In

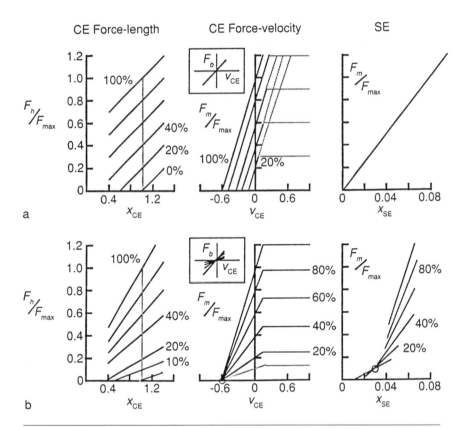

**Figure 12.8** Simplified relations for elements within the basic Hill muscle model structure. (a) Linearized, based on typical physiological operating ranges (note that the CE force-length-velocity relation here is additive as opposed to instantaneous). (b) Bilinearized elements, with each element a multiplicative function of the activation input (until saturation); the $CE_{fl}$ and $CE_{fv}$ relations may be combined multiplicatively or additively, with different effects. CE force-velocity insets: $F_b$ is the viscous force "lost" $(F_m = F_h - F_b)$.

both cases the neurocontrol strategy is optimal with regards to minimizing position tracking error (Seif-Naraghi, 1989). Yet the optimized reference error is considerably greater when employing the linearized model, especially when trying to track the descending ramp.

In general, history suggests that linearization is a curve-fitting approximation that has contributed little to our conceptual understanding of neuromuscular control.

## Insights Into Nonlinear Behavior Via Bilinear Muscle Model Elements

In the previous section, a bilinear model of muscle (Figure 12.8b, p. 266) was used to show the limited conditions under which a muscle "looks linear." Though

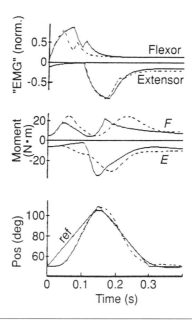

**Figure 12.9** Effects of linearization on optimized tracking movements by an antagonistic muscle-joint elbow model. The performance criterion being optimized is the mean squared error between actual and reference (triangular) trajectories; neural excitation pulse heights and widths are optimized. The optimized model response for the nonlinear model (in solid) is linearized and then reoptimized (in dash).

Hogan (1984) used the bilinear CE force-length relation to establish principles of stiffness-impedance modulation, to this author's knowledge, no one has attempted to formulate a dynamic, bilinear model for muscle. Why consider bilinearized elements? There are two fundamental reasons.

First, the essence of each of the nonlinear relations of muscle mechanics can be captured by bilinear relationships (Figure 12.8b); this simplified approach helps us more clearly visualize the implications of nonlinear properties. For example, reconsider the simulation shown in Figure 12.9, in which muscle properties were linearized based on the operating ranges that occurred over the course of this tracking task. The degradation in performance with the linearized model, even when reoptimized, happens because the optimization algorithm cannot take advantage of activation-dependent variation in damping and stiffness properties to provide more abrupt changes in muscle force-torque levels. This simulation also illustrates perhaps the most important part of the usual CE force-velocity relation: nonlinear properties allow an antagonist effectively to "disappear" and then "reappear" selectively. The commonly used strategy of impedance modulation via coactivation of antagonists provides an especially important case, and optimization solutions often take advantage of this (Winters, Stark, & Seif-Naraghi, 1988). In contrast, for linearized models there is no value in having co-contracting muscles, nor is it predicted in optimization solutions (Seif-Naraghi, 1989; Seif-Naraghi & Winters, 1989a, 1990).

Second, though not as developed as linear theory, bilinear systems theory suggests certain forms for neural feedback control structure that seem compatible with known neurocircuitry. In a bilinear system, in addition to linear terms in the state equations, there are multiplicative terms between a state variable and an input (or another state, in some cases). From our perspective, the control input can be thought of as modifying the (normally constant) parameter in front of the state variable and thus, as *capable of modulating the inherent properties of the dynamic system.* Such structural variation is the sign of an adaptive system. Interestingly, bilinear theory also shows that for systems with saturating inputs-states (as here), a bilinear system may actually be more controllable (Mohler, 1991). Furthermore, the form of such feedback controllers is multiplicative, which (we will see later) provides a reasonable representation of known neurospinal circuitry.

# Another Possibility: Including Neural Elements Within the Model

There is a fundamental problem with adding neural elements to the modeling process: Unlike the musculoskeletal system where the same tissues are used throughout and modeling the mechanical properties of each tissue is feasible, it is not possible to create a tractable model for neurocircuitry. There are many billions of neurons in the human CNS, and even single motoneurons can receive thousands of converging inputs. We know little about the function of the vast majority of CNS neurons. Yet the current trend toward modeling *only* the musculo-skeletal system (with "CNS output" being the model input) may not be the best approach; incorporation of lower-level neural elements may influence higher-level neuromotor movement strategy. The goal of this section is to suggest forms of neural elements that are simple enough to fit within the framework of a larger modeling process and yet detailed enough to capture fundamental neuromusculoskeletal behavior.

From a traditional feedback-control systems perspective, the musculoskeletal system represents the "plant" to be controlled (Figure 12.10). This basic structure is expanded in Figure 12.11 to include a few well documented elements of spinal neurocircuitry. There are pathways from the higher CNS that modulate feedback block parameters, the sign of an adaptive control system. Two approaches toward synthesizing lower neuromotor circuitry with muscle dynamics can be taken: (1) those that do not separate muscle and reflex activity; and (2) those that do.

## Modeling Basic Peripheral Neurosensorimotor Apparatus

**Modeling Motoneuronal Apparatus.** There are two main types of motoneurons (mn): $\alpha$ and $\gamma$ ($\beta$-mn will not be discussed). To a first approximation, $\alpha$- and $\gamma$-mn receive a common central drive, exciting muscle fibers that are structurally in parallel (Figure 12.12). Each muscle is innervated by $\alpha$-mn of a wide variety of sizes (Figure 12.13), with most muscles possessing a range of individual fiber compositions that, from a biomechanics viewpoint, can be broadly classified into

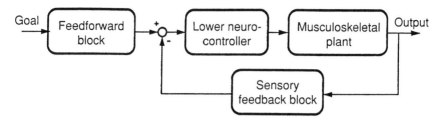

**Figure 12.10** Traditional feedforward-feedback control block diagram structures for neuromusculoskeletal systems.

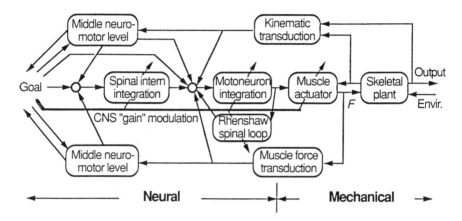

**Figure 12.11** Feedback-feedforward structure that includes basic spinal neurocircuitry connections and emphasizes the diverging and converging flow of information, multiple feedback loops, and CNS gain modulation at various levels (lighter lines crossing through blocks). All signals represent vectors and most also possess significant inherent time delays between blocks.

several types: fast, fatigable fibers innervated by larger $\alpha$-mn, fast, fatigue-resistant fibers, and slow, fatigue-resistant fibers innervated by smaller $\alpha$-mn. For most muscles, there are roughly equal numbers of fast and slow $\alpha$-mn (Winters & Stark, 1988). The larger $\alpha$-mn tend to be the last recruited by the CNS drive to the motoneuron pool (Figure 12.13), have a larger motor unit (i.e., branch to innervate a larger number of muscle fibers), excite faster-contracting muscle fibers, and reach a lower peak firing rate. The $\gamma$-mn tend to be of a size similar to smaller $\alpha$-mn, and are often separated functionally into static ($\gamma_s$) and dynamic ($\gamma_d$) classes (Loeb, 1984). This variation of mn types has caused some investigators to employ separate control for recruitment and firing rate (Hatze, 1981). Others have separated a given muscle into multiple (usually three) structurally parallel muscle elements with different properties (Hemami, 1985). These options, though viable, add to overall model complexity and to the required

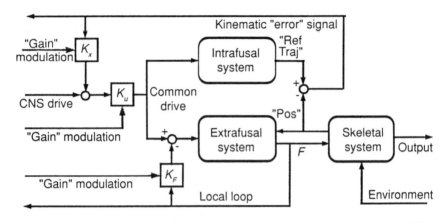

**Figure 12.12**   A neuromuscular model that retains the features of a traditional model reference adaptive control structure (e.g., Landau, 1979). This structural arrangement, which especially applies to γ-static and small-to-medium α-mn, is sometimes called α-γ *linkage* or α-γ *coactivation*. The square blocks (each with two inputs) represent multiplicative elements; the smaller circles represent additive junctions. The extrafusal position signal is a function of bicausal interaction between muscle dynamics and the skeletal system.

number of control inputs. An alternate strategy that implicitly combines both of these observations and assumes ordered recruitment (slow fibers first) involves scaling the CE maximum unloaded velocity as a function of activation (see Figure 12.4, p. 261).

**Modeling Sensory Apparatus.**   The major sensors are the muscle spindles, the Golgi tendon organs, joint angle and contact stress sensors, and environmental contact sensors (pressure, temperature). Biological sensors (transducers) tend to be sensitive both to the absolute magnitude of a specific variable and to its rate of change.

Spindles measure a relative opposed to an absolute muscle length (Figure 12.6-12.14). They often display a hypersensitivity to initial (isokinetic) stretch within a certain region, with further stretch identifying a less sensitive region that may be better approximated by assuming a product (multiplicative) relationship (Houk & Rymer, 1981) or more involved nonlinear, first-order dynamics (Hasan, 1983). The sources of such behavior are not yet resolved, but given the series arrangement between the sensory apparatus, $SE_{IF}$, and $CE_{IF}$ (see Figure 12.14), $CE_{IF}$ yielding may help explain the transition from hypersensitivity to lower sensitivity with lengthening, and the lack of this effect with shortening.

**Transmission Time Delays.**   The speed at which information travels by active spread along neurons ranges from 2 to 120 m/s, the speed being higher for larger nerve axons. Distances from the spinal cord to muscles range from about 5 to 100 cm. Therefore, both sensory and motor transmission times can approach tens of milliseconds, which is significant relative to the temporal dynamics of the

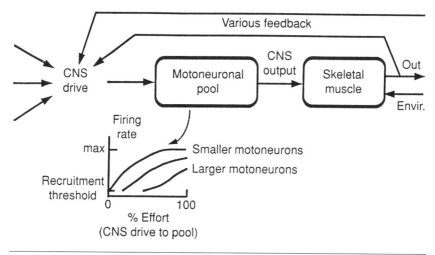

**Figure 12.13**   Basic motoneuronal mapping as relevant to neuromuscular modeling. The motoneurons to a given muscle (i.e., the motor pool) are recruited from small to large as the neural excitation drive increases, because a larger mn has a higher threshold (the size principle). Although represented here by extremes in size and a common neural drive to a motor pool, in fact there tends to be a continuum of axon sizes and muscle fiber fatigue properties, plus the possibility for bypassing the default recruitment scheme for certain types of tasks.

musculoskeletal system for many tasks. Active propagation along muscle tissue (initiated from the neuromuscular junction, or motor endplate) is on the order of 1 to 5 m/s, which for longer muscles produces additional time delays that are significant and, presumably, nonuniform muscle contraction.

**Simpler Muscle-Reflex ("Motor Servo") Neuromuscular Models.**   Lumping muscle and spinal neurocircuitry into a single functional unit has developed to represent input-output data from experiments involving forced perturbations in length-angle or force-torque. These models can be viewed from within the framework of a rack-pinion-spring analogy (Houk & Rymer, 1981; Hasan, Enoka, & Stuart, 1985). Bizzi, Accornero, Chapple, & Hogan (1984) have emphasized stiffness (slope) modulation. In contrast, Feldman (1974) considered the muscle-reflex apparatus to exhibit spring-like behavior in which the spring bias, or threshold, varied with changes in the central drive, thus creating a set of invariant force-length curves, which define a postural state. Within this framework, the EMG is considered an internal signal that varies as a consequence of changing central drives and external perturbations (Feldman et al., 1990); hence, EMG fluctuation with perturbation is predicted. This helps explain (in part) how measured, invariant stiffnesses can be considerably higher than those predicted from muscle CE tension-length (and often, even SE) relations.

Houk (1979) suggested the term *motor servo* for functional muscle-reflex units. He also used the phrase *stiffness regulation* to suggest that asymmetric neural

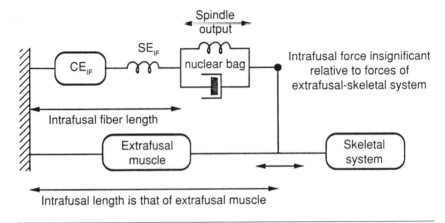

**Figure 12.14**   Illustration showing the muscle spindle sensory transduction process and its structural location within the musculoskeletal system.

excitation may help to balance asymmetric muscle tension-length mechanics to produce approximately linear stiffness levels for a certain motor set, and how spindle and Golgi feedback gains might be modulated in concert to help set certain quasi-static stiffness levels. This work has been extended to nonlinear damping behavior by Wu, Houk, Young, & Miller (1990), who observed that the nonlinear system appeared to be inherently stable for a wide variety of contact conditions.

## Models With Separate Neural and Muscular Elements

Few investigators have attempted to employ individual neural and muscular elements within a large-scale neuromusculoskeletal model. Loeb and Levine (1990) used an approach with a foundation in traditional control systems, modulating feedback gain matrices during cat locomotion by gating different patterns of connectivity, which presumably would be under the control of a central pattern generator that drives motor pools. An alternative approach would be to use fewer controllers per muscle and emphasize a more feedforward-dominated strategy that takes advantage of nonlinear muscle properties and aggressively modulates feedback gains.

# The Challenges of Using Both Feedforward and Feedback

Feedback control theory rests on the concept that the best control input to a plant should be based on maximum information, which includes the feedback of information regarding internal states and overall performance. In the design of feedback control systems, there are three major concerns: (1) maintenance of system stability, (2) capability both to control and to observe the system states, and (3) desirable system dynamic performance.

The existence of loop time-delays on the order of 40 ms reduces the potential for using real-time feedback during dynamic movements. From a stability viewpoint, it becomes essential that feedback gains not be too high. Indeed, this is one of the reasons for the popularity of simulations that have isolated musculoskeletal models with neural excitation (CNS output) inputs. However, EMG records show signs of sensory feedback modulation by both short-loop (stretch reflex) and longer-loop (conditional) reflexes. Thus, at minimum, a feedforward strategy must be modulated conservatively by sensory feedback.

The following assumptions and observations are fundamental to my feedforward-dominated approach:

1. Most tasks are goal-directed, with behavior learned from prior experience.
2. The neuromuscular periphery is purposefully nonlinear, and as a direct consequence, inherently capable of "smart" dynamic performance, especially when interacting with the environment—muscle mechanical properties in particular are aggressively modulated during movement tasks.
3. With sudden perturbation, nonlinear muscle properties provide predictable ongoing stiffness until spinal neurocircuitry (working with sensory information) restores a desired equilibrium.
4. Kinematic feedback gains tend to increase with activation level (Lacquaniti, Licata, & Soechting, 1982); because of nonlinear-muscle mechanical properties, simulations show that the threshold at which such delayed kinematic-feedback gains start to cause unstable system behavior increases nearly linearly with activation (Winters, 1985).
5. This modulation of feedback gains is essentially a multiplicative process (whether viewed as a feedforward signal that modulates feedback gains or a feedback signal that modulates feedforward signals is a matter of perspective).
6. Feedback behavior is more conservative than feedforward behavior and is best viewed as a sculpturing of feedforward signals.
7. A primary use of sensory feedback may be task-specific off-line learning by neural networks.

In summary, the feedforward-dominated modeling approach emphasized here is inherently nonlinear, adaptive, and structured for impedance modulation.

## Neuromuscular Models and the Optimization Process

Optimization theory is a useful tool for exploring goal-directed behavior (Hogan & Winters, 1990). There are four components of the optimization process (Figure 12.15a):

1. A dynamic system to be controlled
2. A scalar performance criterion that defines the movement goal
3. A controller that acts on the system
4. An algorithm that determines the appropriate controller

For 3-D movements, numerical solutions are necessary. Though a wide variety of numerical methods are available, many iterations are generally required to determine a global extrema (usually a minimum) for the performance criterion within the available control space. From this perspective, there may be analogies between the process by which the algorithm converges toward a solution and the processes that might be used by the neuromotor system. We have found that the same numerical optimization algorithm (in our case, one that possesses both random and gradient-like features; Winters, 1991) often converges to solutions more quickly when we use nonlinear instead of linearized muscle models (Seif-Naraghi & Winters, 1990). This must be due to sculpturing of the landscape of the criterion within the control space. Interestingly, neural networks tend to work best with nonlinear systems.

In Figure 12.15b, the optimization algorithm has been incorporated into the neuromotor modeling process. Here it is suggested that during the training process, the mid-level neuromotor structures that are involved in determining appropriate

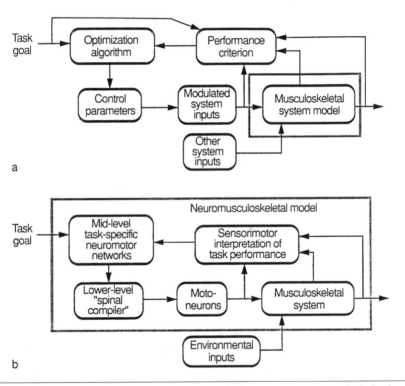

**Figure 12.15**   Conceptual view of the relationship between a neuromusculoskeletal model and the optimization process. (a) Traditional dynamic optimization problem for the musculoskeletal system, showing structural relationship between components. (b) Concept of the optimization algorithm being incorporated within the neuromusculoskel-etal modeling framework as dynamic, task-specific neural networks that drive a "smart" lower neuromuscular system that utilizes sensory information primarily for off-line modulation of future performance.

movement execution strategies, given a goal, represent an advanced form of an optimization algorithm (Winters & Mullins, 1993). Since a separate neural network should not be required when there are subtle changes in scaling within a given class of task (e.g., in movement magnitude or direction, desired speed, or weight of objects, such as cups), the optimizing (*not* optimal) network can be thought of as an input-output process that maps a task goal into a neuromotor execution strategy that unfolds over time. The task goal, specified as a performance criterion, is assumed to come from higher neural structures; note that approaches such as neural networks, fuzzy set control, and expert systems are starting to converge within the literature, and now hybrid approaches seem to represent our best chance for approximating these higher levels associated with organizational planning.

In parallel with task-specific, lower-level neural networks should be a second, more general network that is involved in coordinating overall postural stability and impedance modulation (both between segments and with the environment). By default, this network would drive neural signals back toward base levels of about 10% of maximum. However, based on sampling goals to task-specific neural networks and information from sensors, it could work in parallel to sculpture signals in three interrelated ways: (1) temporarily varying regional coactivation levels, (2) varying feedback gains if the default strategy of tying such gains to muscle activation levels is not appropriate (both spindle and force; including 1a interneurons), and (3) general maintenance of overall postural stability.

Advanced features of neural systems that have yet to be realized by our current optimization algorithms include extensive parallel processing, a capacity for inference (e.g., scaling to new conditions or subtle changes in goals), and neuromotor hardware that is not tuned exclusively for any one task but instead enables many tasks to be done well (though none, perfectly).

# Modeling for Inverse and Forward Optimization Schemes

Optimization techniques are used for two separate reasons in movement biomechanics (Zajac & Winters, 1990): to solve the muscle redundancy (load-sharing) problem, and to generate movement from the perspective of the neuromotor system. The classifications static optimization (SO) and dynamic optimization (DO), respectively, have been used to distinguish between these (Zajac & Gordon, 1989). However, a different classification scheme has recently been proposed (Winters, 1991; Winters & Van der Helm, 1993): inverse optimization (IO) and forward optimization (FO). It is also suggested that the terms *cost function* (*cf*) or *penalty function* be used for the scalar IO criterion, whereas the term *performance criterion* (*pc*) best represents the FO case.

## Inverse Optimization

In Winters (1991), five categories of inverse optimization (IO) are developed for solving the load-sharing problem:

1. inverse heuristic (reductionist) "optimization" approaches ($IO_h$),
2. inverse static optimization ($IO_s$),
3. inverse dynamic optimization ($IO_d$),
4. inverse dynamic integrated optimization ($IO_{di}$), and
5. inverse-forward dynamic optimization ($IFO_d$).

All of these approaches start with *prespecified kinematic knowledge*, which is either assumed or measured experimentally (Figure 12.16a). Usually the "model" for muscle is simply a force generator or a static activation-force curve. However, dynamic models, such as simple forms of the Hill model, can be solved inversely to estimate activation signals (Happee, 1992).

$IO_h$ uses heuristic rules based on physiology and "expert" experience to share load between tissues; typically one set of rules is assumed to be applicable for a variety of tasks. $IO_s$ and $IO_d$ solve optimization problems with cost functions ($J_{cf}$) of the form

$$J_{cf} = \sum_{l=1}^{n_l} \left[ \sum_{m=1}^{n_m} K_{lm} \mid J_{lm}^{p_m} \mid \right] + \sum_{j=1}^{n_j} (K_j \mid J_j^{p_j} \mid) \qquad (12.1)$$

where $J_{lm}$ is the $l$th type of cost subcriteria for muscle $m$ (of $n_m$ muscles) and $J_j$ are joint-based cost subcriteria; $K$ are relative weights (constants); and $p_m$ and $p_j$ are the powers for the respective $J$ (often of magnitude 2, with the magnitude itself having interesting implications; Dul, Johnson, Shiavi, & Townsend, 1984). This scalar cost (or penalty) function is a function only of tissue forces, stress, and so on, and not of kinematics. Often there is a separate $J_{cf}$ for each joint. The difference between $IO_s$ and $IO_d$ is that $IO_s$ involves solving the problem only once for a steady-state posture whereas $IO_d$ involves algebraically solving the inverse dynamic equations of motion (with positions, velocities, and accelerations estimated numerically) and then the cost criterion instant by instant, that is, at each time step. $IO_{di}$ differs from $IO_d$ in that, given the required joint torque time history, the load-sharing strategy as defined by the cost function $J_{cf-i}$ is a function of the entire task time history

$$J_{cf-i} = \int_{t_0}^{t_{max}} J_{cf}\, dt \quad \text{or} \quad \sum_{i=t_0}^{t_{max}} J_{cf_i} \qquad (12.2)$$

where for the right (numerical) case there is a series of time steps $i$ between the start of the task ($t_0$) and the completion of the task ($t_{max}$). The control parameters that are modulated to minimize this function can be parameters that define variable trajectories over time (e.g., muscle force trajectories). A useful (but as yet rarely used) approach is to incorporate length- and velocity-dependent "inverse" muscle models that allow prediction of muscle activation, energy dissipation, stiffness, and so on; such measures could then be incorporated into IO subcriterion. Estimation of mechanical stiffness in particular provides techniques for IO-based methods that address the problem of postural stability, and can predict effects such as antagonistic co-contraction (Van der Helm & Winters, 1994). This method of only considering that class of solutions which are guaranteed to be mechanically stable has recently been called inverse static stable optimization ($IO_{ss}$). "EMG-to-force processing" methods (e.g., Hof, 1990) could also be embedded within this IO framework.

$IFO_d$ assumes that predicted time histories for joint kinematics, net torques, and possibly EMGs can be considered "reference" signal trajectories [$y_{u-ref}(t)$], which may not provide "perfect" knowledge—thus some "slack" is allowable. The overall cost criterion then includes an additional subcriterion that penalizes any deviation from these reference trajectories (Winters, 1991):

$$J_{cf-f} = J_{cf-i} + \sum_{u=1}^{n_u} K_u \left[ \int_{t_0}^{t_{max}} \mid (y_{u-ref}(t) - y_u(t))^{p_u} \mid \right] dt \qquad (12.3)$$

To assure that kinetic behavior remains within the neighborhood of the reference trajectories, the relative weights ($K_u$) penalizing deviations in relevant "output" values ($y_u$) from reference values ($y_{u-ref}$) should be high and the power $p_u$ should be at least 2.0 (notice similarity to a least mean squares approach); the higher the individual $K_u$, the greater the trust in the given signal. If one then makes the reasonable assumption that subtle dynamic coupling changes are insignificant, this approach has the potential to predict smoother muscle force and activation trajectories and yet be relatively computationally efficient—an alternative to traditional forward dynamic approaches such as Seif-Naraghi and Winters (1990) and Yamaguchi (1990), which also use a kinematic reference model.

There are three reasons for the wide use of ID methods: (1) Typically, forces of the distal contact locations between the environment and the body are measured (or there is no contact), so the equations can be solved in an efficient, distal-to-proximal sequence; (2) compared to forward optimization methods (to be described) they are computationally efficient (Bean, Chaffin, & Schultz, 1988); and (3) they provide the required information regarding muscle force and joint contact loading (e.g., Seireg & Arvikar, 1989). However, as discussed elsewhere (Zajac & Gordon, 1989; Zajac & Winters, 1990), these methods provide limited insight into the underlying neuromotor strategy.

## Forward Optimization (FO)

There are two classes of forward optimization: forward static optimization ($FO_s$) and forward dynamic optimization ($FO_d$). $FO_s$ differs dramatically from $IO_s$ in that a kinematic postural configuration is not assumed a priori. Thus, the performance criterion ($J_{pc-s}$) takes the form (Winters, 1991)

$$J_{pc-s} = J_{cf} + K_k J_k \qquad (12.4)$$

where $J_k$ is a kinematically-based, task-related subcriterion. In $FO_s$, muscle forces are determined as a byproduct of a larger, task-based optimization process similar to that which the CNS must solve in choosing a certain posture from various alternatives (Figure 12.16b). This approach has significant applications for 3-D posture and movement, especially for kinematically complex systems such as the torso, neck, and shoulder, which must support significant body mass and transfer loads. It also has theoretical implications regarding postural control and stiffness regulation (Hogan, 1990) and practical implications in rehabilitation and orthopedics where many tasks are essentially quasi-static (Andersson & Winters, 1990).

To help illustrate possible applications, consider the large-scale head-neck system shown in Figure 12.17. Given that a certain head position and orientation is desired, let's develop four different approaches for "setting" a neck posture. One approach prescribes a "reference" posture, perhaps based on radiographic data. In this case, we combine a subcriterion that penalizes deviation from this ideal configuration with one that penalizes muscle activity, such as "muscle stress" (i.e., force per unit of physiological cross-sectional area) and only considers

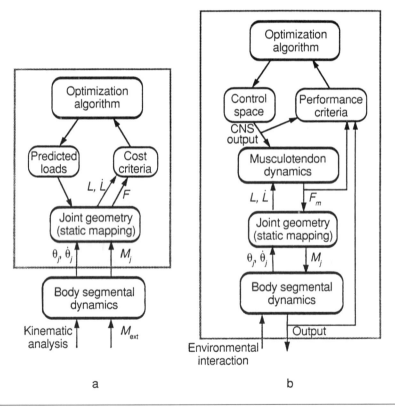

a                                        b

**Figure 12.16**   Typical information flow for various optimization approaches. (a) Inverse methods, showing how the optimization algorithm does not influence body segmental dynamics. (b) Forward methods, emphasizing interactions between elements and a performance criterion that is a function of input, state, and output information and the entire time history of the task.

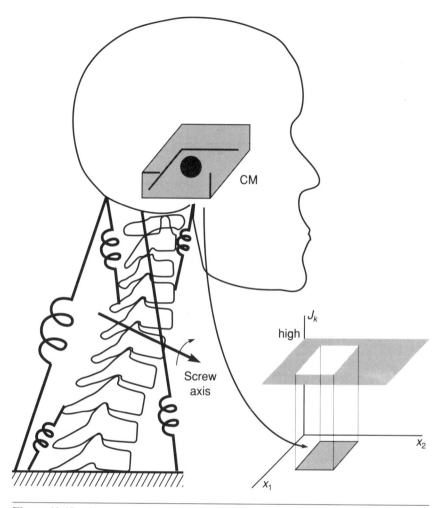

**Figure 12.17**   Conceptual view of the head-neck system where a certain goal-directed head orientation is specified by a kinematic subcriterion in which there is high penalty for deviation outside of an acceptable range of the end-point (head). The kinematics of the neck is not prespecified. Only several representative spring-like muscles are shown.

stable solutions. Another approach specifies only an acceptable orientation range for the most distal segment, that is, the head. Specifying such a range can be done by the use of conditional statements (very sharp boundary for $J_k$ as in Figure 12.17) or by penalizing position deviation to a very high power. This task-based subcriterion can then be combined with muscle-based and stability-based subcriteria, with both the muscle forces and the neck configuration determined as a byproduct of solving the optimization problem.

A third approach again uses the sharp boundary shown in Figure 12.17, now combined with a "subcriterion" that involves finding the minimum potential energy of the system (a scalar) that is also a relative minimum. This is an especially intriguing choice, because the solution of the constrained optimization problem is also a solution to the mechanics problem, because the PE would then increase for any kinematic perturbation from the equilibrium position. (For stability, essentially, for any small perturbation in the PE-kinematic space, the loss in potential energy due to falling masses must be more than made up by the increase in the overall elastic energy stored in muscles and, to a lesser extent, in passive tissues). Because the PE-kinematic space changes shape due to controllable muscle-like springs (Figures 12.6-12.7, pp. 263-264), there is a complex mapping between the criterion-kinematic DOF space and the criterion-control space. By careful specification of $J_{pc-s}$ in which the $J_k$ subcriterion is discontinuous, it is possible to solve both the task-specific optimization problem and the mechanics problem.

Fourth, $FO_s$ solutions could depend on the *combination* of several postural configurations. During horizontal and vertical head-tracking tasks in 10° increments in healthy subjects, the finite screw axis of the head relative to the torso goes through well documented regions of the neck (Winters & Peles, 1990; Winters et al., 1993). By using such screw axes as references, along with penalty subcriteria such as muscle stress, this $FO_s$ approach can make predictions regarding postural configurations for the neck (Daru, 1989).

For $FO_d$, often called *dynamic optimization*, the performance criterion specifies the movement task goal. An optimization algorithm then determines values for the control parameters that will drive the *dynamical* model to an optimal solution. Typically, the control parameters have represented open-loop neurocontroller inputs to a dynamical model of the musculoskeletal system (Lehman & Stark, 1979; Zajac & Winters, 1990). However, the dynamical model can include neurocircuitry, and control parameters can also include feedback gains (Loeb & Levine, 1990; Seif-Naraghi, 1989; Winters & Mullins, 1993).

A generalized performance criterion can be formulated for $FO_d$ problems that is of the form (Zajac & Winters, 1990)

$$J_{pc_d} = J_{pc_k} + J_{pc_{nm}} + J_{pc_{bj}} , \qquad (12.5)$$

where the three terms on the right represent the task-specific kinematic subcriteria, various neuromuscular penalties (muscle stress, force, energy measures, neural "effort", etc.) and any bone-joint penalty, respectively. Each may be quite complex and may include terms that depend on an outcome of the task (e.g., time to get to a target or peak joint loading) and on terms that are integrated over time during the task (e.g., muscle stress).

The $FO_d$ method is elegant and, in many ways, parallels the problem that must be solved by the CNS during goal-directed voluntary movement. However, considerable computational cost is required, especially for 3-D movements involving multilink systems. Thus, in many cases, models must be significantly simplified; as computational power increases, these methods will be more widely used for realistic neuromusculoskeletal systems (Yamaguchi, 1990).

# How Neuromuscular Model Structure
# Influences Research Questions

This chapter has suggested that the neuromuscular modeling process should be influenced by the research questions being asked. However, the opposite is also true. Overly complex models can be disadvantageous because *too much* information can cause underlying principles or sources of basic behavior to be missed. However, using overly simplified neuromuscular models can potentially cause worse problems: misinformation (faulty insights due to inadequate models) or disinformation ("tunnel vision," i.e., information that may keep a research team focused *away* from a path that could lead to improved understanding).

For example, in some cases the limb segment trajectories that occur in response to idealized external perturbations have been reasonably well approximated by linearized second-order inertia-spring-dashpot models. Yet as external inputs (or the instructions to subjects) change, a new set of parameters are needed to fit the new data, and such studies tend to become curve-fitting exercises. This early focus on such modeling efforts may have delayed our subsequent understanding of the important role of certain muscle properties in influencing neuromotor strategies that underlie the interplay between voluntary movement, dynamic interaction with the environment, and impedance modulation (Winters & Stark, 1987).

Another example is the widespread use of inverse static-dynamic analysis not just to estimate muscle forces but also to make implications regarding movement *strategy*. Consider the complex 3-D quasi-static adjustments in postural equilibrium performed while standing or sitting, which represent perhaps the largest class of everyday movements (Hogan & Winters, 1990). The $IO_s$ method starts with an assumed (prespecified) kinematic configuration and the appropriate "free body" diagrams. In reality, however, for 3-D systems with kinematic redundancy many subtly different configurations are possible that will satisfy basic task objectives. The organizational challenge of the CNS is to choose between these possible strategies while maintaining postural stability throughout the entire system (rather than just for a certain "free body"). For instance, co-contraction of antagonistic muscles is not normally predicted by $IO_s$ and is usually considered to be a suboptimal strategy. When employing $FO_s$, co-contraction may be necessary for reasons related to postural stability. Indeed, an impedance field orthogonal to the intended direction of movement may help guide a movement along a path. In summary, inverse methods efficiently document changes in tissue loading (which is valuable) but provide a limited perspective with regard to how such conditions cause changes in task strategy. Kinematic variation and postural stability are normally not explicitly addressed within the IO modeling framework. However, as seen in Winters & Van der Helm (1994) it is possible, for the quasi-static case, to explicitly consider only stable solutions when solving $IO_s$ problems.

# A Suggested Strategy for Choosing
# a Neuromuscular Model

This author's bias is to err initially on the side of a reasonably complex model, then use task-specific sensitivity analysis to reduce the model (Winters, 1990;

Winters & Stark, 1985). Sensitivity analysis allows one to identify any parameters to which the model's behavior is insensitive (Lehman & Stark, 1982). This allows the model properties described by such parameters to be eliminated or simplified with confidence (Figure 12.18).

## APPLICATIONS TO REALISTIC 3-D MOVEMENTS

Neuromotor strategies underlying human movement vary with the class of task. This in turn influences concepts in neuromuscular modeling. The following examples of different classes of human movement tasks illustrate various modeling strategies.

### Eye-Head Coordination in Spatial Tracking Movements

The special feature of these movements is the lack of external applied force. Consequently, movements are made relatively predictably, and eye movements

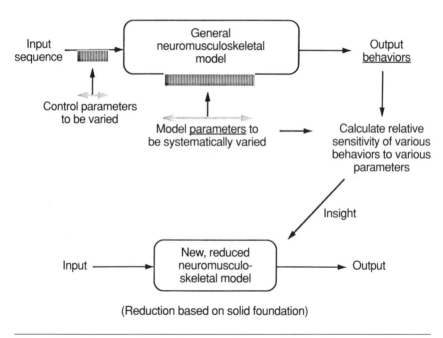

(Reduction based on solid foundation)

**Figure 12.18** Conceptual view of a suggested strategy for obtaining appropriate neuromuscular models. For a given task (or class of tasks), start by using a more realistic model, and systematically check sensitivity of strategic "behaviors" to model parameters. For the same task (or class of tasks), if the system is insensitive to certain model parameters, a model that is reduced based on this knowledge can then be used with confidence for this task (or class of tasks).

in particular are very stereotyped. For fast (saccadic) eye tracking movements, Hill-based muscle models are appropriate (e.g., Clark & Stark, 1975; Lehman & Stark, 1979); for slower movements, even simpler, unicausal linear models can be used. Sensitivity analysis findings show that linearization of SE, CE tension-length, and PE elements is more justified here than in any other region of the body, and CE force-velocity can be linearized piecewise with a higher agonist CE viscosity (Lehman & Stark, 1979; Winters, 1985).

Though "lumped" antagonistic head rotation muscles can be employed for fast tracking movements (Zangemeister, Lehman, & Stark, 1982), such models fail to capture much of the underlying head movement strategies. The cervical column is a complex mechanical structure, with about 40 pairs of muscles, most of which cross multiple joints. This structure supports the head, which has a large mass (see Figure 12.17, p. 279). Using a forward static model of the head-neck system that consisted of 3 joints (lumped to T5-T4, C7-C6, and C3-C2) and 12 muscles, we were surprised to find that certain head orientations—including common upright postures—were mechanically unstable (and not reachable) unless the slopes of the CE force-length relationships over the operating range of consequence were appropriately high (Daru, 1989). We have also found experimentally that the 3-D finite screw axis of the head during vertical, horizontal, and oblique angle-tracking movements varies in direction and location through the neck region (e.g., Winters & Peles, 1990). Though 12-muscle computer models (Daru, 1989; Winters & Van der Helm, 1994) and anthropomorphic models (Liang, 1989) replicate some of these trends, we find that a full range of 3-D head movements in all directions requires a remarkable number of muscles and complex *quasi-static* neuromotor strategies that we as yet do not understand.

For fast horizontal movements, neck EMG activity demonstrates coactivation in muscles that superficially would not seem to be necessary for horizontal movements: This suggests that the impedance is being increased in the vertical direction to help guide the horizontal movement! These observations imply that forward, rather than inverse, musculoskeletal models—both static and dynamic—of the head-neck system are warranted.

What about the nervous system? Neck musculature is well supplied with muscle spindles. Are these only used to assist in head movement control, or do they also help regulate the larger postural control needs of the body? Given the vast neurophysiological literature on the vestibuloocular and vestibulocollic systems (e.g., Peterson & Richmond, 1988), this seems to be an ideal system for exploring neuromuscular models that include neural networks.

# Roles for Arm-Torso Musculature in Arm Movements

From a biocybernetic perspective, the distinguishing feature is the enormous variety of tasks which use the arm. Perhaps this variety explains why many of the controversies in the human movement field involve neuromotor strategies for arm movements (Hogan & Winters, 1990). Upper arm research falls into three classes: (1) simple, goal-directed, fast tracking movements (usually point-to-point; e.g., Gottlieb, Corcos, & Agarwal, 1989), (2) slower, more natural movements investigating strategies underlying posture, equilibrium, and movement; and (3) investigation of practical activities of daily living tasks, usually within the context of rehabilitation engineering.

For fast tracking movements, a nonlinear Hill-based muscle model is usually necessary, especially to simulate antagonistic muscle behavior. Because there usually are no externally applied forces, the SE can often be linearized (or even eliminated). The role of muscles that cross multiple joints requires further investigation; a number of theories related to uses for such muscles have been proposed (Gielen, Ingen Schenau, Tax, & Theeuwen, 1990; Hogan, 1984) that need to be tested for a wider variety of tasks and for muscles with Hill-like muscle properties.

To test and extend posture-equilibrium theories, nonlinear or bilinear models that include both neural *and* muscular components are recommended. CNS controllers for such models require neural drives both for basic movement generation and for modulation of relative stability (e.g., control of impedance coactivation and feedback gain modulation). Neural networks may generate such control signals (Katayama & Kawato, 1993). Denier van der Gon, Coolen, Erkelens, and Jonker (1990) outline approaches in which neural networks (1) create internal representations of movements, (2) learn appropriate reflex responses, and (3) generate sequences of activation patterns.

The shoulder complex and torso are often ignored in upper limb modeling, although shoulder and torso muscles are normally active during voluntary arm movements. Indeed, for fast arm movements while standing, postural-base muscles throughout the torso and lower limb are often active *before* upper arm prime movements, in anticipation of the mechanical coupling that such movements will cause (Bouisset & Zattara, 1990), serving to stiffen proximal structures or to create appropriate anticipatory individual muscle drives. The coordination between the upper limb and more proximal structures, via a five-degree-of-freedom shoulder complex, represents one of the least understood neuromechanical systems of the body. It also provides a potential window for exploring many of the most intriguing questions in neuromotor organization and control, such as relations between posture and movement, maintenance of static and dynamic stability, and coordination between body components during developing movements. Both IO and FO methods are now being actively applied, using state-of-the-art models of the shoulder complex, to study such relations during various goal-directed tasks (Van der Helm, 1991; Van der Helm & Winters, 1994; Winters & Van der Helm, 1993).

## Active Maintenance of Biped Postural Stability During Movements

In the past, mostly inverse methods have been used to analyze these activities. However, forward methods—both static and dynamic—are crucial to development of deeper understanding and perhaps uncovering new organizational principles. For instance, Bergmark (1987) and Crisco and Panjabi (1990) recently showed, using potential energy arguments, that multiarticular rather than uniarticular muscles are mathematically most important for the maintenance of postural stability of the lumbar spine.

In whole-body experiments, musculature throughout the body can contract in anticipation of arm movements (Bouisset & Zattara, 1990) or due to applied perturbations (e.g. Keshner & Allum, 1990). Given the complexity of skeletal dynamic coupling, of muscle properties, and of relevant neural circuitry, there is considerable challenge in maintaining 3-D postural stability of large-scale, inverted pendulum systems while performing various movement tasks (Bergmark, 1987). In tandem with experimental studies, $FO_d$ and $FO_s$ simulations provide a great opportunity for helping identify fundamental neuromotor principles.

## Locomotion and Propulsion

These classes of propulsive tasks differ from those of postural stability in several ways (Hof, 1990; Ingen Schenau, Bobbert, & Soest, 1990; Mungiole & Winters, 1990): (1) The overall task goals are easier to identify; (2) movements tend to be more stereotyped and easier to investigate; (3) a large mass (i.e., the body) is being moved, so joint loading and muscle forces are high; (4) intersegmental dynamics are of great importance; (5) locomotion takes advantage of momentum changes in the system and any temporary instability (e.g., when one or both feet are off the ground); and (6) issues such as SE energy storage-utilization and musculoskeletal power transfer are of greater relative importance (Cavagna & Kaneko, 1977).

For these tasks, the CE-SE structure of the Hill model is necessary. Additionally, because most muscles go through periods of both low and high activation and of both shortening and lengthening, nonlinear musculotendon models are necessary to represent basic behavior. There is bicausal interaction between the CE, SE, and system mass-inertia throughout such tasks (Figure 12.19a); for instance, variable SE stretch allows the muscle CE element to stay within favorable operating ranges (Chapman, 1985; also Figure 12.19b). For propulsive tasks, perhaps the greatest need is the use of sensitivity analysis to better understand the dynamic interaction between elements of the model and the subsequent effects on performance.

## LIMITATIONS AND FUTURE DEVELOPMENTS IN NEUROMUSCULAR MODELING

This chapter has emphasized that a *variety* of neuromuscular models must be pursued, with the type of model dependent on the goal of the research, and, in particular, the type of task under study. It seems unlikely that any one neuromuscular modeling approach will ever be capable of fully representing all neuromuscular system features. Each modeling approach has limitations. In integrating models into 3-D movement analysis, we suggest retaining simplicity of structure, though never at the cost of fundamental nonlinear properties significant to the class of tasks under study. A useful way to determine the simplest muscle model capable of representing the relevant properties is to start with a model that includes salient muscle, skeletal, and perhaps neural properties, and then employ task-specific

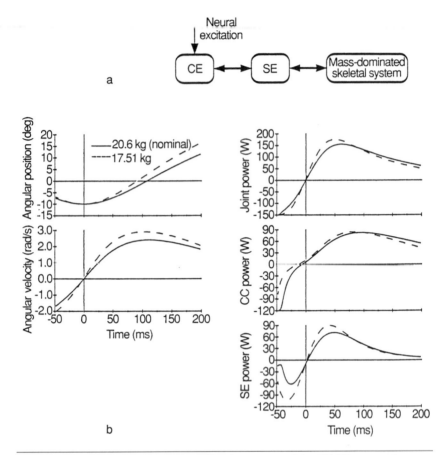

**Figure 12.19** Evolving concepts regarding importance of the CE-SE structural arrangement during goal-directed dynamic movements. (a) Conceptual view, in which CE, SE, and a mass-dominated skeletal framework have dynamic, bicausal interactions. With multiple muscles, the higher neuromotor system can take advantage of this arrangement. In particular, SE, skeletal mechanics, and selective contraction facilitate efficient use of CE properties, whereas the overall arrangement (including multiarticular muscles) can be used to modulate dynamic impedance selectively (e.g., at foot contact). (b) Simulations showing sensitivity of joint, CE, and SE power to mild changes in the magnitude of the mass lifted by the ankle plantar flexors during a "voluntary" stretch-shortening cycle with 75% pretension. A nonlinear antagonistic muscle-joint model for the ankle system (Winters & Stark, 1985) was used. The experimental apparatus described in Mungiole (1991) was also included in the modeling process.

sensitivity analysis to eliminate any unnecessary model features. One then knows a priori the effects of model reduction.

When lower neural dynamics are more widely included in the modeling process, with the specific approach depending on the class of tasks under investigation (e.g., cyclic tasks vs. tracking tasks), the system being controlled will be a *neuro*musculoskeletal system rather than a musculoskeletal system. Midlevel neural "networks," bearing only slight resemblance to currently popular artificial neural networks and more to the evolving dynamical neural networks that are based more closely on known neurocircuitry, will start to replace traditional optimization algorithms, modulating both feedforward and feedback control parameters (Katayama & Kawato, 1993). Basic planning (e.g., choosing appropriate midlevel networks) and refining goals (i.e., performance criteria) require a higher level of neural structure. This will still present a modeling challenge; but interestingly, approaches such as neural networks, fuzzy set control, and expert systems are starting to converge within the literature, and systems of fuzzy neurons, in particular, appear to hold promise for these intermediate levels of decision making and organizational planning.

Finally, it is suggested that forward modeling and forward optimization approaches will take on a greater importance in the future. Most real-life movements involve the entire body, including muscles quite distant from the "prime movers"; we need to represent such interactions with models in which the kinematic configuration is not prespecified. Additionally, as those exploring functional neuromuscular stimulation for gait have found, understanding the strategies that underlie the maintenance of stable 3-D posture is in many ways more difficult than creating basic dynamic movement patterns; to achieve this understanding, we must develop neuromusculoskeletal models that integrate concepts of 3-D posture and equilibrium with general neuromotor strategies for movement planning and execution. The challenges are many, indeed.

# REFERENCES

Alexander, R.M., & Ker, R.F. (1990). The architecture of leg muscles. In J.M. Winters & S.L-Y. Woo (Eds.), *Multiple muscle systems* (pp. 568-577). New York: Springer-Verlag.

Andersson, G.B.J., & Winters, J.M. (1990). Role of muscle in postural tasks: Spinal loading and postural stability. In J.M. Winters & S.L-Y. Woo (Eds.), *Multiple muscle systems* (pp. 377-395). New York: Springer-Verlag.

Bean, J.C., Chaffin, D.B., & Schultz, A.B. (1988). Biomechanical model calculation of muscle contraction forces: A double linear programming method. *Journal of Biomechanics, 21*, 59-66.

Bergmark, A. (1987). *Mechanical stability of the human lumbar spine.* Doctoral dissertation, Lund Institute of Technology, Lund, Sweden.

Bizzi, E., Accornero, N., Chapple, W., & Hogan, N. (1984). Posture control and trajectory formation during arm movement. *Journal of Neurosciences, 4*, 2738-2744.

Bouisset, S., & Zattara, M. (1990). Segmental movement as a perturbation to balance? Facts and concepts. In J.M. Winters & S.L Y. Woo (Eds.), *Multiple muscle systems* (pp. 498-506). New York: Springer-Verlag.

Cavagna, G.A., & Kaneko, M. (1977). Mechanical work and efficiency in level walking and running. *Journal of Physiology, 268*, 467-481.

Chapman, A.E. (1985). The mechanical properties of human muscle. *Exercise and Sport Sciences Reviews, 13*, 443-501.

Clark, M.R., & Stark, L. (1975). Time optimal behavior of human saccadic eye movements. *IEEE Transactions on Automatic Control, AC-20*, 255-272.

Crisco, J.J., & Panjabi, M. (1990). Postural biomechanical stability and gross muscle architecture in the spine. In J.M. Winters & S.L-Y. Woo (Eds.), *Multiple muscle systems* (pp. 438-450). New York: Springer-Verlag.

Daru, K.M. (1989). *Computer simulation and static analysis of the human head, neck, and upper torso*. Master's thesis, Arizona State University, Tempe.

Denier van der Gon, J.J., Coolen, A.C.C., Erkelens, C.J., & Jonker, H.J.J. (1990). Self-organizing neural mechanisms possibly responsible for movement coordination. In J.M. Winters & S.L-Y. Woo (Eds.), *Multiple muscle systems* (pp. 335-342). New York: Springer-Verlag.

Dul, J., Johnson, G.E., Shiavi, R., & Townsend, M.A. (1984). Muscular synergism II: A minimum-fatigue criterion for load-sharing between synergistic muscles. *Journal of Biomechanics, 9*, 674-684.

Edman, K.A.P., Elzinga, G., & Noble, M.I.M. (1978). Enhancement of mechanical performance by stretch during tetanic contractions of vertebrate skeletal muscle fibres. *Journal of Physiology, 280*, 139-155.

Ettema, G.J.C., & Huijing, P.A. (1990). Architecture and elastic properties of the series elastic element of muscle-tendon complex. In J.M. Winters & S.L-Y. Woo (Eds.), *Multiple muscle systems* (pp. 57-68). New York: Springer-Verlag.

Feldman, A.G. (1974). Control of length of a muscle. *Biophysics, 19*, 776-771.

Feldman, A.G., Adamovich, S.V., Ostry, D.J., & Flanagan, J.R. (1990). The origin of electromyograms—Explanations based on the equilibrium point hypothesis. In J.M. Winters & S.L-Y. Woo (Eds.), *Multiple muscle systems* (pp. 195-213). New York: Springer-Verlag.

Gielen, S., Ingen Schenau, G.-J. van, Tax, T., & Theeuwen, M. (1990). The activation of mono- and bi-articular muscles in multi-joint movements. In J.M. Winters & S.L-Y. Woo (Eds.), *Multiple muscle systems* (pp. 302-311). New York: Springer-Verlag.

Gottlieb, G.L., Corcos, D.M., & Agarwal, G.C. (1989). Strategies for the control of single mechanical degree of freedom voluntary movements. *Behavior and Brain Science, 12*, 189-210.

Hannaford, B., & Winters, J.M. (1990). Actuator properties and movement control: Biological and technological models. In J.M. Winters & S.L-Y. Woo (Eds.), *Multiple muscle systems* (pp. 101-120). New York: Springer-Verlag.

Happee, R. (1992). *The control of shoulder muscles during goal directed movements*. Doctoral thesis, Delft University of Technology, The Netherlands.

Hasan, Z. (1983). A model of spindle afferent response to muscle stretch. *Journal of Neurophysiology, 49*, 989-1006.

Hasan, Z., Enoka, R.M., & Stuart, D.G. (1985). The interface between biomechanics and neurophysiology in the study of movement: Some recent approaches. *Exercise and Sport Sciences Reviews*, **13**, 169-234.

Hatze, H. (1977). A myocybernetic control model of skeletal muscle. *Biological Cybernetics*, **25**, 103-119.

Hatze, H. (1981). *Myocybernetic control models of skeletal muscle*. Pretoria: University of South Africa Press.

Hemami, H. (1985). Modeling, control, and simulation of human movement. *CRC Critical Reviews in Biomedical Engineering*, **13**, 1-34.

Hill, A.V. (1938). The heat of shortening and the dynamic constants of muscle. *Proceedings of the Royal Society of London*, **126B**, 136-195.

Hof, A.L. (1990). Effects of muscle elasticity in walking and running. In J.M. Winters & S.L-Y. Woo (Eds.), *Multiple muscle systems* (pp. 182-194). New York: Springer-Verlag.

Hogan, N. (1984). Adaptive control of mechanical impedence by coactivation of antagonistic muscles. *IEEE Transactions on Automatic Control*, **AC-29**, 681-690.

Hogan, N. (1990). Mechanical impedance of single- and multiarticular systems. In J.M. Winters & S.L-Y. Woo (Eds.), *Multiple muscle systems* (pp. 149-164). New York: Springer-Verlag.

Hogan, N., & Winters, J.M. (1990). Principles underlying movement organization: Upper limb. In J.M. Winters & S.L-Y. Woo (Eds.), *Multiple muscle systems* (pp. 182-194). New York: Springer-Verlag.

Houk, J.C. (1979). Regulation of stiffness by skeletomotor reflexes. *Annual Review of Physiology*, **41**, 99-114.

Houk, J.C., & Rymer, Z.W. (1981). Neural control of muscle length and tension. In V.B. Brooks (Ed.), *Handbook of physiology, Sec. 1, Vol. II, The nervous system: Motor control, Part I* (pp. 257-323). Baltimore: Williams & Wilkins.

Huxley, A.F. (1957). Muscle structure and theories of contraction. *Progress in Biophysics and Biophysical Chemistry*, **7**, 257-318.

Ingen Schenau, G.J. van, Bobbert, M.F., & Soest, A.J. van (1990). The unique action of biarticular muscles in leg extensions. In J.M. Winters & S.L-Y. Woo (Eds.), *Multiple muscle systems* (pp. 639-652). New York: Springer-Verlag.

Joyce, G.C., Rack, R.M.H., & Westbury, D.R. (1969). The mechanical properties of cat soleus muscles during controlled lengthening and shortening movements. *Journal of Physiology*, **204**, 461-467.

Katayama, M., & Kawato, M. (1993). Virtual trajectory and stiffness ellipse during multijoint arm movement predicted by neural inverse models. *Biological Cybernetics*, **69**, 353-362.

Keshner, E.A., & Allum, J.H.J. (1990). Muscle activation patterns coordinating postural stability from head to foot. In J.M. Winters & S.L-Y. Woo (Eds.), *Multiple muscle systems* (pp. 481-497). New York: Springer-Verlag.

Lacquaniti, F., Licata, F., & Soechting, J.F. (1982). The mechanical behavior of the human forearm in response to transient perturbations. *Biological Cybernetics*, **44**, 35-46.

Landau, Y.D. (1979). *Adaptive control. The model reference approach*. New York: Marcel Dekker.

Lehman, S., & Stark, L. (1979). Simulation of linear and nonlinear eye movement models. sensitivity analysis and enumeration studies of time optimal control. *Journal of Cybernetics and Information Science, 2*, 21-43.

Lehman, S., & Stark, L. (1982). Three algorithms for interpreting models consisting of ordinary differential equations: Sensitivity coefficients, sensitivity functions, global optimization. *Mathematical Biosciences, 62*, 107-122.

Liang, D. (1989). *Mechanical response of an anthropomorphic head-neck system to external loading and muscle contraction.* Master's thesis, Arizona State University, Tempe.

Loeb, G. (1984). The control and responses of mammalian muscle spindles during normally executed motor tasks. *Exercise and Sport Sciences Reviews, 12*, 157-204.

Loeb, G.E., & Levine, W.S. (1990). Linking musculoskeletal mechanics to sensorimotor neurophysiology. In J.M. Winters & S.L-Y. Woo (Eds.), *Multiple muscle systems* (pp. 165-181). New York: Springer-Verlag.

McMahon, T.A. (1984). *Muscles, reflexes, and locomotion.* Princeton, NJ: Princeton University Press.

Mohler, R.R. (1991). *Nonlinear systems, Vol. 2: Applications to bilinear control.* Englewood Cliffs, NJ: Prentice Hall.

Mungiole, M. (1991). *Factors influencing the mechanical output of the ankle plantarflexor muscles during concentric action, with and without prior stretching.* Doctoral dissertation, Arizona State University, Tempe.

Mungiole, M., & Winters, J.M. (1990). Overview: Influence of muscle on cyclic and propulsive movements involving the lower limb. In J.M. Winters & S.L-Y. Woo (Eds.), Multiple muscle systems (pp. 550-567). New York: Springer-Verlag.

Peterson, B.W., & Richmond, F.J. (1988). *Control of head movement.* New York: Oxford University Press.

Rack, P.M.H., & Westbury, D.R. (1974). The short-range stiffness of active mammalian muscle and its effect on mechanical properties. *Journal of Physiology, 240*, 331-350.

Seif-Naraghi, A.H. (1989). *Predicted optimized neuromuscular control strategies for single-joint goal-directed movements.* Doctoral dissertation, Arizona State University, Tempe.

Seif-Naraghi, A.H., & Winters, J.M. (1989a). Effect of task-specific linearization on musculoskeletal system control strategies. *ASME Biomechanics Symposium,* **AMD-98**, 347-350.

Seif-Naraghi, A.H., & Winters, J.M. (1989b). Changes in musculoskeletal control strategies with loading: Inertial, isotonic, random. *ASME Biomechanics Symposium,* **AMD-98**, 351-354.

Seif-Naraghi, A.H., & Winters, J.M. (1990). Optimized strategies for scaling goal-directed dynamic limb movements. In J.M. Winters & S.L-Y. Woo (Eds.), *Multiple muscle systems* (pp. 312-334). New York: Springer-Verlag.

Seireg, A., & Arvikar, R.J. (1989). *Biomechanical analysis of the musculoskeletal structure for medicine and sport.* New York: Hemisphere.

Van der Helm, F.C.T. (1991). *The shoulder mechanism: A dynamic approach.* Doctoral thesis, Delft University of Technology, The Netherlands.

Van der Helm, F.C.T., & Winters, J.M. (1994). Optimized workspace postures for a large-scale upper limb system: neuro-mechanical mapping and "field" possibilities. In *Proceedings of the 13th Southern Biomedical Engineering Conference*. Washington, D.C.: University of the District of Columbia.

Winters, J.M. (1985). *Generalized analysis and design of antagonistic muscle models: Effect of nonlinear properties on the control of human movement.* Doctoral dissertation, University of California, Berkeley.

Winters, J.M. (1990). Hill-based muscle models: A systems engineering perspective. In J.M. Winters & S.L-Y. Woo (Eds.), *Multiple muscle systems* (pp. 69-93). New York: Springer-Verlag.

Winters, J.M. (1991). Optimized strategies for goal-directed human movements. In J. Menon (Ed.), *Trends in biological cybernetics* (pp. 13-25). Sreekanteswaram, India: Council of Scientific Research Integration.

Winters, J.M., & Mullins, P.A. (1993). Synthesized neural/biomechanical models used for realistic 3-D tasks are more likely to provide insights into human movement strategies (commentary). *Behavior and Brain Science,* **15**, 805-807.

Winters, J.M., Osterbauer, P., Peles, J.D., Derickson, K., Debur, K., & Fuhr, A. (1993). 3-D head axis of rotation during tracking movements: A tool for assessing neuro-mechanical neck function. *Spine,* **18**, 1178-1185.

Winters, J.M., & Peles, J.D. (1990). Neck muscle activity and 3-D head kinematics during quasi-static and dynamic tracking movements. In J.M. Winters & S.L-Y. Woo (Eds.), *Multiple muscle systems* (pp. 461-480). New York: Springer-Verlag.

Winters, J.M., & Stark, L. (1985). Analysis of fundamental movement patterns through the use of in-depth antagonistic muscle models. *IEEE Transactions on Biomedical Engineering,* **BMER-32**, 826-839.

Winters, J.M., & Stark, L. (1987). Muscle models: What is gained and what is lost by varying model complexity. *Biological Cybernetics,* **55**, 403-420.

Winters, J.M., & Stark, L. (1988). Simulated mechanical properties of synergistic muscles involved in movements of a variety of human joints. *Journal of Biomechanics,* **12**, 1027-1042.

Winters, J.M., Stark, L., & Seif-Naraghi, A.H. (1988). An analysis of the sources of muscle-joint system impedence. *Journal of Biomechanics,* **12**, 1011-1025.

Winters, J.M., & Van der Helm, F.C.T. (1993). Comparing simulation approaches for the shoulder: inverse static, inverse dynamic, forward static, forward dynamic. In *Proceedings of the 15th Annual International Conference of the IEEE Engineering in Medicine and Biology and Society* (pp. 1153-1154). Piscataway, NJ: IEEE.

Winters, J.M., & Van der Helm, F.C.T. (1994). Relations between stability, redundancy, and optimization for postural neuro-mechanical systems: principles. In *Proceedings of the 13th Southern Biomedical Engineering Conference*. Washington, D.C.: University of the District of Columbia.

Wu, C-H., Houk, J.C., Young, K-Y., & Miller, L.E. (1990). Nonlinear damping of limb motion. In J.M. Winters & S.L-Y. Woo (Eds.), *Multiple muscle systems* (pp. 214-235). New York: Springer-Verlag.

Yamaguchi, G.T. (1990). Performing whole-body simulations of gait with 3-D, dynamic musculoskeletal models. In J.M. Winters & S.L-Y. Woo (Eds.), *Multiple muscle systems* (pp. 663-679). New York: Springer-Verlag.

Zahalak, G.I. (1990). Modeling muscle mechanics (and energetics). In J.M. Winters & S.L-Y. Woo (Eds.), *Multiple muscle systems* (pp. 1-23). New York: Springer-Verlag.

Zajac, F., & Gordon, M.E. (1989). Determining muscle's force and action in multiarticular movement. *Exercise and Sport Sciences Reviews,* **17**, 187-230.

Zajac, F., & Winters, J.M. (1990). Modeling musculoskeletal movement systems: Joint and body-segment dynamics, musculotendinous actuation, and neuro-muscular control. In J.M. Winters & S.L-Y. Woo (Eds.), *Multiple muscle systems* (pp. 121-148). New York: Springer-Verlag.

Zangemeister, W.H., Lehman, S., & Stark, L. (1982). Simulation of head move-ment trajectories: Model and fit to main sequence. *Biological Cybernetics,* **41**, 19-32.

# Part III

---

# Implementation and Scope of Three-Dimensional Analysis

---

# Chapter 13

# Musculoskeletal Applications of Three-Dimensional Analysis

*Michael W. Whittle*

Most of the early research in three-dimensional biological measurement had little practical application, although the authors often expressed hopes that improvements in clinical care might result from their methods of measurement (Whittle, 1987). Happily, that situation has now changed, and there are inarguable clinical benefits and the expectation of an increasing number of useful applications in other clinical areas in the future.

The clinical world has been slow to realize the benefits of three-dimensional measurement for a number of reasons. Probably the most important is that many clinicians are unfamiliar with the technical process and are perhaps a little afraid of it. The high cost of modern analytical systems has also deterred many who would like to explore the clinical use of biomechanical data. The most progress has been made in centers where there is an active collaboration between engineers, who can make the measurements and calculate the results, and clinicians, who are able to apply the resulting data to the clinical situation.

The clinical area that has shown the greatest practical benefit from 3-D measurement is gait analysis, in particular when applied to the detailed diagnosis and treatment planning of cerebral palsy. This chapter reviews this application, as well as applications in prosthetics and orthotics, upper limb motion, spinal motion, joint mechanics, and sporting activities.

## GAIT ANALYSIS

After more than 100 years of technical development, gait analysis has now become a useful clinical tool. The techniques of three-dimensional kinetic and kinematic analysis can provide a detailed biomechanical description of normal and pathological gait, but the equipment remains very expensive. Where such equipment is not available, valuable clinical information can still be obtained using simpler methods (New York University, 1986). An electronic bibliography on gait analysis has been published by Vaughan, Besser, Sussman, and Bowsher (1992), and a textbook on the subject has been written by Whittle (1991).

# Normal Gait

Far more has been published on normal gait than on the potentially more worthwhile subject of pathological gait. There are, however, two good reasons to study normal gait—to provide a basis for understanding pathological gait and to provide normative data against which pathological data can be judged. Though the scientific literature contains a great deal of data on normal subjects, this mostly relates to the simpler measures such as the general gait parameters, the timing of the gait cycle, and the sagittal plane joint angles (Murray, 1967; Perry, 1974). Normal ranges have not yet been fully defined for some of the more complex measures that cannot be obtained directly, such as the angles of joints whose axes move relative to the plane of progression, the joint moments, or the limb segment energy transfers. Because many parameters vary with the sex, age, and height of the subject and with the speed of walking, it may be necessary to measure hundreds of subjects to adequately define the normal range.

## Gait Pattern Description

Scientific gait analysis started with the photographic measurements made by Marey and Muybridge in the 1870s. The use of cinephotography dominated the subject for 100 years, until the first practical optoelectronic systems were introduced in the 1970s. Although expensive, the accuracy, speed, and convenience of modern television-computer systems (Whittle, 1982) has now made them the standard measurement tool in the majority of gait analysis facilities.

The scientific study of gait involves the identification of a number of events occurring during the gait cycle, which is defined as the interval of time between any of the repetitive events of walking, typically between one foot contacting the ground and the same foot contacting the ground again. The gait cycle is divided, for each foot, into a *stance phase* (when the foot is on the ground) and a *swing phase* (when it is not; see Figure 13.1). In the normal individual, the stance phase commences when the heel contacts the ground. This is followed by *foot-flat*, when the rest of the foot comes down. The time at which the swing-phase foot passes the stance-phase foot is *mid stance*. It is followed by *heel-off*, when the heel lifts and the body weight is taken by the forefoot. The stance phase ends at *toe-off*, when the remainder of the foot leaves the ground. The swing phase is divided into an acceleration phase and a deceleration phase, with *mid swing*, which corresponds to mid stance, being the time at which the two feet are side by side. The most comprehensive description of normal walking is that given by Inman, Ralston, and Todd (1981). The gait cycle has also been described in some detail by Whittle (1991) and Perry (1992).

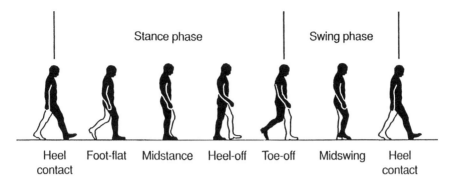

**Figure 13.1** Positions of the legs during a single gait cycle from right heel contact to right heel contact.
*Note*: From *Gait Analysis: An Introduction*, by M. Whittle, 1991, Oxford: Butterworth-Heinemann. Copyright 1991 by Michael W. Whittle. Reprinted by permission.

Because the foot is on the ground for longer than it is in the air, there is an overlap between the stance phases on the two sides—the *double-support time*, when both feet are on the ground at the same time (Figure 13.2). The *single-support time* for one foot is the time when only that foot is on the ground—during the swing phase of the other foot.

From the age at which walking begins, the gait pattern matures rapidly at first, then more slowly, until the adult pattern is reached essentially at the age of 4 years (Sutherland, Olshen, Biden, & Wyatt, 1988). From this age onwards, there are changes in stride length and cadence, which relate to growth, but there is little change in the underlying pattern. In old age, there is a slowing down of gait, but most of the observed changes relate to the speed of walking, rather than to any specific abnormalities of old age. However, the incidence of pathology affecting gait becomes much higher with advancing age, and thus many old people walk with an abnormal gait pattern (Cunha, 1988; Murray, Kory, & Clarkson, 1969).

## General Gait Parameters and Joint Angles

The successive positions of the feet on the ground define a number of gait parameters (Figure 13.3). The number of separate steps taken in a certain period of time (typically, 1 minute) is called the *cadence*. The distance the body as a whole moves forward in a given time can be used to calculate the *walking velocity*.

The *step length* is the distance by which each foot, in turn, advances in front of the other one. Two step lengths, added together, make the *stride length*, which is the distance by which either foot moves forward during the gait cycle. Other foot placement parameters are the *walking base*, which is the side-to-side distance between the line of the two feet, and the *angle of toe-out* (or less commonly, *toe-in*), which is measured between the midline of the foot and the direction of the walk.

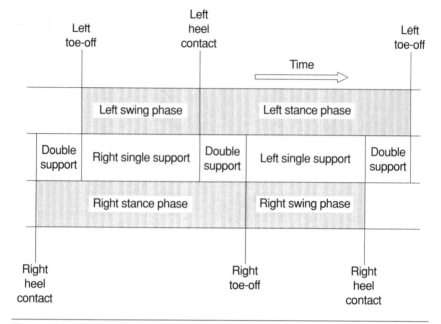

**Figure 13.2**   Timing of single and double support during a single gait cycle from right heel contact to right heel contact.
*Note*: From *Gait Analysis: An Introduction*, by M. Whittle, 1991, Oxford: Butterworth-Heinemann. Copyright 1991 by Michael W. Whittle. Reprinted by permission.

If suitable limb markers are used and the necessary software is available, 3-D kinematic systems can measure many of the characteristics of the gait cycle by following the movements of the trunk and the two feet. Such measurements include the general gait parameters (cadence, stride length, and velocity), the step length, the walking base, the toe-out angle, the timing of the single- and double-support phases, and the timing of the events of foot-flat and heel-off.

Initially, the data from 3-D kinematic systems were used primarily to produce stick figure diagrams of the positions of the limbs during walking (Figure 13.4). Although visually appealing, this type of presentation is of limited value in interpreting the results of gait analysis. Of far greater use are plots of joint angles and other parameters related to the function of the muscles and joints. Three-dimensional kinematic systems make measurements of the angles of motion about the major joints in three dimensions. The greatest range of motion occurs in the sagittal plane, and data from this plane are the most often used (Figure 13.5), but both coronal and transverse plane motions may have significance, particularly in studies of pathological gait. Unfortunately, the clinical definitions of 3-D joint motion (flexion-extension, abduction-adduction, and internal-external rotation) are imprecise in scientific terms and thus are difficult or impossible to reconcile completely with measured 3-D joint rotations. Despite many attempts to devise systems of axes that correspond to the clinical terminology, there is little agreement on how this can best be achieved.

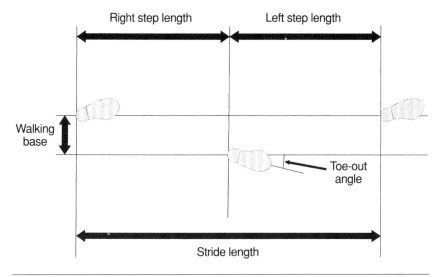

**Figure 13.3**   Terms used to describe foot placement on the ground.
*Note*: From *Gait Analysis: An Introduction*, by M. Whittle, 1991, Oxford: Butterworth-Heine-mann. Copyright 1991 by Michael W. Whittle. Reprinted by permission.

## Intersegmental Resultant Forces and Moments

A kinematic system gives only part of the data required for a detailed biomechani-cal analysis of human movement. The further information needed to measure musculoskeletal forces is provided by a kinetic system. This most commonly consists of one or more force platforms, which measure the ground reaction force during walking and other activities. If both kinetic and kinematic systems are used, the resulting 3-D data, in a common coordinate system, can provide the input for mathematical models, to derive joint moments, joint forces, and the details of energy transfer between limb segments (Bresler & Frankel, 1950; Cavagna & Margaria, 1966; Paul, 1966). Measurements of forces, moments, and energy have proved to be particularly valuable research tools and have also shown their value in many clinical conditions.

Figure 13.6 shows the internal moments of force generated about the hip, knee, and ankle joints in the sagittal plane during walking (Winter, 1987). The existence of an internal moment does not necessarily imply muscular contraction, because internal moments may also be generated by "passive" structures, such as liga-ments. However, multiplying the moment of force by the angular velocity about the joint gives the power generated or absorbed, which relates more directly to muscular activity and to the storage of energy (Winter, Quanbury, & Reimer, 1976).

Mathematical modeling can provide an estimate of joint forces, but in most cases, there exists a situation of indeterminacy, so that it is generally impossible to derive a unique, exact solution to the large number of simultaneous equations involved. Thus, it is necessary to make simplifying assumptions, which may or

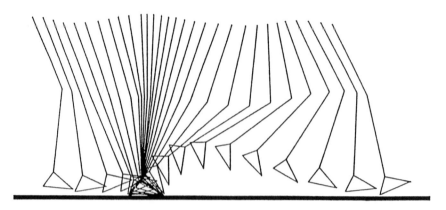

**Figure 13.4**   Position of right leg at 40-ms intervals during a little more than one gait cycle.

*Note*: From *Gait Analysis: An Introduction*, by M. Whittle, 1991, Oxford: Butterworth-Heinemann. Copyright 1991 by Michael W. Whittle. Reprinted by permission.

may not be correct. For instance, it may be assumed that some form of optimization exists or that there is no cocontraction of antagonistic muscles. In some cases, this uncertainty about muscle activity can be reduced, for example, by using electromyography to determine whether a particular muscle is contracting.

Although the study of limb mechanics has concentrated on the sagittal plane, there is now an increasing realization that for a complete picture, it is necessary to make 3-D measurements. In the knee, for example, there is a differential loading between the medial and lateral compartments that depends largely on the coronal-plane knee moment (Goodfellow & O'Connor, 1978). The extent to which arthritic symptoms can be relieved by high tibial osteotomy now can be predicted by making 3-D measurements of the joint angle and moment.

## Muscle Activity

Although it is not a three-dimensional measurement, an important aspect of gait analysis, particularly in a clinical setting, is the measurement of the electrical activity of the muscles by means of electromyography (EMG). Good reviews of the subject were given by Basmajian (1974) and Shiavi (1985). Figure 13.7 shows the typical EMG activity of six muscle groups during the gait cycle (Inman et al., 1981; Perry, 1974; Whittle, 1991). Most commercial 3-D gait analysis systems are able to record EMG data through an analog-to-digital converter while kinematic and force platform data are being acquired. EMG data are normally recorded at a higher sampling frequency than kinematic data. The analytical software must be able to cope with this and to synchronize the two types of data.

The most precise form of EMG recording involves inserting a fine wire into the muscle to be measured, which almost eliminates interference from other sources of electrical activity. However, this is uncomfortable, particularly when there is a lot of muscle movement, such as in walking. Thus, for gait analysis,

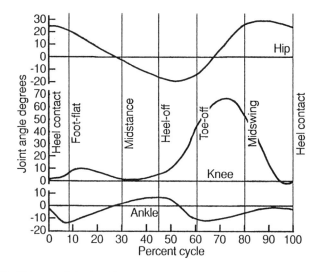

**Figure 13.5**   Angles of hip (flexion positive), knee (flexion positive), and ankle (dorsi-flexion positive) during a single gait cycle.
*Note*: From *Gait Analysis: An Introduction*, by M. Whittle, 1991, Oxford: Butterworth-Heinemann. Copyright 1991 by Michael W. Whittle. Reprinted by permission.

it is more usual to use the less accurate but more comfortable surface electrodes, unless there is a particular need for the detail provided by fine-wire measurements.

One of the disadvantages of EMG is that it is only semiquantitative. Muscles are able to develop tension while they are shortening (concentric contraction), maintaining length (isometric contraction), or lengthening (eccentric contraction). The relationship between electrical activity and muscle tension differs for these three types of contraction, as well as being nonlinear with marked hysteresis and varying with the recent contraction history of the muscle (Perry & Bekey, 1981). Many attempts have been made over the years to define the EMG-force relationship, but this has proved impossible, except under extremely artificial laboratory conditions.

On the other hand, EMG is helpful in gait analysis for providing an explanation of those gait abnormalities observed by a kinematic system. Ultimately, it is the muscles that are responsible for gait, and examining the EMG makes it possible to identify the "guilty motor patterns" in a pathological gait (Winter, 1985).

## Work, Energy, and Power

One of the major characteristics of many pathological gaits is that they are inefficient in terms of energy consumption and cause rapid fatigue in the patient. Analysis of the energy consumption involved in walking is a routine measurement in many gait laboratories, and its reduction is used as one indication of successful treatment (Waters, Lunsford, Perry, & Byrd, 1988). The usual means of measuring metabolic energy consumption is by collecting the expired air, using a face mask

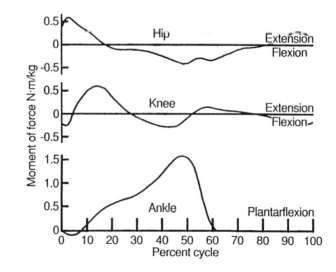

**Figure 13.6**  Internal moments about hip, knee, and ankle joints, in newton-meters per kilogram body weight. Positive moments are generated by hip extensors, knee extensors, and ankle plantarflexors.

*Note*: From *Gait Analysis: An Introduction*, by M. Whittle, 1991, Oxford: Butterworth-Heinemann. Copyright 1991 by Michael W. Whittle. Reprinted by permission.

or mouthpiece, and calculating the oxygen used and the carbon dioxide produced during a given activity. It is, however, theoretically possible to make the same type of measurement using a combined kinetic-kinematic system, calculating the changes in energy of each body segment during the walking cycle. This type of estimation (the *estimated external work* of walking) also requires the masses, centers of gravity, and moments of inertia of the body segments. However, there is not a direct correspondence between metabolic activity and physical work. Muscles use metabolic energy even when performing an eccentric contraction, which in physical terms is negative work. For this reason, the results of such an analysis are unlikely to correspond exactly to expired gas measurements, although they may still give clinically useful information (Gage, Fabian, Hicks, & Tashman, 1984).

Perhaps of more value is the analysis of the changes in energy in the segments of the limbs during walking. The normal gait cycle is very efficient in terms of energy usage, because of a number of optimizations and energy exchanges between the limb segments. In pathological gait, these processes frequently break down, and the cause of the overall increased energy consumption can be deduced from the changes that occur within the limbs.

## Pathological Gait

The study of pathological gait may be carried out for a number of reasons, of which the most important are clinical decision making, assessment and documentation, and clinical research.

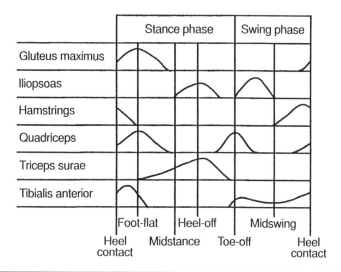

**Figure 13.7**    Typical activity of major muscle groups during the gait cycle.
*Note*: From *Gait Analysis: An Introduction*, by M. Whittle, 1991, Oxford: Butterworth-Heinemann. Copyright 1991 by Michael W. Whittle. Reprinted by permission.

## Clinical Decision Making

All clinical decision making, whether it involves a gait disorder or not, relies on three types of information—a description of the nature and history of the problem (obtained by interview from the patient or caretakers), a physical examination that concentrates on the particular areas likely to be affected, and several laboratory tests or special investigations. In this context, gait analysis should be thought of as a special investigation, the results of which must be considered along with the history and examination as well as any other investigations that have been made, such as X-rays or blood biochemistry (Rose, 1983). Gait analysis is particularly valuable in making therapeutic decisions, because it can yield information on the motor deficit that underlies a visible gait disorder.

**Cerebral Palsy.**    One of the most important clinical applications of gait analysis at present is in planning treatment for children affected by cerebral palsy. This topic is considered in chapter 16.

**Stroke.**    The clinical syndrome called *stroke* is a result of the brain damage that occurs after an interruption of the cerebral blood supply. It often includes difficulty with walking (Perry, 1969), due to paralysis affecting one side of the body (hemiplegia or hemiparesis). Although gait analysis has not yet clearly shown a benefit in the management of stroke, a number of research studies have suggested that an accurate assessment of the gait disorder may improve the decision-making process and lead to an improved treatment outcome.

**Excessive Fatigue.**    In normal walking, the body uses six different mechanisms (the *determinants of gait*) to reduce energy consumption (Inman et al., 1981; Saunders, Inman & Eberhart, 1953):

1. Pelvic rotation
2. Pelvic tilt
3. Knee flexion in stance phase
4. Ankle mechanism
5. Foot mechanism
6. Lateral displacement of the body

In pathological gait, these mechanisms frequently break down, leading to an increased energy consumption and more rapid fatigue. Gait analysis may be used to detect the loss of these mechanisms, and gait training may be of use in attempting to restore them.

**Parkinsonism.**    The medical treatment of parkinsonism has undergone a revolution in recent years, with the introduction of the drug L-Dopa. Because it is not always easy to establish the correct dosage for a particular patient, gait analysis has been used to monitor the patient's performance, in order to more properly titrate the dosage (Klenerman, Dobbs, Weller, Leeman, & Nicholson, 1988).

**Orthopaedic Conditions.**    Gait analysis may be beneficial in treating some orthopaedic conditions. When a patient has arthritis that affects several joints and it is decided to perform a number of joint replacements, it is not always obvious which joint should be corrected first. Pain or lack of function in one joint frequently causes an alteration in the gait pattern, which in turn puts increased stress on a second joint. If this second joint is replaced before the nonfunctional joint, there is a chance of early failure, due to the adverse mechanical environment. Gait analysis has been used to detect this type of effect and to indicate the optimum order for joint replacement (Rose, 1983). Another orthopaedic condition that may benefit from preoperative assessment by gait analysis is unicompartmental arthritis of the knee. Realigning the joint by means of high tibial osteotomy may give good relief of symptoms and may postpone by some years the need for knee joint replacement. However, the outcome of this procedure is extremely variable. Gait analysis, which concentrates on the knee angle and moment in the coronal plane, gives a good indication of the likely results of the operation (Prodromos, Andriacchi, & Galante, 1985; Wang, Kuo, Andriacchi, & Galante, 1990).

## Assessment and Documentation

Gait analysis is often used to assess and document the current condition of a patient, even when the data are not used directly as part of the clinical decision-making process. Such assessment may be performed only once, or it may be done repeatedly to monitor changes over a period of time.

Assessment on a single occasion may be useful to confirm a functional diagnosis. For example, it may be used to distinguish between inadequate anterior tibial activity and triceps surae overactivity in a child who walks with a plantarflexed foot. The strength and activity of these muscles during walking may differ markedly from their state when tested manually, and gait analysis may be needed to make an accurate diagnosis. The measurement of muscle power output is particularly useful in this type of study. Gait analysis on a single occasion may also be used to ensure that there are no other gait abnormalities that might have been missed in the course of the routine clinical examination.

Repeated gait analysis over a period of time may be valuable in two ways— either to document the natural course of a disease or to monitor the effects of treatment. Whether a patient's condition is improving or worsening, objective evidence on the rate of change helps the clinician provide a prognosis, which may either assist in treatment planning or indicate when a particular stage in the condition is likely to be reached. For example, gait analysis may provide an estimate of how long it will be before a boy with Duchenne muscular dystrophy becomes unable to walk, or it may give an idea of the likely final outcome for someone recovering from paralysis due to spinal cord injury.

When used to monitor the effects of treatment, gait analysis provides an accurate, objective means of measuring function, which may be of value in a number of conditions and treatments. It may be used to titrate drug dosage in parkinsonism, or to decide whether nonsteroidal anti-inflammatory drugs are adequately controlling a patient's rheumatoid arthritis. It may also be used before and after a surgical procedure, to measure the extent to which surgery has been beneficial.

Litigation is another area in which gait analysis may make an important contribution (Whittle, 1991). By providing objective data, it can remove some of the uncertainty in claims for damages when someone's ability to walk has been affected. It may also be of value in claims of negligence against medical practitioners, by recording a patient's condition before treatment and showing that the best possible information was taken into account when planning treatment.

## Research

Even greater than the impact of gait analysis on clinical practice has been its impact on research. The research applications of gait analysis fall into two categories—clinical and fundamental.

**Clinical Research.**    Clinical research is designed to answer questions about disease processes or therapeutic methods that can be expected to lead directly to improvements in treatment. There is obviously an overlap between the use of gait analysis in clinical practice and its use in clinical research, although in the former case the patient whose gait is being examined may expect to benefit directly, whereas in the latter case, only future patients can benefit. The features of gait analysis that make it valuable as a research tool are its objectivity and its ability to measure things that cannot be felt by the patient or seen by an observer. Gait analysis may be particularly valuable in providing evidence to confirm (or refute) the efficacy of a particular treatment. Sometimes, gait analysis will fail to demonstrate either a benefit or a lack of benefit from a particular form of treatment, because of inadequate statistical power and precision in the experiment. Alternatively, the degree of musculoskeletal "challenge" may be inadequate to make the distinction evident. For example, concerning the use of orthoses to protect the joints during sporting activities, gait analysis may be unable to either demonstrate or refute any benefit of such devices.

**Fundamental Research.**   Fundamental scientific research differs from clinical research in that it is carried out without a clear idea of the benefits that might result. The methodology of 3-D kinetic and kinematic measurement provides a powerful research technique in several fields of investigation. Although a great deal is already known about the walking process, it continues to be studied, particularly by those interested in motor control. Much has still to be learned about the coordination of the many muscles that act during the gait cycle and the response of the system to perturbation. In the discipline of biomechanics, mathematical modeling of the muscles, bones, and joints requires input of 3-D kinetic and kinematic data, and the models allow predictions that can be tested by further data collection. As well as being of value in investigating walking, 3-D measurement is used in fundamental research on a range of other human activities, including spinal motion and athletic activities.

# PROSTHETICS AND ORTHOTICS

In the design and testing of prosthetic limbs, accurate 3-D measurement is of great value in three critical areas—socket design (Boone & Burgess, 1989), foot mechanisms (Wagner, Seinko, & Susan, 1987), and knee mechanisms (Murray, Mollinger, Sepic, Gardner, & Linder, 1983).

Orthoses, which are external supports for some part of the body, generally have enjoyed much less success than prosthetic limbs, and 3-D measurement systems have been used to identify their shortcomings (Lehmann, Condon, Price, & deLateur, 1987; Perry, 1974).

# UPPER LIMBS

Three-dimensional measurement systems are used to study the upper limbs, as well as the lower. Upper limb motion is of particular interest in the field of rehabilitation, when dysfunction of the arm makes it impossible for a patient to perform some of the necessary activities of daily living and treatment is aimed at restoring these lost functions. Studies have been made of complicated 3-D movements, such as that of the shoulder joint and scapula (Engin & Peindl, 1987). Because the movements of upper limbs are usually fairly small, they may be studied using simple electromechanical devices, such as the vector stereograph (Grew & Harris, 1979).

# SPINE

In the musculoskeletal applications of 3-D measurement, the study of the spine ranks second only in importance to gait analysis. The two major areas of interest in this field are back shape (which is outside the scope of this chapter) and spinal motion (Parnianpour, Nordin, & Kahanovits, 1988; Thurston, Whittle, & Stokes, 1981).

# JOINT MECHANICS

Three-dimensional measurement has been used to study the kinesiology of natural joints under different loading conditions, both in isolated preparations and in vivo. However, the measurement of joint motion using skin markers is fairly inaccurate, and few "heroic" studies have been performed in which pins have been inserted into the adjacent bones (Gregerson & Lucas, 1967; Lafortune & Cavanagh, 1985). Another type of 3-D measurement of joint functioning is the instrumented clinical examination, in which the motion of the limbs is monitored as the examiner puts the joints through a range of motion (Steiner, Brown, & Zarins, 1990).

The detailed biomechanical information provided by 3-D measurement systems and the associated mathematical modeling has had a major impact on the design of prosthetic joints (Rose, 1983). Three-dimensional measurement systems also have been used to study the in vivo biomechanics of different prostheses and surgical procedures (Jefferson & Whittle, 1989).

# SUMMARY

Gait analysis, using three-dimensional kinetic and kinematic measurement systems, is now an established technique of major importance in the clinical management of certain conditions, particularly cerebral palsy. Both gait analysis and other types of 3-D measurement have shown themselves to be valuable tools for both clinical and fundamental research on a variety of musculoskeletal disorders.

# REFERENCES

Basmajian, J.V. (1974). *Muscles alive: Their functions revealed by electromyography*. Baltimore: Williams & Wilkins.

Boone, D.A., & Burgess, E.M. (1989). Automated fabrication of mobility aids: Clinical demonstrations of the UCL computer-aided socket design system. *Journal of Prosthetics and Orthotics*, **1**, 187-190.

Bresler, B., & Frankel, J.P. (1950). The forces and moments in the leg during level walking. *American Society of Mechanical Engineers Transactions*, **72**, 27-36.

Cavagna, G.A., & Margaria, R. (1966). Mechanics of walking. *Journal of Applied Physiology*, **21**, 271-278.

Cunha, U.V. (1988). Differential diagnosis of gait disorders in the elderly. *Geriatrics*, **43**, 33-42.

Engin, A.E., & Peindl, R.D. (1987). On the biomechanics of the human shoulder complex—1: Kinematics for determination of the shoulder complex sinus. *Journal of Biomechanics*, **20**, 103-117.

Gage, J.R., Fabian, D., Hicks, R., & Tashman, S. (1984). Pre- and postoperative gait analysis in patients with spastic diplegia: A preliminary report. *Journal of Pediatric Orthopedics*, **4**, 715-725.

Goodfellow, J., & O'Connor, J. (1978). The mechanics of the knee and prosthesis design. *Journal of Bone & Joint Surgery*, **60B**, 358-369,

Gregerson, G.G., & Lucas, D.B. (1967). An in vivo study of the axial rotation of the human thoracolumbar spine. *Journal of Bone & Joint Surgery*, **49A**, 247-262.

Grew, N.D., & Harris, J.D. (1979). A method of studying human body shape and movement—The "vector stereograph." *Engineering in Medicine*, **8**, 115-118.

Inman, V.T., Ralston, H.J., & Todd, F. (1981). *Human walking*. Baltimore: Williams & Wilkins.

Jefferson, R.J., & Whittle, M.W. (1989). Biomechanical assessment of unicompartmental knee arthroplasty, total condylar arthroplasty, and tibial osteotomy. *Clinical Biomechanics*, **4**, 232-242.

Klenerman, L., Dobbs, R.J., Weller, C., Leeman, A.L., & Nicholson, P.W. (1988). Bringing gait analysis out of the laboratory and into the clinic. *Age & Ageing*, **17**, 397-400.

Lafortune, M.A., & Cavanagh, P.R. (1985). The measurement of normal knee joint motion during walking using intracortical pins. In M. Whittle & D. Harris (Eds.), *Biomechanical measurement in orthopaedic practice* (pp. 234-243). Oxford: Clarendon Press.

Lehmann, J.F., Condon, S.M., Price, R., & deLateur, B.J. (1987). Gait abnormalities in hemiplegia: Their correction by ankle-foot orthoses. *Archives of Physical Medicine & Rehabilitation*, **68**, 763-771.

Murray, M.P. (1967). Gait as a total pattern of movement. *American Journal of Physical Medicine*, **46**, 290-333.

Murray, M.P., Kory, R.C., & Clarkson, B.H. (1969). Walking patterns in healthy old men. *Journal of Gerontology*, **24**, 169-178.

Murray, M.P., Mollinger, L.A., Sepic, S.B., Gardner, G.M., & Linder, M.T. (1983). Gait patterns in above-knee amputee patients: hydraulic swing control vs. constant-friction knee components. *Archives of Physical Medicine & Rehabilitation*, **64**, 339-345.

New York University. (1986). *Lower limb orthotics*. New York: Prosthetics and Orthotics, NYU Postgraduate Medical School.

Parnianpour, M., Nordin, M., & Kahanovits, N. (1988). The triaxial coupling of torque generation of trunk muscles during isometric exertions and the effect of fatiguing isoinertial movements on the motor output and movement patterns. *Spine*, **13**, 982-992.

Paul, J.P. (1966). Forces transmitted by joints in the human body. *Proceedings of the Institute of Mechanical Engineers*, **181**, 8-15.

Perry, J. (1969). The mechanics of walking in hemiplegia. *Clinical Orthopaedics & Related Research*, **63**, 23-31.

Perry, J. (1974). Kinesiology of lower extremity bracing. *Clinical Orthopaedics & Related Research*, **102**, 18-31.

Perry, J. (1992). *Gait analysis: Normal and pathological function*. Thorofare, NJ: Slack.

Perry, J., & Bekey, G. (1981). EMG-force relationships in skeletal muscle. *CRC Critical Reviews in Bioengineering*, **7**, 1-22.

Prodromos, C.C., Andriacchi, T.P., & Galante, J.O. (1985). A relationship between gait and clinical changes following high tibial osteotomy. *Journal of Bone & Joint Surgery*, **67A**, 1188-1194.

Rose, G.K. (1983). Clinical gait assessment: A personal view. *Journal of Medical Engineering & Technology*, **7**, 273-279.

Saunders, J.B.D.M., Inman, V.T., & Eberhart, H.S. (1953). The major determinants in normal and pathological gait. *Journal of Bone & Joint Surgery*, **35A**, 543-558.

Shiavi, R. (1985). Electromyographic patterns in adult locomotion: A comprehensive review. *Journal of Rehabilitative Research & Development*, **22**, 85-98.

Steiner, M.E., Brown, C., & Zarins, B. (1990). Measurement of anterior-posterior displacement of the knee: A comparison of the results with instrumented devices and clinical examination. *Journal of Bone & Joint Surgery*, **72A**, 1307-1315.

Sutherland, D.H., Olshen, R.A., Biden, E.N., & Wyatt, M.P. (1988). *The development of mature walking*. London: Mac Keith Press.

Thurston, A.J., Whittle, M.W., & Stokes, I.A.F. (1981). Spinal and pelvic movement during walking—A new method of study. *Engineering in Medicine*, **10**, 219-222.

Vaughan, C.L., Besser, M.P., Sussman, M.D., & Bowsher, K.A. (1992). *Biomechanics of human gait: An electronic bibliography*. Champaign, IL: Human Kinetics.

Wagner, J., Seinko, S., & Susan, T. (1987). Motion analysis of SACH vs. Flex-Foot in moderately active below-knee amputees. *Clinical Prosthetics & Orthotics*, **11**, 55-62.

Wang, J.W., Kuo, K.N., Andriacchi, T.P., & Galante, J.O. (1990). The influence of walking mechanics and time on the results of proximal tibial osteotomy. *Journal of Bone & Joint Surgery*, **72A**, 905-909.

Waters, R.L., Lunsford, B.R., Perry, J., & Byrd, R. (1988). Energy-speed relationship of walking: Standard tables. *Journal of Orthopaedic Research*, **6**, 215-222.

Whittle, M.W. (1982). Calibration and performance of a three-dimensional television system for kinematic analysis. *Journal of Biomechanics*, **15**, 185-196.

Whittle, M.W. (1987). Gait analysis—Its usefulness or otherwise. *British Journal of Rheumatology*, **26**(Abs), 86.

Whittle, M.W. (1991). *Gait analysis: An introduction*. Oxford: Butterworth-Heinemann.

Winter, D.A., Quanbury, A.O., & Reimer, G.D. (1976). Analysis of instantaneous energy of normal gait. *Journal of Biomechanics*, **9**, 253-257.

Winter, D.A. (1985). Concerning the scientific basis for the diagnosis of pathological gait and for rehabilitation protocols. *Physiotherapy Canada*, **37**, 245-252.

Winter, D.A. (1987). *The biomechanics and motor control of human gait*. Waterloo, ON: University of Waterloo Press.

# Chapter 14

# Applications in Ergonomics

*Steven A. Lavender*
*Sudhakar L. Rajulu*

The principles described in the previous chapters are also applicable to the field of industrial ergonomics. The term *ergonomics* stems from the Greek expressions *ergon*, or work, and *nomos*, a system of laws. More specifically, ergonomics refers to the study of work through the application of mechanical principles or laws. The objective of an ergonomic approach is the improved interfacing between people and their environment. Although this includes aspects such as noise, lighting, and ventilation, current ergonomic efforts within industrial environments focus on the biomechanical and cognitive performance requirements. In essence, the objective of an industrial ergonomics program could be narrowly stated to be the reduction of occupational musculoskeletal injuries in terms of both incidence and cost.

With the rapid increase in the cost of health care seen in the last decade, most industries have begun to focus on ways in which these costs can be contained. Many industries have employed disability management programs to get employees quickly back into a productive capacity after injury. However, the most cost-effective approach is through preventive medicine. The purpose of ergonomics is to design a workplace and work procedures such that injuries do not result. Ideally, this is implemented during the initial design of the workplace. In existing work environments, the prevention of occupational injuries must be achieved mainly through changes in job design and work methods.

Several factors appear in the epidemiological literature to be potential contributors to the development of musculoskeletal injuries in the work environment. These factors, shown in Figure 14.1, include body motions, workplace factors, anthropometric factors, and psychosocial factors. For example, the workplace factors contain many of the variables measured in traditional ergonomic evaluations, such as the weights lifted, the frequency of lifting, the height from which the lift is performed, the repetition rate, and the reach distances.

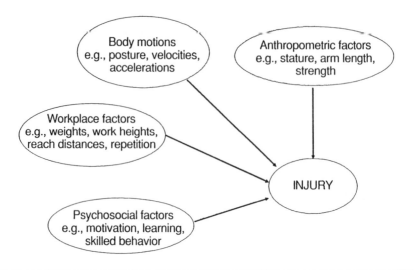

**Figure 14.1**   Factors that contribute to the development of occupational injuries.

In many work environments the constraints on human motion are determined by the workplace. This means that a particular task, by the nature of its design, results in the employee performing the task using a consistent sequence of motions. Under these circumstances, where changes to the work environment would result in changes in the motions necessary to complete the task, the model in Figure 14.2 might be more applicable. In this situation, studying the motions is central to the ergonomic assessment. Jobs in which the tasks are not repetitive and the motions are quite variable should be analyzed in terms of the other factors shown in Figure 14.1 that could be responsible for the injury statistics. The focus of this chapter is on the analysis of workplaces where the motions and postural changes are an integral component in the performance of the occupational task.

As postural demands of the workplace move the body out of an upright, relaxed posture, internal loads on the muscles and joints are increased. This results from the increase in the horizontal moment arm distance from the center of gravity of a limb or the torso to the corresponding joint's axis of rotation. Therefore, even static postural deviations without the application of an external load require an increase in the muscle force necessary to stabilize the posture. When the body is viewed as a collection of levers and mechanical linkages, it becomes clear that the musculoskeletal system generally works at a mechanical disadvantage (Basmajian, 1976). In other terms, the muscles operate over very small lever arms relative to the application point of external loads, or the center of mass of the corresponding body segments. As a result the muscles are the primary source of mechanical loading on the body's joints. Under conditions in which body motion is required, the situation is accentuated. Additional muscle forces are generated that must be large enough to overcome the inertia of the body part(s) and accelerate the body segments into motion. Additional muscles are recruited

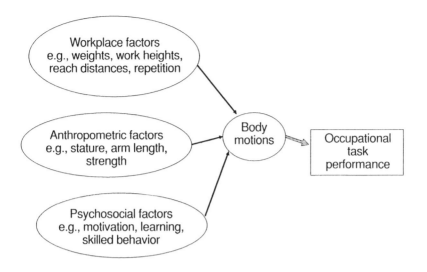

**Figure 14.2**    Factors hypothesized to control body motions in highly repetitive tasks.

to supply force for the purposes of controlling and terminating the motions. When the motions are highly repetitive, this repeated mechanical loading has been related to the onset of cumulative trauma disorders (Armstrong, 1983; Chaffin & Andersson, 1991; Kumar, 1990).

Cumulative trauma represents the gradual wear and tear of structures within the musculoskeletal system. Conversely, an injury with a sudden onset is usually referred to as an acute trauma injury and is usually discussed in terms of accidents and safety issues. In cumulative trauma, body structures become increasingly painful over a period of time, either through repetitive motion or repetitive force application. Ultimately, this process leads to disability, hence, the ergonomic concern.

The current techniques for measuring human motion allow the sampling of many characteristics of joint motion that are important in ergonomic analyses. The complexity of these analyses are dependent on the joint(s) of interest. For example, in the knee and elbow joints there is generally a primary axis about which rotation occurs. However, in more complex joints like the wrist, the shoulder, or the back, the motion takes place about two or more axes. Therefore, in conducting 3-D ergonomic assessments that include joint motions, the complexity will depend in part on the joints being monitored.

Several motion characteristics must be considered when performing ergonomic analyses. These include not just the position of the joint, but also (a) the range of rotation, (b) the velocity at which the joint rotates, and (c) the acceleration of the joint. This chapter seeks to convey the advantages and disadvantages of the techniques available for measuring the motion of body joints in the workplace. Though an analysis of the body motion required by a particular task should enhance the job evaluation process, movements are just one aspect of an ergonomic evaluation. Therefore, this chapter also describes the integration of motion data with the data describing the other aspects of ergonomic assessments. Further, this chapter discusses how this information can be used in the development of ergonomic interventions. Finally, the limitations of kinematic data in the interpretation of an ergonomic analysis are explained.

# MEASURING TECHNIQUES

Though several types of systems have been developed to quantify 3-D motion, there are two types of measuring devices commonly used in the workplace: electromechanical goniometers and video-based motion analysis systems. Both systems can provide joint motion data in terms of angular displacement. This section discusses each of these systems in terms of their basic principles, the supporting hardware necessary for data collection, and some of the considerations in setting up the systems at the worksite.

## Electromechanical Goniometers

Motion monitors that use electrical devices to monitor changes in motion are generally called electromechanical goniometers. Many variations of electrogoniometers exist, but the simplest use electrical potentiometers (or sensors), typically one for each axis of rotation, to measure the angle of joint rotation with respect to a reference position. Changes in joint angle produce changes in the potentiometer resistance, which in turn produce changes in the output voltage. By calibrating the output voltages against known angles of rotation, the signal from the potentiometers can be used to measure the joint positions in the planes of interest. Because the structural aspects of the body's joints differ from one to another, the goniometer's design will vary depending on the joint being studied. All designs require that the goniometers be accurately placed over the joint's axis of rotation. These monitors are usually secured in place using specially designed harnesses and straps. Figure 14.3 shows a goniometer-based system for measuring wrist postures and motion.

## Hardware Requirements for Goniometer Data Collection

At the data collection site, the instruments are typically powered by small lightweight batteries. These supply the voltages across the potentiometers. The signals from each potentiometer are fed, via cable, to an analog-to-digital converter (A/D system) and into a personal computer. The computer and the A/D hardware must be able to sample data at a rate between 60 and 300 samples a second. The selected rate depends on the maximum velocity at which the monitored joint can rotate. For example, wrist motion should be sampled at 300 Hz, whereas torso motion can be adequately described using sampling rates of 100 Hz or even less (Priemer, 1991).

Typically, data is transported from the instrument to the data collection computer via a cable, which must be long enough to allow the worker to move freely.

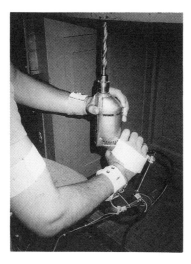

**Figure 14.3** The use of wrist electrogoniometer system for quantifying wrist postures and motions in an assembly task.
*Note.* Photo courtesy of Dr. Richard Schoenmarklin.

Alternatively, a radio telemetry system can be used to transmit the data. The collected data are analyzed by appropriate software at the data-processing site. Some researchers use portable tape recorders, which can be worn by the subjects for the duration of the test (Snijders, van Riel, & Nordin, 1987). Though this method is convenient and inexpensive, it does not allow instantaneous checks of data quality during the test or the immediate evaluation of the task components.

## Set-Up Requirements for Electrogoniometer Testing at the Worksite

Most electrogoniometers need a calibration check before use to ensure that the relationship between the voltage and the angle of joint rotation is within the factory-installed calibration limits. A calibration device is normally provided with the goniometer to accomplish this task. Placement of the instrument on the subject also requires a trained person who can locate the joints from anatomical landmarks. If a radio telemetry system is used for data transport, radio antennas should be properly located around the worker so as to minimize electrical noise and maximize the signal strength.

## Video-Based Motion Analysis Systems

Three-dimensional motion analysis based on video-imaging techniques requires the use of at least two cameras to capture different views of the body motion.

Most systems require the placement of light-emitting or light-reflecting markers over landmarks, such as a joint's center of rotation. Other systems do not use markers but require the user to locate the body's joints manually during data analysis.

Computational algorithms have been created that allow 3-D analysis with arbitrary camera placements. Additional cameras may be used when the body is expected to move outside the view of the primary cameras. The cameras must be synchronized so that images can be coordinated with each other on playback.

The video input from each camera is analyzed individually and then a composite image is produced by combining the synchronized video input. Joint angular position is then derived from the composite image with reference to a specific joint or location within the body. Several joint motions can be monitored simultaneously, provided that all the joints in question are within the view of the cameras.

## Hardware Requirements at the Worksite

At the worksite, two or more cameras with videorecorders are employed to collect the video data. A synchronizer is required to ensure that the cameras are synchronized. A simple method is to use an audio or light signal to start data collection. The cameras should have zoom capability so that the field of view can be adjusted to maximize the resolution of the markers (while keeping the subject in view). Finally, sturdy camera stands are necessary to provide stable images free from camera motion.

Either light-emitting (active) or light-reflecting (passive) markers are necessary for the detection of body segments in the video data. Passive markers have the advantage of keeping the subjects free from encumbering wires. However, the contrast between markers and the background illumination must be maintained. Often the ambient lighting determines what type of marker may be used, and the intensity of the marker illuminations will have to be adjusted to maintain adequate contrast. Poor contrast will generate spurious signals or noise when the video data is processed.

## Set-Up Requirements for Video-Based Systems at the Worksite

The worksite being assessed must be thoroughly evaluated for obstructions to each camera's view. In general, any unnecessary materials surrounding the worker should be removed. Markers must be positioned on the subject over each joint's center of rotation. Where possible, the lighting conditions should be adjusted to levels that will enhance the contrast between markers and the background. (Passive markers can be highlighted using directional lighting.) Measurements of the worksite should be taken to allow for scaling of the video image.

## JOINT MOTION DATA REDUCTION

When electromechanical goniometers are used, the joint position is readily available in the form of voltages. First, the calibration data are used to obtain the angular displacement of the joint. Then the instantaneous velocities and accelerations are calculated using either numerical differentiation or other transformation techniques.

In video-based motion analysis systems, the joint position data can be obtained only after each videotape is processed and all multiple-camera views combined. This can require a considerable amount of time, depending on the type of task performed. The amount of processing time also varies depending on whether joints are indicated by markers or manually located during the video analysis. (Manual selection of the joint centers is typically more time consuming than active or passive marker methods.) The data are combined to obtain the composite image and the joint positions in a 3-D coordinate system. Once the position data are obtained, computation of velocity and acceleration is relatively quick.

# ADVANTAGES AND DISADVANTAGES OF GONIOMETER- AND VIDEO-BASED SYSTEMS

These two alternative methods both have certain limitations, which must be considered when selecting a technique for ergonomic evaluations. Some of these limitations can be overcome by choosing or controlling the variables in the data-collection environment; others are more difficult to control.

## Goniometer-Based Systems

Goniometer systems are useful in studying the motions of a single joint and are generally more accurate than video-based systems. By using a telemetry system or a long cable to transmit data, it is possible to record motions even when the worker has to perform tasks at various work stations. Because the instruments stay with the person, it is easier to monitor motions in places where camera placement would be difficult. Furthermore, it is not necessary to rearrange the workplace to accommodate a goniometer system. The set-up time for goniometers is less than that of video-based motion analysis systems. Joint motion characteristics, such as position, velocity, and acceleration, can be readily obtained at the worksite itself. This is a tremendous benefit when one is providing on-site ergonomic evaluations.

Goniometers require accurate placement such that their centers follow the joint center in parallel. Thus, they are usually strapped or taped over the joint, which may impose limitations on the joint motion. This is a potential source of bias in the motion data collected with electrogoniometers. Firm attachment is required to avoid relative-motion artifacts. Care should also be exercised to prevent any unnecessary physical discomfort that may arise while the subject is wearing the instrument.

Certain joints are better measured by using goniometers than by using video-based techniques. For example, when monitoring wrist motions, hand movement is likely to obstruct the markers used in a video-based system. However, each joint in the body requires specially designed goniometers. This often requires complex designs to capture the rotations about the necessary axes. For example, detecting all axes of rotation in the shoulder presents an especially difficult problem for the design of goniometer systems.

Other liabilities involved with use of goniometers include the need for regular calibration checks, the complexity of measuring motion in multiple joints simultaneously, and the cables or telemetry required for data transmission. In general, goniometers require frequent calibration checks and careful handling. Yet industrial environments often present harsh conditions for operating and maintaining scientific instruments.

When studying the motions in multiple joints, the use of multiple goniometers becomes a limiting factor. The additional instruments will increase the hardware requirements in terms of A/D and computational capacities, the setup time, and the complexities of data collection and analysis.

At times, the long cable used for transmitting data creates unnecessary complications. The cable can become entangled with the materials around the worker, or the worker may perform the task in an unusual manner while attempting to accommodate the cable. Telemetry is a viable alternative; but in industrial settings, telemetry systems may be affected by the electrical background noise and interference from multipath signals.

## Video-Based Motion Analysis Systems

Unlike goniometer systems, which can monitor only one joint at a time, video-based systems can observe several joints simultaneously. This is particularly advantageous when studying whole-body motion. An additional advantage of video-based systems, especially those that employ passive markers, is that they produce little interference with a worker's motions or tasks. Whereas in goniometer methods, the data collection hardware must be attached to the subject, video-based systems require that only small markers be applied. This eliminates any resistance-to-movement or weight concerns that affect some goniometer designs.

Video analysis systems have the additional advantage of being able to superimpose the results of motion onto the actual video images. Such a presentation is beneficial in terms of illustrating ergonomic problems to the workforce. Also, because videotaping includes other aspects of the workplace, the ergonomist is provided with additional information for the assessment of the workplace. This is not possible with goniometry systems, unless cameras are used in conjunction with a goniometer system.

There are several disadvantages to using video-based motion analysis systems. First and foremost, they require several cameras and sufficient open space to observe the motion. Controlled work environments with open areas around the worker are more suitable for video analysis than are assembly areas or cluttered workstations. Many industrial settings, for example, automobile assembly plants and meat-packing plants, have extremely limited space surrounding the workers. This makes unobstructed viewing from all directions nearly impossible. Second, workers often move about while performing their tasks. During the filming process, such movements often obscure markers from at least one of the cameras.

Further difficulties arise if a worker performs tasks at various stations. In such situations, reorienting the cameras will be time consuming and difficult. Realistically speaking, video motion analysis is best suited for laboratory conditions, that is, where the subject's surroundings are under the investigator's control.

Passive, or light-reflecting, markers, while allowing the subject the most freedom of movement, impose further problems for video systems. Lighting conditions in industrial workplaces are rarely under the control of the investigator. Generally, there is a poor contrast between the markers and the surrounding background. Without proper contrast levels, video analysis software is unable to differentiate markers from the background. Even with proper lighting, reflective surfaces in the work environment may generate spurious signals, which interfere with those coming from the markers.

Most video analysis systems can analyze data only up to 120 Hz. Many joints rotate at a speed greater than 120°/s; hence, the resolution of the system will be far inferior relative to goniometry systems, in which the sampling frequency is controlled by the A/D device and can vary up to several kilohertz.

Another disadvantage of video analysis is the data-processing time, which may be as much as several hours when obtaining 3-D joint positions and rotations. In contrast, goniometer-based systems generally require less time (normally a few minutes) for data processing.

Finally, workers often distrust and are distracted by the presence of video cameras. The use of video systems may generate poor compliance with the ergonomic assessment. Furthermore, the data obtained may not be representative of the manner in which the tasks are normally performed.

# CONCERNS ABOUT OBTAINING ACCURATE ERGONOMIC ASSESSMENTS WITH MOTION DETECTION DEVICES

A critical issue in ergonomic assessments is the definition of the components in the work cycle. Even in repetitive work environments, job cycles usually consist of several tasks. When assessing a worksite, an ergonomist first must identify the normal tasks. Visual inspection of the tasks usually indicates which tasks affect the body part of interest. Then it is necessary to develop a representative sample of the activities associated with a particular task. For example, if an individual lifts items off a pallet, measurements should be obtained from the full range of work heights and reach distances (see Figure 14.4). Along with the motion data, these workplace dimensions—specifically, the magnitude of the load being handled, estimates of the moment arms about the joints of interest, the heights of the work, and the frequency of each task—must be recorded for observation.

This raises the issue of the number of samples necessary to describe the work activity adequately. In general, the more variability in the motions, the more samples needed. For very simple, repetitive tasks, at least 5 to 10 samples should be collected. However, if there is variation in the task (e.g., loads are lifted from

**Figure 14.4**   Using the Lumbar Motion Monitor developed at Ohio State University to quantify back motions during a repetitive lifting task.

different levels), then multiple samples must be obtained from each level. With at least 5 to 10 observations a level, an average and a standard deviation can be computed for each of the motion variables studied. The standard deviation quantifies the variability in the observations. Averaging many observations within a level reduces the error in the mean for each level, thus making the underlying patterns more apparent.

Whenever possible, one should sample several individuals performing the same job in order to assess the variability between employees. This variability in the motions, as well as variability along anthropometric dimensions, must be considered when developing ergonomic solutions.

The duration of the data collection process depends on the cycle times for the particular tasks. Subjects should wear the monitoring equipment for a short period before actual data collection to allow them to acclimatize to the presence of the video equipment or the feel of the goniometer system and the presence of the investigator.

While participating in an ergonomic assessment, people often receive extra attention from their coworkers. Whenever possible, the number of observers should be minimized so as to reduce what psychologists call the Hawthorne effect, or the altering of one's work behaviors while being observed (Oborne, 1982).

## INTERPRETATION AND INTEGRATION OF DATA FROM MULTIPLE SOURCES

Because the data collected during ergonomic evaluation comes from several sources, the ergonomist is faced with the task of integrating the data to assess which variables are responsible for potential ergonomic hazards. These variables, grouped into the factors shown in Figure 14.1 (see p. 312), must be reviewed for their potential contribution to the problem being investigated. The contribution of joint motions depends on the type of work performed. For example, if the task is relatively static, only the position data may be of importance. However, for more dynamic tasks, the higher-order motion components (i.e., velocities and accelerations) become important.

Figure 14.5 presents a flow chart that can be used in the interpretation of the data collected during an ergonomic assessment. After inspection of the worksite, the investigator must decide whether ergonomic hazards exist. If no hazards appear to be present, but a worksite problem has been identified from injury reports or from high employee turnover, psychosocial factors may need to be considered (see Bigos et al., 1991). Assuming there are biomechanical problems, the next question centers on whether the postures or the motions are the cause of concern. This is evaluated by examining whether there is deviation from the joint's relaxed neutral posture. If there is no such deviation, then the ergonomic analysis should focus on the workplace and anthropometric factors that contribute to the forces acting on the joints under investigation. If postures do deviate from a neutral orientation, the focus of the analysis depends on whether the task observed is primarily static or primarily dynamic. For mainly static tasks, the focus should be on the work postures. For mainly dynamic tasks, the focus must include the quantification of postural extremes, the range of motion, the velocity, and the acceleration in each plane of motion. (The interpretation of the dynamic analysis is much more difficult.) Though some motion must occur during most assembly and material handling tasks, the investigator must evaluate the potential risks of the motions observed. Furthermore, if motion is eliminated to the point where static work postures are maintained, other ergonomic problems (e.g., muscle fatigue) will predominate.

Electrogoniometric techniques have been used by Marras and Schoenmarklin (1993) for quantifying wrist motions in individuals who perform industrial tasks that have either high or low historical risks of carpal tunnel syndrome. Comparisons between the extreme postures and ranges of motion showed little difference based on risk grouping. However, examination of the *velocities* showed greater separation between these groups. The accelerations of the wrist in the flexion-extension and radial-ulnar planes showed the greatest differentiation between the high- and low-risk groups. Similar results were observed with regard to forearm rotation or the pronation-supination of the hand. Logistic regression models indicated that the acceleration in the flexion-extension plane was the strongest predictor of risk-group membership with an odds ratio of 6 : 1 (Marras, 1992).

When evaluating the motions in each plane, an important criterion is the smoothness of the motions. The muscle forces acting on joints increase as motions change. In general, smaller velocities with smooth acceleration patterns should lead to smaller muscle forces and smaller mechanical loads acting on the joints. Counting the zero crossings in a joint's acceleration data is a means of evaluating the smoothness of a motion.

For static and dynamic tasks, the postural and motion data (if applicable) must be integrated with the workplace and anthropometric measures. Marras et al.

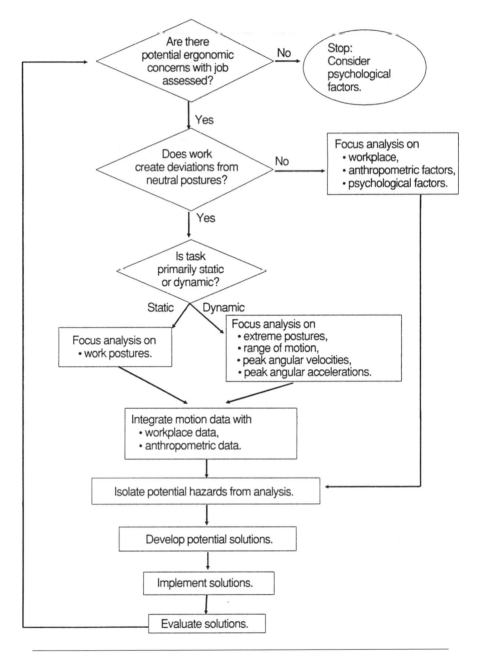

**Figure 14.5**   A flow chart illustrating the decision process used when work postures and body motions are thought to contribute to workplace injuries.

(1993) used an electrogoniometer to quantify spine motion in repetitive material-handling jobs. For each of the 400 jobs in which trunk motions were measured, the historical risk of back injury was obtained from accident records. Jobs in which there were no back injuries and no turnover were categorized "low risk." High-risk jobs had back-injury rates of 12 or more incidents for every 200,000 hours of exposure. Workplace variables, such as the rate at which lifts were performed, the weights lifted, the moment arms between the objects lifted and the spine, the height at which each lift originated and terminated, and each participating employee's anthropometry, were measured. Through multiple logistic regression modeling, these investigators were able to develop a five-factor model with odds of better than 10 to 1. The model incorporated three measures of trunk motion and two workplace measures. The peak forward flexion of the torso, the peak side-bending velocity, and the average twisting velocity comprised the trunk motion components. The lift rate (lifts per hour) and the peak moment created by the object handled (object weight × distance between object center and the spine) comprised the workplace components of the model. Figure 14.6 shows a job in the beverage industry that has been compared to the model. Each of the scales in the figure represent one of the five variables in the model; a bar indicates the observed value for the particular task. The model quickly indicates which variables pose the greatest risk of injury. For example, the markings show the measures obtained from a product transfer task in which beverage cases were moved from a pallet below knee level to another pallet at approximately waist level. Clearly, the rate at which the lifting was performed and the amount of forward bending were very high in this task. An overall probability value, represented by the dashed arrow, can be obtained by averaging the probability values for each variable obtained from the scale below (even extreme values cannot exceed 100%). This overall probability indicates that the sampled task resembles the high-risk tasks on which the model is based. But more importantly, the model indicates the key variables that must be addressed by ergonomic intervention.

Though anthropometric variables did not enter into the model created by Marras et al. (1993), other researchers have found that taller individuals tend to experience greater mechanical loading of the spine than do their shorter counterparts (Magnusson et al., 1990). Such factors must be considered when interpreting motion data to isolate potential ergonomic hazards. Based on this type of analysis, potential ergonomic solutions can be proposed.

# APPLICATION OF MOTION ANALYSIS IN DESIGN OF SOLUTIONS TO ERGONOMIC PROBLEMS

Ergonomic solutions could vary in form and complexity from reorienting incoming parts on a production line to redesigning the entire workstation. For example, if a particular assembly task requires employees to combine forearm rotation with wrist flexion, the key task elements responsible for the motions must be identified. Perhaps by altering the orientation of the parts or the sequence of assembly steps the joint motions at the workstation can be reduced.

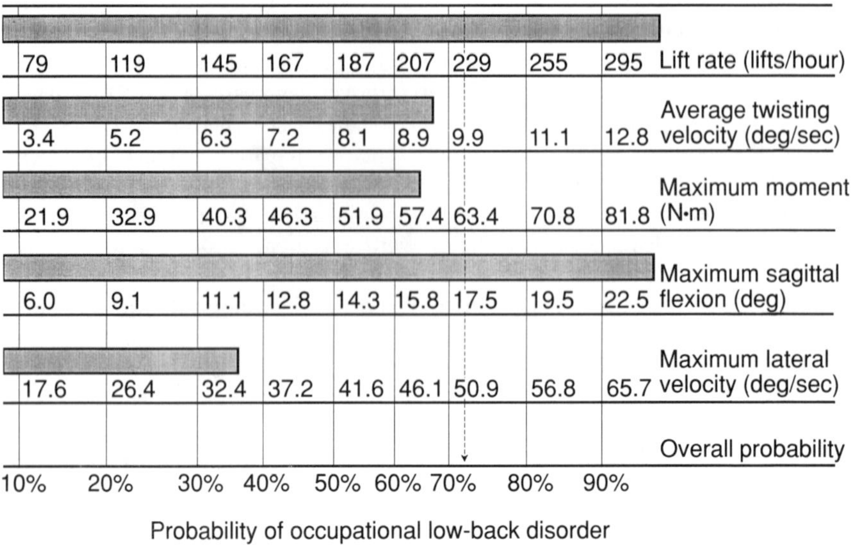

| 79 | 119 | 145 | 167 | 187 | 207 | 229 | 255 | 295 | Lift rate (lifts/hour) |
|----|-----|-----|-----|-----|-----|-----|-----|-----|------------------------|
| 3.4 | 5.2 | 6.3 | 7.2 | 8.1 | 8.9 | 9.9 | 11.1 | 12.8 | Average twisting velocity (deg/sec) |
| 21.9 | 32.9 | 40.3 | 46.3 | 51.9 | 57.4 | 63.4 | 70.8 | 81.8 | Maximum moment (N·m) |
| 6.0 | 9.1 | 11.1 | 12.8 | 14.3 | 15.8 | 17.5 | 19.5 | 22.5 | Maximum sagittal flexion (deg) |
| 17.6 | 26.4 | 32.4 | 37.2 | 41.6 | 46.1 | 50.9 | 56.8 | 65.7 | Maximum lateral velocity (deg/sec) |
|  |  |  |  |  |  |  |  |  | Overall probability |

10%    20%    30%  40%    50% 60% 70%    80%    90%

Probability of occupational low-back disorder

**Figure 14.6**  The multivariate model developed by Marras et al. (1993) that predicts back-injury risk-group membership based on trunk motion variables and workplace measures. The horizontal bars indicate values for each of the five variables obtained from data collected in a warehousing job in the beverage industry. The overall probability that this job resembles the high-risk jobs (incidence rate greater than 12 per 200,000 hours of exposure) on which the model is based is represented by the vertical dashed arrow.

Ergonomic assessments that focus on low-back disorders must integrate the factors that generate motions with those that generate mechanical loading of the spine. Workplaces that require lifting, for example, include motion in the back and external load application, both of which increase the loads on the spine. Furthermore, workplaces that require trunk motions in the transverse plane (twisting) *and* the frontal plane (lateral bending) have also been associated with increased low-back injury rates; such motion combinations should be minimized (Andersson, 1991; Frymoyer et al., 1983). The actual ergonomic intervention to be employed is situation dependent. For example, potential ergonomic solutions for situations in which back injuries are prevalent include the use of pallet lifts, hoists, articulated arms, conveyor systems, adjustable work platforms, tilt bins, reach extenders, and turntables. In essence, these solutions address the workplace and anthropometric factors described in Figure 14.2 (see p. 313) by reducing the trunk motion and the biomechanical loading of the spine. The use of these ''engineering controls'' is considered the most effective intervention method for reducing the incidence of low-back disorders (Snook, Campanelli, & Hart, 1978). Snook (1987) suggests that although worker selection based on strength testing has been shown to be effective, it should only be used in jobs that are not amenable to the use of engineering controls. A more complete discussion of possible interventions can be found in various ergonomic texts (see Chaffin & Andersson, 1991; Grandjean, 1969; Kodak, 1982; Tichauer, 1978).

Once the solutions are developed and implemented, they should be evaluated using methodologies similar to those used in the initial assessment. For simple solutions, workstation mock-ups can be constructed or existing workstations temporarily altered to incorporate the proposed intervention. This way interventions can be tested before full-cost implementation. Furthermore, employee feedback concerning the proposed interventions is especially important. Without the workers' acceptance, the intervention may not be used or may not be used properly. Moreover, the evaluation should address whether the original problem has been solved, whether other problems continue to exist, and whether new problems have been created through the intervention.

# FUTURE TRENDS IN THE APPLICATION OF HUMAN MOVEMENT ANALYSIS TO ERGONOMICS

The techniques for quantifying human motion in the workplace are still being developed. Ergonomics programs in industry are becoming increasingly sophisticated, in part driven by the demands of many organizations for good, quantitative descriptions of the ergonomic problems before they fund intervention programs. The current tools for measuring trunk and wrist-forearm motions, combined with models that predict risk-group membership, often provide the information that management needs to make funding decisions. The models allow ergonomists to quickly evaluate proposed design changes in terms of their anticipated effect on the critical variables. Where this cannot be determined theoretically, simulations can be set up in which actual motions are measured. Such evaluations should determine whether the proposed modifications will be successful, without having to wait for injury statistics to accumulate.

Eventually, these models will be incorporated into software that can highlight problem areas as the motion data are being analyzed. At this point, dynamic biomechanical models could further interpret the data to estimate the mechanical joint loading under the conditions observed. Data bases that contain population strength capabilities, cataloged by motion variables, could also be integrated with motion data collected at the worksite. Together, such analyses will provide the ergonomist with the information necessary to understand the expected capabilities and limitations of any operator under various measured working conditions.

# REFERENCES

Andersson, G.B.J. (1991). The epidemiology of spinal disorders. In J.W. Frymoyer (Ed.), *The adult spine: Principles and practice* (pp. 107-146). New York: Ravens Press.

Armstrong, T.J. (1983). An ergonomics guide to carpal tunnel syndrome. *AIHA Ergonomic Guide Series*. Akron, OH: American Industrial Hygiene Assn.

Basmajian, J.V. (1976). *Primary anatomy* (7th ed). Baltimore: Williams & Wilkins.

Bigos, S.J., Battie, M.C., Spengler, D.M., Fisher, L.D., Fordyce, W.E., Hansson, T.H., Nachemson, A.L., & Wortley, M.D. (1991). A prospective study of work perceptions and psychosocial factors affecting the report of back injury. *Spine,* **16**, 1-5.

Chaffin, D.B., & Andersson, G.B.J. (1991). *Occupational biomechanics.* New York: Wiley.

Frymoyer, J.W., Pope, M.H., Clements, J.H., Wilder, D.G., MacPherson, B., & Ashikaga, T. (1983). Risk factors in low-back pain. *Journal of Bone & Joint Surgery,* **65-A**, 213-218.

Grandjean, E. (1969). *Fitting the task to the man.* London: Taylor & Francis.

Kodak. (1982). *Ergonomic design for people at work: Volume 1.* New York: Van Nostrand Reinhold.

Kumar, S. (1990). Cumulative load as a risk factor for back pain. *Spine,* **15**, 1311-1316.

Magnusson, M., Granqvist, M., Jonson, R., Lindell, V., Lundberg, U., Wallin, L., & Hansson, T. (1990). The loads on the lumbar spine during work at an assembly line: The risks for fatigue injuries of vertebral bodies. *Spine,* **15**, 774-779.

Marras, W.S. (1992). Toward an understanding of dynamic variables in ergonomics. *Occupational Medicine: State of the Art Reviews,* **7**, 655-677.

Marras, W.S., Lavender, S.A., Leurgans, S.E., Rajulu, S.L., Allread, W.G., Fathallah, F.A., & Ferguson, S.A. (1993). The role of dynamic three-dimensional trunk motion in occupationally-related low back disorders: The effects of workplace factors, trunk position, and trunk motion characteristics on risk of injury. *Spine,* **18**, 617-628.

Marras, W.S., & Schoenmarklin, R.W. (1993). Wrist motions in industry. *Ergonomics,* **36**, 341-351.

Oborne, D.J. (1982). *Ergonomics at work.* Chichester: Wiley.

Priemer, R. (1991). *Introductory signal processing.* Singapore: World Scientific.

Snijders, C.J., van Riel, M.P.J.M., & Nordin, M. (1987). Continuous measurements of spine movements in normal work situations over periods of 8 hours or more. *Ergonomics,* **30**, 639-653.

Snook, S.H. (1978). A study of three preventive approaches to low back injury. *Journal of Occupational Medicine,* **20**, 478-481.

Snook, S.H., Campanelli, R.A., & Hart, J.W. (1987). Comparison of different approaches for the prevention of low back pain. In *Ergonomic interventions to prevent musculoskeletal injuries in industry* (pp. 57-72). Chelsea, MI: Lewis.

Tichauer, E.R. (1978). *The biomechanical basis of ergonomics: Anatomy applied to the design of work situations.* New York: Wiley.

# Chapter 15

# Application of Three-Dimensional Analysis to Sports

*Alain Durey*
*R. Journeaux*

It is essentially by the use of predictive biomechanical models that tridimensional techniques have been applied and successfully used in sport applications. The observation and description of time-space parameters have been a constant preoccupation of coaches. Scientists and coaches exploit the technical possibilities of 3-D analysis to improve the athlete's performance. However, to study on-site performances without disturbing the ongoing competitions or the training program, new technical and methodological approaches are required. By means of a few examples, we will illustrate the potential of a 3-D quantitative analysis—the methodological constraints and, in this perspective, the present technical limitations and needed improvements—by analyzing, in a teaching framework, basic tennis strokes, followed by examples in waterskiing, pole vaulting, and table tennis.

## BASIC TENNIS STROKES

Are there basic strokes in tennis? Do the terms topspin, backspin, slice, and chop have a physical meaning? Is it possible to describe and define a player's repeatable strokes, and are they characterized by similar spatiotemporal patterns? Can these be described with precision? De Kermadec (FFT/INSEP & De Kermadec, 1985, 1986) attempted to answer some of these questions using annotated stick diagrams in his film studies of the French Open tennis tournaments. Using two 16-mm cameras at 100 frames per second with a 1/1,000 s time exposure, he filmed the world's top players during their matches at Roland Garros (1984 to 1987) to determine typical standard strokes for each player. Each view was then digitized, recorded, and restored in stick diagram form. From the two views, a pseudo-top view was drawn. The purpose here was mainly to attempt a classification of the movements and the stable morphokinetic sequences for teaching and training purposes.

The behavior of the tennis ball after the stroke has already been modeled by Durey (1987). The model was useful in establishing the necessary conditions for "special effect" balls. For example, to produce ball rotation, the velocity vector of the tennis racket's mesh must not be perpendicular to the plane of the racket frame. The model makes it possible to calculate the kinematic parameters of the ball and the mesh after impact, on the basis of the values of these parameters before impact. For a particular ball effect, the velocity and the orientation of the racket requires morphokinematic organization and control by the player. How is this pattern organization managed by the players to produce similar ball effects? Each rotation producing an effect is defined by the direction of the rotation vector, which has 3-D components. Standard ball rotations were defined by Durey and De Kermadec (1984) and are represented in a plane perpendicularly oriented with respect to the ball velocity vector as shown in Figure 15.1. To produce a specific ball effect, the racket must have a well defined velocity and orientation. For an underspin effect, as in Figure 15.2c, the racket must be moving downward with the racket frame plane slightly oriented upward. For a topspin effect, Figure 15.2b, the racket has an upward movement with the racket frame plane oriented downward. To produce a ball without spin the movement of the racket should be in a direction perpendicular to the racket frame plane, Figure 15.2a. When the players' strokes are grouped according to these standard effects both constants

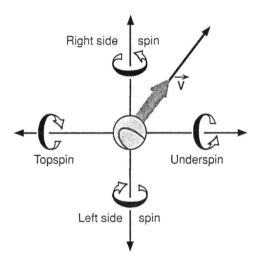

**Figure 15.1**   Definition of standard ball rotations in a plane perpendicularly oriented with respect to the ball's velocity vector: Right-side spin rotation vector angle, 0°; top-spin rotation vector angle, 90°; left-side spin rotation vector angle; 180°; underspin rotation vector angle; 270°.

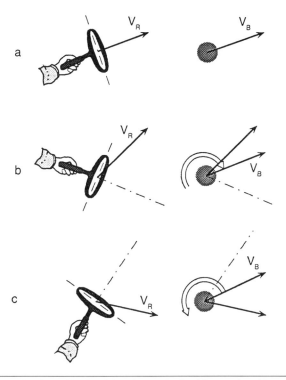

**Figure 15.2**   Production of balls with spin: (a) Ball without spin, (b) topspin rotation, and (c) underspin rotation ($V_R$, racket velocity before impact; $V_B$, ball velocity after impact).

and variants can be defined for the same type of effect. The flat shot of Leconte, Figure 15.3a, is unique; it can be called a "whip shot." The ball is taken up high and hit horizontally after a vertical excursion of the racket. On the contrary, McEnroe, Figure 15.3b, develops his stroke on a single, slightly inclined plane, which produces almost no rotation. All the stick diagrams presented in this paper follow the temporal pattern given in Table 15.1. Real serves can be described in terms of standard serves: right-hand slice, left-hand slice, underspin, topspin, without spin. A serve said to be "right-hand sliced" is essentially a serve with right-side spin. Generally, a player's second serve is a combination of side spin and top spin (the rotation axis is about 45° in the plane of the Figure 15.1). The slice serve was first fully exploited by McEnroe. Yet even when he makes a flat serve, he uses the same organizational pattern; only the end of the stroke is modified by a change in the pronation and endorotation of the arm. McEnroe has an atypical pattern of organization in which the tennis racket evolves into two orthogonal planes. Noah's serve, Figure 15.4, a-c, is a good example of a serve carried out along a single principal plane, in the same way many world-class players execute the serve. The three views of the analysis make it possible to see the movement aspect that cannot be seen on the movie frame. They also

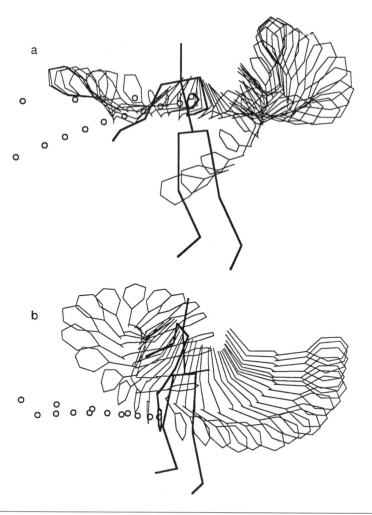

**Figure 15.3** Two sorts of forehand without spin: (a) H. Leconte, profile view; and (b) J. McEnroe, profile view.

enable us to make relevant and reliable comparisons insofar as the views (top, profile, and back) are thus standardized: for instance, the position of the ball at the moment of impact in comparison with the front foot position. The point of impact depends on the kind of service, therefore the ball throw must be appropriate.

Our observations are comparable to those of Van Ghelüwe and Hebbelinck (1985). However, their results are limited to laboratory conditions involving three players. Our approach represents actual competition situations and is therefore more global and takes into consideration movement patterns as well as tactical aspects in relation to the ball trajectory models (Durey, 1987). This model can

**Table 15.1   Time Intervals Between Two Consecutive Racket Positions**

| Images | | $\Delta t$ (ms) |
|---|---|---|
| 0 to 5 | correspond to all the frames of the film | 10 |
| 5 to 10 | correspond to every second frame | 20 |
| 10 to 15 | correspond to every third frame | 30 |
| 15 to 20 | correspond to every fourth frame | 40 |
| 0 to −5 | correspond to all the frames of the film *before* impact | 10 |
| −5 to −10 | correspond to every second frame | 20 |
| −10 to −15 | correspond to every third frame | 30 |
| −15 to −20 | correspond to every fourth frame | 40 |
| −20 to −25 | correspond to every fifth frame | 50 |

*Note.* The position represented by the stick figure corresponds to the first frame *after* impact (reference position).

simulate actual trajectories and even create new ones. One can thus estimate the advantage of a lifted passing shot over one without spin, Figure 15.5a, as well as that of an underspin drop shot compared to one without spin, Figure 15.5b.

The game of tennis is constantly evolving, and players continually strive to achieve better strokes. Today, new solutions have been found (e.g., the two-hand racket hold used by Courier, Agassi, Seles, and Sanchez in the finals competition at Roland Garros in 1991). These solutions are not only the result of tactical choice and biomechanical constraints, they are also largely dependent on changes in the game itself. Three-dimensional analysis is used in this study to determine the parameters of shocks and ball trajectory models and to describe and compare the strokes of the best world-class players.

# WATERSKIING

The second example deals with waterskiing. Here, the different phases of the jump (cut, ramp, and flight) have been modeled to find the ideal conditions for a jump. The model for each phase is based on aerodynamics, hydrodynamics, and impact mechanics. The aim is to develop a global model, taking into account the interaction of the different phases of the jump, rather than to refine each phase model.

The jumps of the ski master competition held in 1990 at Marseille-Marignane were filmed using six video cameras. The cameras were positioned in such a way as to cover all three phases of a ski jump. Film taken from two crane-mounted cameras made it possible to recalculate the topviews of each skier's "cut" trajectory. The skier position in the horizontal plane enables describing and characterizing the cut with times $t_1$ and $t_2$ and distances $D_1$ and $D_2$, Figure 15.6. In reference to the different skiers, $t_1$ times lie between −1 s and −1.4 s,

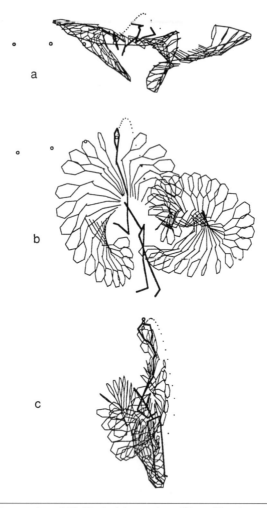

**Figure 15.4** First service of Y. Noah: (a) top view, (b) profile view, and (c) back view.

$D_1$ distances lie between 19 m and 25 m, $t_2$ times lie between −3 s and −4 s, and $D_2$ distances lie between 32 m and 45 m. Two other cameras gave us a 3-D image of the skier on the ramp. The speed of the skiers' centers of mass at take-off varied between 88 km · h$^{-1}$ and 117 km · h$^{-1}$. Surprisingly, the highest speeds were not associated with the highest jumps. At take-off, the velocity vector angles of the skiers' centers of mass, with respect to the horizontal, ranged from 9° to 17.5° (Figure 15.7). Again, the largest angles were not related to the best performances. Nevertheless, if the angle was less than 11°, the jump was short. The skis formed an angle of 10° to 70° with respect to the velocity vector taken at the skier's center of mass during the flight. Although the angle varied

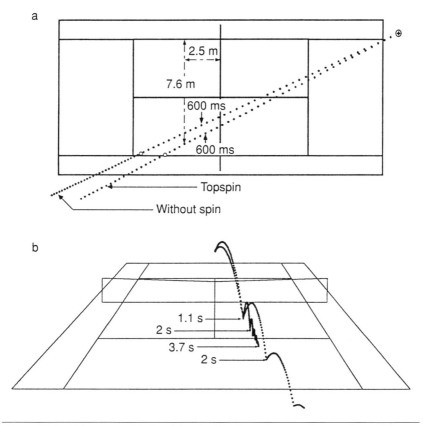

**Figure 15.5**  Advantage of balls with spin: (a) The trajectory of a topspin passing shot can be lower and more exterior than a passing shot without spin. (b) An underspin drop shot has a shorter rebound than a drop shot without spin.

considerably between jumps, it remained fairly constant during most of the aerial phase. The angle varied during the first few meters after take-off, but stabilized afterwards. An angle of 60° or greater resulted in a poor performance.

In the flight model, the aerodynamic forces in the skis were assumed to be similar to those applied to an airplane wing; by assuming that the angle between the ski and the velocity vector of the skier's center of mass was constant during the flight, the model adequately predicted the flight distance. The aerodynamic forces exerted on the water-skier's body were neglected.

Using this model, other flight conditions were simulated. The optimal ski angle with respect to the velocity was found to be 30°, regardless of the magnitude of the speed. The estimated flight distance is 54 m if the angle is 30°, and 47 m if it is doubled (60°). Therefore the best performance is obtained by increasing both the velocity and its angle.

The camera positions provided adequate information for the cut phase and for the ramp phase. The 3-D analysis of the flight phase was complicated by the

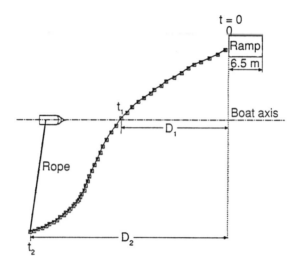

**Figure 15.6**   Topview of water-skier's "cut" trajectory.

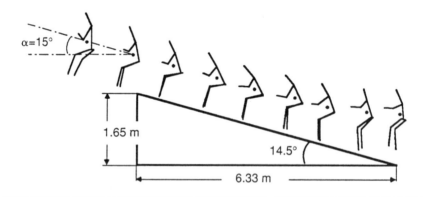

**Figure 15.7**   Center of mass velocity vector angle with respect to the horizontal.

calibration procedures related to the nature of the rather large area in question, namely 60 m long, 20 m wide, and 6 m high in relation to the water level. The models were validated by actual data collected during international competitions and were used to estimate the optimal values for the best performances.

## POLE VAULT MODEL

In this work, the theoretical models of Hubbard (1980) and Griner (1984) were used. These models, including our improvements, have been validated by results obtained from a 3-D analysis of actual jumps. The major difficulty lies in the amount of space required to execute a complete jump, in which there is a running portion combined with a high vertical vault. Thus, two sets of cameras should be used: the first to follow the general 3-D trajectory of the pole vaulter and the second to focus on the athlete's movement. A panning camera might simplify the camera setup but would complicate the 3-D reconstruction procedure.

A pole vault team coach must decide early on major training orientations to be implemented over several months, the results of which can only be appreciated much later. He cannot afford to make a mistake. The consequences would be too damaging to the athletes' short-lived careers. Therefore, the French national pole vault team coach saw the value of sport biomechanics in the development of a pole vault model that would enable him to simulate jumps on a computer. The model he asked us to develop makes it possible to determine the best individual performance that can be achieved, based on the morphological characteristics of the jumper, the lever arm of the pole, the stiffness of the pole, running speed, and take-off. The simulation can indicate possible improvements in the choice of pole materials, the ideal conditions for take-off, and how the athlete might better perform the jump. (Model validation was performed through a 3-D analysis of jumps filmed during the training camp of the French national team.)

In a pole vault model, joint reaction forces and moments, as well as ground reaction forces, need to be considered. Additionally, the elastic properties of the pole must be included. Thus, Hubbard's model (1980) for the pole vaulter was improved by the inclusion of damping factors at each joint and then combined to the pole model of Griner (1984). First, the pole was modeled as a segment of variable lengths but without any mass. This approximation was justified, because the inertia of the pole is small compared to that of the vaulter. This implies that the greater part of kinetic energy can be assigned to the vaulter.

The three segments of the vaulter's body are the arms, the head-torso combination, and the legs. Deconstructing the arms into two segments seemed unnecessary, because the arm's kinetic energy is small. A submodel for the legs was more relevant, because the velocity of such segments can be high in some parts of the jump. Nevertheless, the great number of parameters involved in the simple model was such that complete control of them was very difficult. A thorough analysis of this reduced model is the first step to take before any further refinements are attempted.

Each segment is characterized by the orientation, $\theta_i$, and the lengths, $l_i$. These lengths are constants for the three segments of the vaulter's body. The center of mass, $G_i$, of each segment of the body is at a distance, $D_i$, from the distal part of the segment. The three segments of the vaulter's body are characterized by the three masses, $m_i$, and by the three moments of inertia, $J_i$, for axes passing through $G_i$ and perpendicular to the plane of motion.

The interactions between the segments involve forces due to the masses of the segments and the voluntary control torques exerted by the vaulter at each articulation. A longitudinal force, $R_1$, and a torque, $C_1$, are applied to Segment 1, the pole, by Segment 2, the arms. A force, $R_2$, and a torque, $C_2$, are applied to Segment 2 by Segment 3. A force, $R_3$, and a torque, $C_3$, are applied to Segment 3 by Segment 4. Each force is the composition of $R_{iL}$ (longitudinal) and $R_{iT}$ (Transverse), with respect to Segment $i$ (see Figure 15.8).

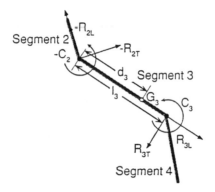

**Figure 15.8**  Interactions on Segment 3, Head-Torso: $-R_{2L}$, $-R_{2T}$, $C_2$—forces and torque on the shoulders; $R_{3L}$, $R_{3T}$, $C_3$—forces and torque on the hips; $G_3$—center of mass of the segment; $l_3$—length of the segment; $d_3$—distance from the center of mass of the segment to the proximal articulation.

Applying dynamic relations to the three segments of the vaulter leads to nine differential relations (Table 15.2). The numerical resolution of these relations is performed on a computer using the classical Runge-Kutta algorithm, fourth order. The other nine unknown variables are computed from the nine linear equations given in Table 15.2 according to the matricial relation **MH = K**. The component $R_{1L}$ is deduced from the pole submodel as a function of $l_1$ and $C_1$ at each step.

Before the computer simulations are carried out, the model needs to be validated by means of experimental input data and then the predicted values compared to the measured ones. Three synchronized high-speed cameras were used to film 30 jumps.

The numerical solution requires input parameters for the given initial conditions (see Figure 15.9). The lengths of the vaulter's segments are measured on film, and the mass, the center-of-mass position, and the moments of inertia are calculated according to Winter (1979). The length of the pole, L, was measured before each vault; the stiffness of the pole, B, was measured by the manufacturer. The initial conditions at takeoff were measured from the first camera closeup of the vaulter. We measured the angle and the magnitude of the velocity and the positions and the angular speed of the segments, while the fourth segment, the legs, was defined by the position of the center of mass of the two legs. After film digitization, the 3-D reconstructions made it possible to evaluate and compare the jump, as well as the vaulters, by transformation of the experimental data (camera film) into data to be compared to the biomechanical model. The five characteristic positions are takeoff ($t = 0$), legs and trunk in line ($t = 0.25$ s), knees at hand level ($t = 0.70$ s), body extended vertically ($t = 0.95$ s), and body clearing the bar ($t = 1.35$ s).

Figure 15.10a gives rough data from the digitization of a 100 frame-per-second film (the profile camera) and Figure 15.10b, the view computed from the profile camera view in the same positions. This calculation places the movement in a

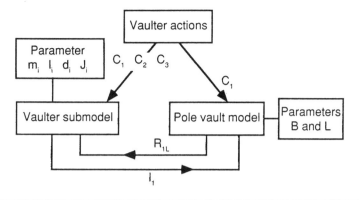

**Figure 15.9**  Interaction of the models for operating numerical solutions.

vertical plane (without perspective). Figure 15.10c shows reduction of the values to three segments: first segment, higher hand–middle of the shoulders; second segment, middle of the shoulders–middle of the hips; third segment, middle of the hips–center of mass of the legs. The adjustment of the model consisted of finding the parameter values, controlled vaulter actions, and damping to make the simulated vault coincide with the real one. The direction of the exerted torques and their temporal organization during the vault were then determined a priori. The data from Hubbard (1980) were used, the values adjusted to fit the real vault to the simulated one. Each torque—$C_1$, $C_2$, and $C_3$—is thus composed of successive time-based exponential functions. The viscous damping coefficients were adjusted at the same time. The comparison criterion was based on the superimposition of both the simulated and the real stick diagrams (Figure 15.11a). The precision of our simulations depended on the limits of the pole submodel, which assumes that no energy is lost and thus leads to higher vaults than the real ones. Three-segment representation of the simulated values ($Dt = 100$ ms) with a perfectly elastic pole. The vault is optimized by the choice of the torques applied to articulations. Thus, the simulated vault is 1 m higher than the real one. From a methodological point of view, it would now be necessary to work in simulation with the data from the inverse dynamic analysis (McGinnis, 1983) on the same basic data. We might be then able to take a 10-segment model for the pole vaulter similar to the one used by Dapena and Braff (1985). It would also be relevant to compare the values of the reaction forces and moments applied by the pole-and-vaulter system to the vaulting box with the values measured in the vaulting box (Vaslin, Couëtard, & Cid, 1993).

When pole stiffness increases, the trajectory of the center of mass is translated toward the running track, yet the height of the vault remains unchanged. For example, a pole having a stiffness of 1,950 Nm yields a trajectory summit that is translated 23 cm horizontally toward the running track compared to the summit yielded by a pole stiffness of 1,900 Nm. The maximum height of the center of mass is higher with increasing velocity. For example, a $V_c$ of 7.5 ms$^{-1}$ gives a maximum height of 5.5 m, whereas a $V_c$ of 8 ms$^{-1}$ yields a maximum height of

**Table 15.2  Elements of Matrix M and Vector H**

| $\ddot{\theta}_1$ | $\ddot{\theta}_2$ | $\ddot{\theta}_3$ | $\ddot{\theta}_4$ | $\ddot{l}_1$ | $R_{2T}$ | $R_{2L}$ | $R_{3T}$ | $R_{3L}$ |
|---|---|---|---|---|---|---|---|---|
| | $J_2$ | $J_3$ | $J_4$ | | $l_2 - d_2$ | | $l_3 - d_3$ | |
| | | | | | $d_3\cos(\theta_3 - \theta_2)$ | $-d_3\sin(\theta_3 - \theta_2)$ | $d_4\cos(\theta_4 - \theta_3)$ | $-d_4\sin(\theta_4 - \theta_3)$ |
| $-m_2 l_1\sin\theta_1$ | $-m_2 d_2\sin\theta_2$ | | | $m_2\cos\theta_1$ | $-\sin\theta_2$ | $\cos\theta_2$ | | |
| $m_2 l_1\cos\theta_1$ | $m_2 d_2\cos\theta_2$ | | | $m_2\sin\theta_1$ | $\cos\theta_2$ | $\sin\theta_2$ | | |
| $-m_3 l_1\sin\theta_1$ | $-m_3 l_2\sin\theta_2$ | $-m_3 d_3\sin\theta_3$ | | $m_3\cos\theta_1$ | $\sin\theta_2$ | $-\cos\theta_2$ | $-\sin\theta_3$ | $\cos\theta_3$ |
| $m_3 l_1\cos\theta_1$ | $m_3 l_2\cos\theta_2$ | $m_3 d_3\cos\theta_3$ | | $m_3\sin\theta_2$ | $-\cos\theta_2$ | $-\sin\theta_2$ | $\cos\theta_3$ | $\sin\theta_3$ |
| $-m_4 l_1\sin\theta_1$ | $-m_4 l_2\sin\theta_2$ | $-m_4 l_3\sin\theta_3$ | $-m_4 d_4\sin\theta_4$ | $m_4\cos\theta_1$ | | | $\sin\theta_3$ | $-\cos\theta_3$ |
| $m_4 l_1\cos\theta_1$ | $m_4 l_2\cos\theta_2$ | $m_4 l_3\cos\theta_3$ | $m_4 d_4\cos\theta_4$ | $m_4\sin\theta_1$ | | | $-\cos\theta_3$ | $-\sin\theta_3$ |

Matrix M

$$d_2R_{1L}\sin(\theta_2 - \theta_1) - d_2R_{1T}\cos(\theta_2 - \theta_1) + C_1 - C_2 - (-a_1\dot\theta_2)$$

$$C_2 - C_3 - a_2\dot\theta_3$$

$$C_3 - a_3\dot\theta_4$$

$$R_{1L}\cos\theta_1 - R_{1T}\sin\theta_1 + m_2l_2\cos\theta_1\dot\theta_1^2 + m_2d_2\cos\theta_2\dot\theta_2^2 + 2m_2\sin\theta_1\dot\theta_1\dot l_1 + m_2g$$

$$R_{1L}\sin\theta_1 + R_{1T}\cos\theta_1 + m_2l_1\sin\theta_1\dot\theta_1^2 + m_2d_2\sin\theta_2\dot\theta_2^2 - 2m_2\cos\theta_2\dot\theta_1\dot l_1$$

**Vector H**

$$m_3\sum_{i=1}^{2} l_i\cos\theta_i\dot\theta_i^2 + m_3d_3\cos\theta_3\dot\theta_3^2 + 2m_3\sin\theta_1\dot\theta_1\dot l_1 + m_3g$$

$$m_3\sum_{i=1}^{2} l_i\sin\theta_i\dot\theta_i^2 + m_3d_3\sin\theta_3\dot\theta_3^2 - 2m_3\cos\theta_1\dot\theta_1\dot l_1$$

$$m_4\sum_{i=1}^{3} l_i\cos\theta_i\dot\theta_i^2 + m_4d_4\cos\theta_4\dot\theta_4^2 + 2m_4\sin\theta_1\dot\theta_1\dot l_1 + m_4g$$

$$m_4\sum_{i=1}^{3} l_i\sin\theta_i\dot\theta_i^2 + m_4d_4\sin\theta_4\dot\theta_4^2 - 2m_4\cos\theta_1\dot\theta_1\dot l_1$$

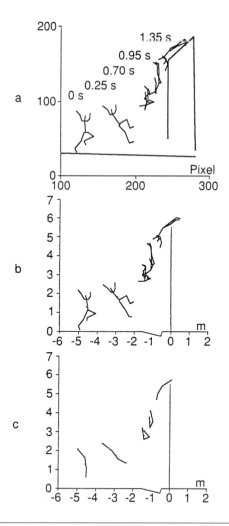

**Figure 15.10**   (a-c) Three views of a 5.50-m pole vault executed by T. Vigneron.

5.8 m. If the angle between the chord of the pole and the horizontal plane is reduced, the maximum height of the trajectory is increased and translated toward the running track. For example, with L = 5 m, $V_c$ = 8.8 ms$^{-1}$, and a chord angle of 28.6°, the summit of the parabola of the vaulter's center of mass has a vertical coordinate of 6.71 m and a horizontal coordinate of 0.44 m, whereas an angle of 27.7° gives a parabola summit for the vaulter's center of mass at 6.77 m vertical and 0.18 m horizontal. The trunk must be as vertical as possible when the vaulter rotates 180° to face the crossbar; the maximum height can be increased by about 10 cm if the vaulter manages to set the hips vertically over the head. This requires an important controlled shoulder torque. When the pole flexion increases, the vault is translated to the jumping pit and the vaulter must produce

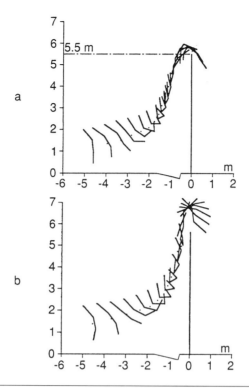

**Figure 15.11**   (a and b) Optimization of T. Vigneron's 5.50-m pole vault.

a greater effort in the shoulders to get an optimized vault. If this lever arm is decreased, the vault is lower and is translated toward the running track. For example, if the lever arm is decreased from 4.72 m to 4.67 m, the height attained is only 5.90 m rather than 6.10 m, resulting in a translation of 10 cm rather than 24 cm. The simulation results show that all athletes can further improve their performances either by maximizing the values of the initial parameters or by modifying the timing of their movement patterns.

## SPATIOTEMPORAL REGISTRATION IN TABLE TENNIS

Three-dimensional analysis has been used in table tennis to relate the different phases of the motor pattern of the athlete to the ball's position along its trajectory. The 3-D data capturing system must fulfill the mechanical considerations associated with the specific strokes, especially those with ball effects (Durey & Orfeuill, 1989), as well as the visual information (Ramanantsoa, Ripoll, & Durey, 1991; Ripoll, 1989). The underlying hypothesis is the existence of stable relations between the athlete's movements and the position of the ball.

Within this spatiotemporal framework, the athlete may be in a position to reduce the number of possible strokes that he can choose from. Indeed, some strokes take less time than others, and these would be used in a situation where a quick response is required. In this case, the player must follow the rhythm imposed by the adversary limiting him or her to certain types of responses or strategies. The game continues as long as this relationship is not perturbed. When one of the players gains the advantage, some phases are shortened to the point where one of the players can no longer respond, and the point is lost.

The French table tennis team was filmed during in-house competitions by means of two 16-mm high-speed cameras set at 100 Hz and having a time exposure of 1/1000 s. A video camera located above the table tracked the ball's trajectory. (Results are shown for the matches played by J.P. Gatien.)

The trajectory of the top of the paddle can be used to characterize the different movement phases that are stable with respect to the ball's trajectory (Figure 15.12). The player's 3-D movement occurs in four phases (Figure 15.13, a-c).

The first phase, Images 1 to 5, corresponds to the ball's displacement from the player's stroke to the net (Figure 15.12, points a and b). After the stroke, the paddle continues its upward movement to a point where it is almost at a stop at Image 5.

In the second phase, Images 5 to 26, the ball moves from the net, forward b, to the net backward c (Figure 15.12). There is a downward movement of the paddle along an axis perpendicular to the table (Figure 15.13c). Its back-and-forth progress involves a motion of the whole body. At the end of this phase, the movement stabilizes itself, and the paddle lies in a plane almost parallel to the table (Figure 15.13b).

In the third phase, Images 26 to 38, the ball moves from the net backward, c, to a point, d, after the bounce (Figure 15.12). The position of this point is a function of the fourth phase duration. The back-and-forth motion is followed by a backswing. It consists of moving the paddle to a final, lateral backhand (or backward forehand) point, while the ball returns to the net up to a point just after the first rebound. This point is dependent on the next and last phase. For

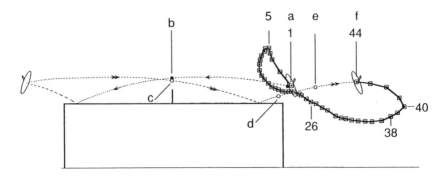

**Figure 15.12**   Synchronization of the relations between the movement phases and ball trajectory.

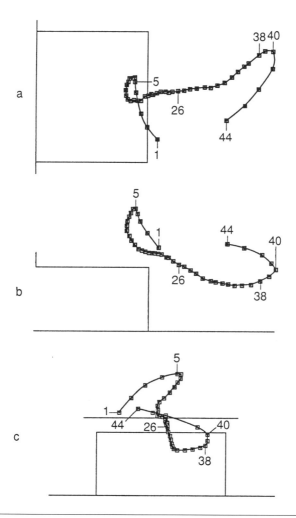

**Figure 15.13**   Racket head trajectory between two forehand strokes executed by J.P. Gatien: (a) Top view, (b) profile view, and (c) back view. The time span of the forehand propulsion phase is invariable (80 ms).

a backhand, the change of direction of the upper portion of the paddle is located at the beginning of this phase (Figure 15.13b). This change of direction is also observed during a forehand, although it is less marked (Figure 15.13a).

In the fourth and last phase, Images 38 to 44, the ball moves from a point, d, after the bounce to the new stroke, f (Figure 15.12). The downswing can be divided into two parts. Initially, the paddle moves in preparation for the stroke followed by the propelling action, which lasts 80 ms regardless of the type of stroke, the distance to the table, or the technique. When a player needs to react quickly, it is always the preparation period of this phase that is modulated (Images

38-40). These observations are consistent for all backhands and forehands, regardless of the opponents.

## CURRENT LIMITATIONS OF AND FUTURE HOPES FOR 3-D ANALYSIS OF SPORT ACTIVITIES

Spatiotemporal representations of movement, or stick diagrams and kinematic concepts to describe them, have evolved since the development of photochromotography and dynamography by Marey and Demeny (1885). We can distinguish two principal uses of 3-D techniques and methods to analyze human movement in sports and elite performances: a graphical approach for descriptive study and a modeling approach to predict performance through computer simulation. The 3-D representation of sport movement is a fundamental step in understanding it and yields objective, quantitative information. For example, Dapena (1988a, 1988b) has detailed the Fosbury flop; Van Gheluwe and Hebbelinck (1985) have studied the tennis serve. Models and computer simulations were used to find invariant parameters to classify complex movements into basic strokes. The kinematic variables (positions, rotations, etc.) were initially considered. For example, Deporte, Van Gheluwe, and Hebbelinck (1990) have studied the temporal evolution of the angles that characterize the pronation and endorotation during a tennis serve. The next step is to generate hypotheses or theories of human movement from 3-D data. Such an attempt has been made in the spatiotemporal modulation of table tennis, in which the underlying hypothesis relating the displacement of the top of the paddle to the ball's trajectory is taken from previous studies based on sport neuroscience.

In seeking invariant parameters related to sport activities and performances, several authors address a more general problem related to the validity of the results. Grainger, Norman, Winter, and Bobet (1983) looked at the reproducibility of the measurement of biomechanical variables calculated from film data on four subjects taken on two different days. Taking a statistical approach, Takei (1989) presented a deterministic model of the vaulting horse; his objective was to determine which mechanical variables are best correlated with a judge's marks. Even though this was based on a planar analysis, one can imagine a similar 3-D approach for many gymnastic performances.

The following models result from empirical analysis using physical elements. Predictive models are highly sought, because they allow simulating the desired movements and eventually lead to the optimization of the performances. To do this, the human body must be modeled as a mechanical system consisting of articulated rigid links. The number of links can be relatively high. The model limitation is not the number of links, nor the number of equations or the numerical techniques to be solved, but rather the number of parameters needed to correctly define the movement. Dapena (1989) developed a 15-link segment model in his optimization of a high jump; he modified certain parameters during the flight phase of the jump to optimize the performance. Yeadon (1990a, 1990b, 1990c) and associates (Yeadon, Atha, & Hales, 1990) used an 11-rigid-segment model with 17 degrees of freedom to describe the aerial movements of the human body.

Neglecting air resistance, the aerial rotations of the human body are described by the conservation of angular momentum. The accuracy of the model is estimated by comparing the simulated somersault's tilt and twist angles with the corresponding values obtained from film data of nine twisting somersaults. The largest errors are due to the digitization process rather than the anthropometric measurements. Yeadon's work was limited to aerial movements.

Others (Neal & Wilson, 1985) developed models that take into account external forces other than body weight; these studies were usually limited to three or four segments. Vaughan (1981) estimated the forces and moments at the wrist during a golf swing, whereas Schleihauf, Gray, and Derose (1983) used a hydrodynamic model to estimate the forces acting on the arm of a swimmer. Finally, Hubbard (1980) and Durey (1987) incorporated the interaction between the athlete and his equipment into movement analysis.

Today, three-dimensional analysis benefits the coach by allowing the more rigorous study of sport activities and elite performances. In addition to providing a better understanding of complex movements, 3-D analysis orients training choices through predictive models. It yields indices and parameters for monitoring and controlling the athlete's progress. The interactive videos also make it possible to improve training for coaches, referees, and judges. If, in the near future, certain long awaited technical breakthroughs are achieved, we can expect greater precision and accuracy in 3-D measurements, refining our predictive models, and even real-time 3-D representations during competition for the expert eyes of judges as well as for the pleasure of spectators.

## ACKNOWLEDGMENTS

We would like to note the support of the following organizations: Ministère de la Jeunesse et des Sports, Paris; Institut National des Sports et de l'Education Physique (INSEP), Paris; Laboratoire de Neuroscience du Sport de l'INSEP; Service Audiovisuel de l'INSEP; Fédération Française de Tennis, Paris; Fédération Française de Tennis de table, Paris; Fédération Française de ski nautique, Paris; Fédération Française d'Athlétisme, Paris; Les sociétés Adidas et 3IT; G. De Kermadec (FFT); and O. Faugeras and G. Toscani (INRIA). Thanks also to the athletes and coaches who contributed to this work: T. Vigneron, F. Salbert, P. Quinon, M. Houvion, J.M. Cau, J.P. Gatien, B. Parietti, P. Birocheau, X. Saiké, M. Gadal, and F. Orfeuil.

## REFERENCES

Dapena, J. (1988a). Biomechanical analysis of the Fosbury flop, Part 1. *Track Technique*, **104**, 3307-3317.
Dapena, J. (1988b). Biomechanical analysis of the Fosbury flop, Part 2. *Track Technique*, **105**, 3343-3350.
Dapena, J. (1989). Simulation of modified human airborne movements. *Journal of Biomechanics*, **14**, 81-89.

Dapena, J., & Braff, T. (1985). A two-dimensional simulation method for the prediction of movements in pole vaulting. In D.A. Winter, R.W. Norman, R.P. Wells, K.C. Hayes, & A.E. Patla (Eds.), *Biomechanics IX-B* (pp. 458-463). Champaign, IL: Human Kinetics.

Deporte, E., Van Gheluwe, B., & Hebbelinck, M. (1990). A three-dimensional cinematographical analysis of arm and racket at impact in tennis. In N. Berne & A. Cappozzo (Eds.), *Biomechanics of human movement: Application in rehabilitation, sports and ergonomics* (pp. 460-467). OH: Bertec.

Durey, A. (1987). Vers des activités didactiques de mise au point de modèles de physique avec des micro-ordinateurs. Exemple: trajectoires, frappes et rebonds de balles en rotation [Toward didactic activities to perfect physical models using micro-computers. Example: Trajectories, strokes, and bounces of rotating balls]. *Thèse d'état en sciences physiques*. Université Paris 7.

Durey, A., & De Kermadec, G. (1984). Tous les coups du tennis et leurs effets [All the tennis strokes and the effects of spins]. *Science et vie Hors Série, Le sport au quotidien*, 44-69.

Durey, A., & Orfeuil, F. (1989). Spins and trajectories in table tennis. Table Tennis Scientist's Conference, Rome.

FFT/INSEP (Producer), & De Kermadec, G. (Director). (1985). *Roland Garros avec Martina Navratilova* [Film]. Paris: INSEP.

FFT/INSEP (Producer), & De Kermadec, G. (Director). (1986). *Roland Garros avec John McEnroe* [Film]. Paris: INSEP.

Grainger, J., Norman, R., Winter, D., & Bobet, J. (1983). Day to day reproducibility of selected biomechanical variables calculated from film data. In Hideji Matsui & Kando Kobayashi (Eds.), *Biomechanics IVB* (pp. 1238-1247). Champaign, IL: Human Kinetics.

Griner, G.M. (1984). A parametric solution to the elastic pole-vaulting pole problem. *Journal of Applied Mechanics*, **51**, 409-414.

Hubbard, M. (1980). Dynamics of the pole vault. *Journal of Biomechanics*, **13** (11), 965-976.

Marey, M.M., & Demeny, G. (1885). Locomotion humaine, mécanisme du saut [Human motion, jumping mechanism]. *Comptes rendus des séances de l'académie des sciences*. Paris: Gauthier Villars.

McGinnis, P.M. (1983). The inverse dynamics problem in pole vaulting. *Medicine & Science in Sports & Exercise*, **15**, 112.

McGinnis, P.M. (1987). Performance-limiting factors in the pole vault. *Medicine & Science in Sports & Exercise*, **19**, 518.

McGinnis, P.M, & Bergman, L.A. (1986). An inverse dynamic analysis of the pole vault. *International Journal of Sport Biomechanics*, **2**, 186-201.

Neal, R.J., & Wilson, B.D. (1985). 3-D kinematics and kinetics of the golf swing. *International Journal of Sport Biomechanics*, **1**, 221-232.

Ramanantsoa, M.M., Ripoll, H., & Durey, A. (1991). Analyse du mouvement et neurosciences: bilan et reflexions de recherches [Analysis of movement and neuroscience results and research notes]. *Actes du colloque national, Methodologie et étude du mouvement en sport, ergonomie et clinique*.

Ripoll, H. (1989). Uncertainty and visual strategies in table tennis. *Perceptual and Motor Skills*, **1989c**, (68), 507-512.

Schleihauf, R.E., Gray, L., & Derose, J. (1983). Three-dimensional analysis of hand propulsion in the sprint front crawl stroke. In A.P. Hollander, P. Huijing, & G. deGroot (Eds.), *International Series on Sport Sciences, Vol. 14* (pp 173-191).

Takei, Y. (1989). Techniques used by elite male gymnasts performing a handspring vault at the 1987 Pan American games. *International Journal of Sport Biomechanics*, **5**, 1-25.

Van Gheluwe, B., De Ruysscher, I., & Graenhals, J. (1987). Pronation and endorotation of the racket arm in a tennis serve. In J. Bengt (Ed.). *Biomechanics VIB* (pp. 667-673). Champaign, IL: Human Kinetics.

Van Gheluwe, B., & Hebbelinck, M. (1985). The kinematics of the service movement in tennis: A three-dimensional cinematographic approach. In D.A. Winter, R.W. Norman, R.P. Wells, K.C. Hayes, & A.E. Patla (Eds.), *Biomechanics, XIB* (pp. 521-526). Champaign, IL: Human Kinetics.

Vaslin, P., Couëtard, Y., & Cid, M. ( 1993, July). Three-dimensional dynamic analyses of the pole vault. Paper presented at International Society of Biomechanics Congress, Paris.

Vaughan, C.L.A. (1981). A three-dimensional analysis of the forces and torques applied by a golfer during the downswing. In A. Morecki, K. Fidelus, K. Kedzior, & A. Wit (Eds.). *Biomechanics IIIB* (pp. 325-331). Baltimore: University Park Press.

Winter, D.A. (1979). *Biomechanics of human movement.* Toronto: Wiley.

Yeadon, M.R. (1990a). The simulation of aerial movement—I. The determination of orientation angles from film data. *Journal of Biomechanics*, **23** (1), 59-66.

Yeadon, M.R. (1990b). The simulation of aerial movement—II. A mathematical inertia model of the human body. *Journal of Biomechanics*, **23** (1), 67-74.

Yeadon, M.R. (1990c). The simulation of aerial movement—III. The determination of the angular momentum of the human body. *Journal of Biomechanics*, **23** (1), 75-83.

Yeadon, M.R., Atha, J., & Hales, F.D. (1990). The simulation of aerial movement—IV. A computer simulation model. *Journal of Biomechanics*, **23**(1), 85-89.

# Chapter 16

# Clinical Gait Analysis: Application to Management of Cerebral Palsy

*James R. Gage*
*Steven E. Koop*

Assessment of gait probably began with Eadweard Muybridge, a photographer who deserves to be considered the father of modern gait analysis. He was hired by Leland Stanford, the founder of Stanford University, to satisfy a bet. Stanford, a great trotting horse enthusiast, had wagered with a friend that there were periods of time when his trotter, Occident, had all four hooves off the ground. The bet was accepted contingent on the fact that it could be proven absolutely. Muybridge, who was completely funded in this endeavor by Leland Stanford, accepted the challenge. The project took 5 years to complete (1872-1877) and required faster photographic emulsions than were in existence at that time. Once Muybridge had developed emulsions that could stop action at 1/1,000 of a second, he set up a series of cameras with trip wires on Stanford's race track and succeeded in obtaining a picture of Occident with all four hooves off the ground. This was the beginning of both modern motion analysis and the motion picture industry.

Although there were other pioneers of motion analysis in Europe and North America, Dr. Vern Inman, a professor of orthopaedics at the University of California at Berkeley, deserves credit for its initial application in clinical use. After World War II many American servicemen returned home as amputees. Dr. Inman, with his colleagues in engineering and physiology, became involved in lower limb prosthetics research. This work was funded by the Committee on Prosthetic Devices, which later became the Advisory Committee on Artificial

Limbs under the National Academy of Sciences. Eventually this work led to the formation of the Biomechanics Laboratory at the University of California in San Francisco and Berkeley. Two of Dr. Inman's orthopaedic residents, Dr. Jacquelin Perry and Dr. David Sutherland, became pioneers in gait analysis in their own right and are still actively involved in research in gait and cerebral palsy.

Gait analysis allows us to monitor the activity of locomotion. There are three major reasons why an investigator might wish to evaluate gait:

1. To improve understanding of how the locomotion system operates (investigative research)
2. To improve activity that is already normal, or above average (improved athletic performance)
3. To bring abnormal or suboptimal performance closer to normal standards (assisting medical treatment decisions)

The parameters and constraints of the analysis system must vary depending on the task at hand (see Table 16.1). For example, if one wishes to investigate the gait of a child with cerebral palsy, the equipment attached to the patient must not encumber walking in a way that modifies the usual gait pattern. Additionally, the environment should be quiet and subdued in order to minimize distraction or excitability, which will also modify tone and, consequently, gait. In this situation walking speed is slow enough that the sampling frame speed and field of view are not critical. For the high-performance athlete, a different set of criteria applies. The athlete also must not be encumbered. A very wide range of motions may occur, and these motions must be sampled rapidly and over a wide field of view. Although maximum flexibility of the testing system is desirable, even with today's technology it is still necessary to target the design of the facility to the specific need for which it is intended.

Most of the active work in clinical gait analysis in North America today is being done in conjunction with work in cerebral palsy or other neuromuscular conditions. Several good clinical systems are now commercially available that are fairly "user friendly," and more will be on the market in the near future. There will be a broader application in sports medicine in the future, but such efforts are presently still in their infancy for two reasons: Physicians, coaches, and trainers who work in these fields still have not become sufficiently aware

**Table 16.1   A Comparison Between Gait Analysis Systems and the Subject to be Tested**

|  | High-performance athlete | Cerebral palsy child |
| --- | --- | --- |
| Marker system | Active (Easy identification) | Passive (less encumbering) |
| Speed | Rapid (200 HZ or more) | Moderate (30 or 60 Hz) |
| Field of view | Wide | Moderate |

of the benefits; and high-speed, portable, flexible automatic systems for outdoor use are not yet being commercially manufactured. Because our own work is almost entirely in the neuromuscular field, we will discuss the current and possible future applications of gait analysis mainly from this perspective.

# OVERVIEW OF 3-D ANALYSIS OF CEREBRAL PALSY AND RELATED CONDITIONS

Although gait analysis has applications for amputee gait, stroke, spinal cord injury, myelomeningocele, and some muscle diseases, the major use of kinesiology laboratories at centers such as Newington Children's Hospital (Newington, Connecticut) and Gillette Children's Hospital (St. Paul, Minnesota) is the evaluation and treatment of children with cerebral palsy. Cerebral palsy is not a disease of muscle (Ziv, Blackburn, Rang, & Koreska, 1984). It is a neurologic disorder that affects the brain in a wide variety of ways, including a partial loss of control of the locomotor process manifested by

- abnormal muscle tone, which varies with position or movement,
- a propensity to develop muscle contractures and bone deformity with growth,
- loss of selective, voluntary control of muscles,
- dependence on primitive reflexes to accomplish ambulation, and
- diminished overall body balance.

The gait abnormalities that occur in cerebral palsy are a combination of the deviations imposed by the neurologic deficit and the secondary adaptations the child uses to circumvent them. Identification of gait abnormalities using only observational gait analysis is difficult. Determining, in all cases, whether a gait deviation is a primary or secondary (coping) abnormality is probably not possible. Thus, in a neuromuscular condition such as cerebral palsy, gait analysis is used to determine precisely the abnormalities of gait and to accurately assess the outcome of a particular treatment or intervention.

It is apparent that, given our present state of knowledge, we are limited in our ability to modify the problem. To some extent we can modify abnormal muscle tone, and to a greater extent we can avoid or relieve muscle contractures and bone deformity that occur with growth. We cannot modify the loss of selective muscle control, the use of primitive reflexes, or poor body balance. Therefore a logical approach to treatment would be to address what is remediable (abnormal muscle tone, contracture, and bone deformity) and to accept what cannot be altered. Currently, the best means to permanently reduce muscle tone is a surgical procedure known as selective dorsal rhizotomy. This surgery consists of bilateral, partial, surgical division of the dorsal (sensory) nerve roots from the upper lumbar segments to the upper sacral segments. The rootlets to be divided are determined by electrical stimulation of each rootlet with simultaneous EMG monitoring of at least four bilateral dermatomes, as well as visual observation of patient muscle response. This procedure, popularized in North America by Dr. Warwick Peacock

(Peacock, Arens, & Berman, 1987), a neurosurgeon, does not yet have long-term outcome results but has significantly improved the lives of some children. Preoperative and postoperative gait analysis is being used to document the outcome of this surgical procedure.

Bone deformity and muscle contractures unresponsive to physical therapy must be corrected by orthopaedic surgery. In spastic cerebral palsy, surgical lengthening of tendons or muscles is probably the most effective way of restoring balance once static contracture of the muscle has developed. Unfortunately, the lengthened muscle is also weakened. The power a muscle generates can be increased by improving its mechanical advantage, by increasing the length of the lever arm upon which it acts, or through transfer of muscle insertions. In children with cerebral palsy, this should be done only after a great deal of thought and analysis, because the results are often unpredictable at best. In general, lengthening is not necessary for isometric muscles (stabilizers) because they are usually weak in cerebral palsy and rarely contracted in children who have walking ability. In most cases eccentric muscles (decelerators and shock absorbers) can be lengthened without significant loss of function, although the resultant weakness can allow antagonist muscles to become unbalanced, relatively more powerful, and eventually contracted. Lengthening is a major problem for concentric muscles (accelerators) because these muscles are necessary to initiate movement of the part upon which they are acting. Individuals with cerebral palsy often do not have adequate acceleration, so the weakness that results from muscle lengthening can reduce function measurably. A prime example of this dilemma is the hip flexion contracture often present in cerebral palsy. The iliopsoas is the prime hip flexor, and a lengthening of this muscle sufficient to correct contracture may weaken the muscle to the point that the patient has difficulty initiating hip flexion. It is obvious that specific information regarding the degree of contracture and the source of power for movement must be present if serious iatrogenic (physician-caused) errors are to be avoided.

Physical examination and observation are generally inadequate to determine specific treatment. Thus, in the past, usually only one muscle group was lengthened at a time. A child then went through an extended period of therapy and, once the outcome of that surgery was known, would undergo a similar surgery for another muscle group. Treatment was carried out in this fashion because the likelihood of a treatment error was greater if multiple muscle groups were addressed during the course of a single surgery. Rang, Silver, and de la Garza (1986) and Bleck (1987) have cogently argued that the overall result of surgery is much better if all contracted muscle groups are lengthened simultaneously rather than by staging the procedures. We agree with this concept. Not only is the morbidity lessened by accomplishing all surgery during the course of a single procedure, but much better function is possible by simultaneously balancing all major lower extremity joints. This is particularly true because many of the muscles that require lengthening are biarticulate, that is, they cross two joints. Surgical lengthening of such a muscle to correct imbalance at any one joint is likely to create imbalance at the joint above or below. For example, if the hamstring muscles are lengthened in a patient with knee flexion contractures, the result is likely to be better extension at the knee but an increased flexion contracture at the hip. This is because the hamstring muscles, in addition to being knee flexors, are also hip extensors. Thus hamstring lengthening also reduces hip extensor power, and in the face of spastic hip flexors, the hips will develop a more fixed flexion deformity. It is useful to think of the back, hip, knee, and ankle as four weights on the corners of a suspended balance board (Figure 16.1). Unless weight is subtracted or added *evenly* at all four corners, the board will tip. Unfortunately, the more muscles that are lengthened at one time, the more likely the possibility of making a judgment error. Therefore the muscle imbalance must be precisely defined. Also, primary abnormalities must be differentiated from adaptive, or coping, mechanisms. An example of this is circumduction gait in a child who does not have sufficient knee flexion for foot clearance during swing. This is a simple example, but other coping mechanisms can be extremely subtle and difficult to detect. If the surgeon focuses on the coping mechanism rather than the primary abnormality, the patient will often be worsened by the procedure rather than helped. It is our personal feeling that the differentiation between primary and secondary abnormalities often cannot be accomplished without dynamic gait analysis. Thus, in summary, if surgery is aimed at the imbalance between agonists and antagonists,

- restoration of balance will best be accomplished if all joints are rebalanced simultaneously;
- the imbalance must be precisely defined, and primary abnormalities must be differentiated from adaptive mechanisms; and
- in most cases, gait analysis will be required to define the problem and will require 3-D kinematics, dynamic electromyography, and kinetics (moments and powers).

Cerebral palsy is classified in three ways:

1. By the type of muscular tone that exists
2. By the body part involved (topographic)
3. By the level of independent function achieved by the patient

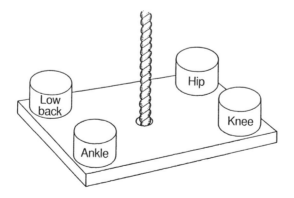

**Figure 16.1**  A demonstration of the necessary balance between muscle forces at all joints in the lower extremity.

For simplicity we will discuss only spastic cerebral palsy, the type for which surgical treatment is the most successful. In the topographic classification there are three major subgroups. Involvement limited predominantly to one half of the body (left or right side) is called *hemiplegia*. Involvement confined largely to the lower extremities is called *diplegia*. Involvement of all four limbs is called *quadriplegia*. Only 22% of children with spastic quadriplegia are able to walk, whereas nearly all children with diplegia or hemiplegia are ambulatory.

The "typical" gait of a child with diplegia or quadriplegia is characterized by flexion, adduction, internal rotation at the hip, and flexion at the knee. During stance the feet are usually flat, with the forefoot externally rotated on the hindfoot. This is often associated with a true bony deformity and external tibial torsion, the combination of which may serve to neutralize the internal femoral torsion and result in a foot progression angle that is neutral or only mildly internal. Initial foot contact with the floor is usually in the forefoot. Sometimes this is due to a contracted heelcord with a resultant plantar flexion deformity at the ankle, although frequently the ankle is neutral and the toe-toe gait comes about because of the knee flexion, which is present throughout stance phase. The net result of either flat-foot deformity or toe-toe gait is some degree of instability of the foot in stance. Torsional malalignment of the long bones may interfere with the normal foot progression angle in stance and contribute to stance-phase instability.

In swing phase, close observation of the knee will usually reveal that flexion-extension range is significantly reduced. This, coupled with excessive pelvic drop and ankle equinus, creates difficulty with clearance of the foot in swing. When the foot contacts the floor during swing, it is twisted internally or externally. In diplegia and quadriplegia, the lateral foot-stabilizing muscles (peroneals) predominate over those on the medial side (tibialis posterior and toe flexors), and the foot is prepositioned to rotate externally. In hemiplegia, the converse is true, and the rotation is internal. In both cases, the effect over time of this torsion on growing bones is to increase the deformity. In terminal swing, the restriction of full knee extension produces a short step and abnormal prepositioning of the foot, which sets the stage for instability in stance as soon as the next gait cycle begins.

Oxygen consumption studies of children with cerebral palsy indicate that the oxygen expense during walking is significantly higher than in children with normal gait. When related to a specific walking velocity, oxygen consumption may be two to five times the normal expenditure. By comparison, the energy required to climb stairs is approximately two times that required for normal walking on level ground.

Thus the primary abnormalities of gait that are usually present in individuals with cerebral palsy are

- loss of adequate limb acceleration,
- loss of stability in stance,
- difficulty clearing the foot in swing,
- inappropriate pre-positioning of the foot in terminal swing,
- inadequate step length, and
- excessive energy consumption.

These abnormalities manifest themselves to different degrees, depending on the severity and pattern of involvement. Because asymmetry is common in diplegia and quadriplegia, the primary abnormalities of gait may not be identical, or even present, on both sides.

# TREATMENT PRINCIPLES AND STRATEGY

The goal of orthopaedic surgery for these children is to correct the primary abnormalities of gait to the greatest degree possible. However, as mentioned previously, the individual with a primary abnormality of gait may employ some type of coping response. For example, a child with inadequate knee flexion who has difficulties with foot clearance in swing may circumvent the problem by circumducting a swing limb, employing hip abduction, and vaulting on the stance limb. Such coping responses create *secondary* abnormalities of gait that are virtually impossible to identify without some type of gait analysis. Even with kinematic gait analysis, they are often very subtle and difficult to detect without a great deal of experience. The calculation of joint moments and powers significantly improves the ability of gait analysis to separate these problems. In general, primary abnormalities of gait are most pronounced in joints over which the individual has insufficient selective control. By comparison, coping responses are generated through secondary abnormalities of gait at joints over which the person has the *most* selective control. Essentially, the problem is to identify the primary gait abnormalities, determine their causes, devise a solution for each, and then correct those that are correctable.

Accelerator muscles provide the power that is needed for mobility, and, if possible, their function should be preserved. One of the real benefits of kinetics (i.e., joint moment and power curves) is that it allows identification of the source and magnitude of the force that is being used to generate propulsion. In normal individuals, approximately 50% of the force for propulsion comes from the gastrocsoleus complex, 30% from the hip extensors, and the remaining 20% from the hip flexors. In children with cerebral palsy whom we have studied using gait analysis, this situation is reversed, and the bulk of the power required for walking comes from the hip flexors and extensors. Furthermore, these children utilize their hamstrings to augment hip extensor power. The surgeon may be required to determine whether the combination of iliopsoas surgical recession and hamstring lengthening required to maintain joint balance is justified when reduction in power may also negatively affect ambulation. Because individuals with cerebral palsy usually walk more slowly than their normal counterparts, pure decelerators can usually be safely lengthened. Unfortunately, many muscles, such as the soleus, work both eccentrically (deceleration) and concentrically (acceleration) during the course of a single gait cycle.

Cerebral palsy appears to affect principally the biarticular muscles, such as the iliopsoas, hamstrings, rectus femoris, and gastrocsoleus. This is probably because the strength and timing of these muscles during normal gait has to be much more precise than those muscles that span only one joint, and because the period of their action is often very brief. Lengthening of a compound muscle tendon will lengthen both the monoarticular and biarticular components of the unit, whereas intramuscular lengthening allows the surgeon to address the specific muscle that is overacting. Intramuscular lengthening has the further benefit of weakening the muscle, which is the source of power, rather than lengthening the tendon, which is simply the cable that attaches the power source to its point of action. If tendon lengthening is done, the muscle initially will be weakened because of displacement on its length-strength curve. The inherent spasticity in the muscle, and the steady growth of the bones to which it is attached, will allow the muscle to regain its former strength and result in recurrence of the original problem. Although the joint crossed by such a muscle will exhibit increased range of motion, the muscle fibers will not be subjected to stretch because lengthening occurred in the tendon component. Stretch is an important component of muscle growth. This contributes to contracture and the overall recurrence of muscle spasticity. Another way to address abnormal action of a compound muscle is to simplify its action by means of tendon transfer. Two examples of this, which we use frequently, are transfer of the distal end of the semitendinosis to the femur, and transfer of the distal end of the rectus femoris posteriorly to one of the knee flexors. In the former case, the action of the muscle is simplified by converting it to a monoarticular muscle. This preserves the desired action of this muscle as a hip extensor, while eliminating it as a knee flexor. In the latter case, the needed action of the muscle as a hip flexor is preserved, but its action as a knee extensor in swing is converted to that of knee flexion, which enhances clearance of the foot. Thus, function can often be enhanced if the action of the biarticular muscles can be simplified or modified.

Muscles always work as part of a force-couple to produce a moment around a joint (Figure 16.2). For example, the lever on which the triceps surae acts to extend the knee in midstance can be lost in one of two ways: The normal foot progression angle can be distorted because of malrotation of a long bone or because of breakdown of the foot. Either deformity interferes with the integrity of the foot as a rigid lever. The plantar flexion-knee extension couple becomes inadequate regardless of the fact that the triceps surae has normal power. Deformity of bone can alter muscle function, and this important concept is often overlooked when treatment is planned.

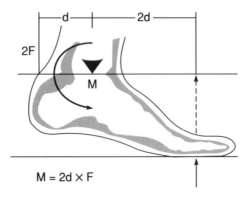

**Figure 16.2**   Moments created at the ankle joint by internal forces, such as muscle, and external forces, such as ground reaction force.

# PATIENT ASSESSMENT

Given this understanding of the nature of cerebral palsy, and the general strategy for treatment, we need to ask the question, How does gait analysis help determine the specific treatment? In addition to the static physical examination and the observation of gait, many dynamic parameters can be measured. Measurement techniques include videotape, linear measurements, kinematics, dynamic electromyography, kinetics, and energy assessment (Figure 16.3).

Videotape is extremely useful in gait assessment. It is relatively inexpensive and readily available. Good quality videotape can be repeatedly reviewed in slow motion, which allows far more accuracy than simple observation of gait. For example, it is possible to distinguish some primary from secondary deviations of gait using videotape and to estimate the kinematics of the knee with fair accuracy. However, videotape does not permit adequate evaluation of the relationship between the trunk and hip, particularly in a subject who is moderately overweight. In addition, unless the videotape is focused closely on the foot and ankle, accurate assessment of the dynamics of ankle motion and foot placement are poor.

Simple gait measurements such as step length, velocity, and cadence can be accomplished with a measured walkway and a stopwatch. This may provide

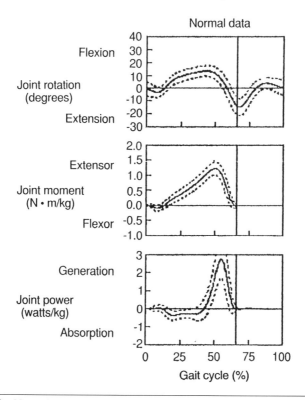

**Figure 16.3**   Normal ankle kinematics and kinetics.

a rough estimate of pre- and postoperative status but are extremely limited measurements of improvement. They are nonspecific in that they are not helpful in determining the cause of gait abnormalities or the special effects of surgery. Other direct measurements include 3-D kinematics, which can describe segment motion simultaneously for both lower extremities in the sagittal, frontal, and transverse planes. However, they are essentially descriptive measurements and give little or no information about the cause of the particular movement observed.

Dynamic electromyography provides objective evidence of muscle activity and can be useful in isolating the cause of a particular abnormality. Because the internal moment produced by muscle forces is always acting to balance the external moment produced by the ground reaction and initial forces, dynamic electromyography can tell us only *when* the muscle is acting; it does not give accurate measurement of the intensity of the muscle activity. Lastly, it cannot discern between primary and secondary abnormalities unless used in conjunction with joint moments. If an individual is walking in an abnormal manner, the ground reaction forces would be altered, and muscles that are under normal voluntary control will be forced to act in a manner that will control the abnormal moment. Their timing and intensity of action will be altered accordingly. Electromyography will report the presence of activity but does not reveal why the muscle is active. In general, dynamic electromyography is not useful in postoperative assessments of treatment outcome.

Kinetics has been defined as the study of moving bodies and the forces that act to produce the observed motion. Kinetics (joint moments and powers) provide our best opportunity to determine exactly how the abnormalities in the control system are affecting locomotion. (A case example is presented to demonstrate how kinematics and kinetics are used in this way.) Unfortunately, kinetics have several major limitations. An extremely accurate 3-D motion analysis system is required, with simultaneous sampling of ground reaction forces through a force plate system. This requires a sophisticated gait analysis facility and a subject who is capable of walking without assistance and of placing a single foot strike on the force plate. The use of upper-extremity aids (crutches and walkers) transfers weight bearing to the aids and invalidates the data, whereas more than one foot strike on the force plate provides additional data that cannot be separated out. Additionally one needs to estimate the mass moments of inertia of the limb segments (Dempster, 1955). Also, joint centers must be estimated from surface marker locations. Nonetheless, despite their limitations, kinetics may prove to be the most useful measurement because they provide information about the *cause* of deviations. Kinetics are the only objective method currently at our disposal that will distinguish primary from secondary abnormalities.

In order to fully assess our treatment interventions, accurate and efficient methods of measuring energy must yet be developed. Both the patient and the surgeon are interested in improving the efficiency of gait, as well as its appearance. Ultimately, energy measurement may be the best single index of the success or failure of treatment. Currently, there are several ways to measure energy expenditure. These include oxygen consumption, carbon dioxide generation, limb segment analysis, and power curves of individual joints. There are major limitations for each of these methods. Oxygen consumption and carbon dioxide generation, though relatively easy to measure, require methodology that is laborious, inappropriate for use during walking (for children with cerebral palsy), and expensive. More recent technology, using cart-based breath-by-breath measurements, has dramatically improved our ability to measure oxygen consumption and carbon dioxide output during walking. Limb segment analysis and calculations of power curves for individual joints are difficult and time-consuming and require complex algorithms based on assumptions that have not yet been validated fully. The correlation between mathematical models and calculations of energy expense and the objective data from oxygen consumption and carbon dioxide output is poorly understood, an area for future research. Very little is known about the energy expense of walking for children with cerebral palsy, and virtually nothing is known about the effect of medical and surgical treatments on this.

Now that treatment principles and strategy have been discussed, we will attempt to demonstrate the usefulness of gait analysis in clinical decision making by means of a clinical example.

# TREATMENT EXAMPLE

JPG is an 8-year-old child with right spastic hemiplegia cerebral palsy. He has had no previous surgical treatment. He walks using a toe-toe gait on the right with excessive internal rotation of the hemiplegic leg. At the beginning of the gait cycle, at the time of initial contact of the foot with the floor, his knee is excessively flexed. As he progresses in stance, the excessive knee extension moment, which is generated by the plantar-flexed foot, drives the leg posteriorly so that the knee reaches nearly full extension by terminal stance. During swing phase, knee flexion is close to normal, yet he does not extend his knee fully during physical examination.

The physical examination also reveals tight hamstrings and tight gastrocsoleus musculature on the right side. The Duncan-Ely stretch test (an indication of rectus femoris overactivity) is mildly positive. In addition, there is an excessive external tibial torsion on the right, measuring approximately 40°. The remainder of the physical examination of the lower extremities is within normal limits.

Kinematic assessment of the frontal plane is essentially normal. In the transverse plane, JPG demonstrates excessive external foot rotation, consistent with the degree of external tibial torsion noted during the physical examination. The sagittal plane kinematics of the right side are shown with a normal comparison (Figure 16.4). Note that the modulation (pattern of movement) of the ankle graph is abnormal, but the heelcord does not appear to be contracted because full dorsiflexion is achieved. The knee graph is close to normal except for the loss of extension in terminal stance. The graph of the hip motion is essentially normal, and the pelvis demonstrates excessive lordosis in late stance.

Dynamic electromyography reveals continuous activity of the right rectus femoris in swing phase. Additionally, there is prolonged activity of the hamstrings in stance and swing and early activity of the gastrocsoleus (gastrocsoleus activity starts in late swing, rather than in early stance, as is normally expected).

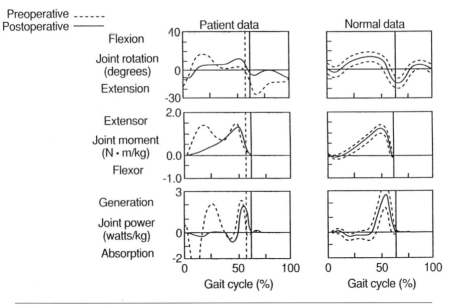

**Figure 16.4**   A comparison of preoperative and postoperative patient kinematics and kinetics, and normal data.

Kinetics of the right ankle reveal a biphasic moment with two distinct bursts of power. The first occurs in midstance. Normally, this is a time of eccentric muscle activity in the soleus muscle, with energy absorption noted. Instead, JPG exhibits a burst of power generation. The second occurs in terminal stance. This is normally a time of concentric gastrocsoleus muscle activity (''the push-off''), with generation of power, and an appropriate burst of power occurs at that time. Integration of the power curve shows that the total power generated during both bursts is 0.44 joules/kg. Normal power generation would be 0.21 ± 0.9 joules/ kg during a single power burst in terminal stance.

On the basis of the clinical and kinesiological examinations, JPG's abnormalities of gait consist of

- contracture of the gastrocnemius portion of the gastrocsoleus complex (triceps surae),
- hamstring contracture with dynamic cospasticity of the rectus femoris during swing phase, and
- excessive external tibial torsion with the resultant external foot-progression angle.

This fits the pattern of Type-3 hemiplegia (Winters, Hicks, & Gage, 1987). On the basis of this analysis, the following surgery was performed on the right leg:

- Baker-style tendoAchilles lengthening (gastrocnemius only)
- Hamstring lengthening (medial and lateral)
- Rectus femoris release (distal end)
- Tibial derotation osteotomy (supramalleolar)

After JPG's recovery from surgery, gait analysis was repeated and the postoperative outcome compared to the preoperative analysis. (The linear measurements, kinematics, and kinetics are shown in Figure 16.5.) The kinematics and kinetics were normalized. JPG is walking slightly slower (110 cm/s vs. 123 cm/s, preoperatively), but he no longer has the abnormal burst of energy in midstance. The calculated work performed at the ankle has been reduced from 0.44 joules/kg to 0.12 joules/kg, which is less than one third of the preoperative value.

Thus preoperative gait analysis enables us to define precisely the patient's problems, devise an appropriate treatment plan, and treat all muscle groups simultaneously. Postoperative analysis allows us to correlate outcome with the treatment given and thus to reduce future treatment errors. Essentially, gait analysis in clinical treatment allows practical application of the scientific method. Facts are accumulated, hypotheses are made to explain the relationships among the observed facts, and the hypothetical outcome can be compared to actual surgical results.

## THE FUTURE—WHERE IS IT GOING?

The technology of gait analysis is moving rapidly, requiring changes in physician attitudes. There is still a widespread perception among orthopaedic surgeons that clinical examination and observation of gait are adequate to determine treatment. Yet human gait is very complex, and a thorough understanding of it demands familiarity with the basic principles of biomechanics and the technology used to measure gait. Some professionals' reluctance to use gait analysis may be due to the amount of time and effort necessary to accomplish this and the necessity for teamwork among the disciplines involved. Other imaging-technology advances in medicine, such as magnetic resonance imaging, do not require such a significant effort on the part of those who merely wish to use the technology. However, we are certain that, given the steady advance of technology and our continued efforts to document the benefits of that technology and to teach it to our colleagues, gait analysis will soon be a routine part of the evaluation of *both* the elite athlete and the physically impaired adult or child.

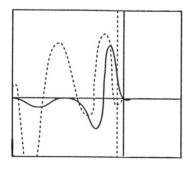

|  | Preoperative | Postoperative |
|---|---|---|
| Concentric energy (joules/kg) | 0.44 | 0.12 |
| Eccentric energy (joules/kg) | -0.39 | -0.13 |

**Figure 16.5**   Ankle power assessment, preoperatively and postoperatively.

Although no one can foresee future events, current directions suggest that

- technologic advances will continue rapidly, and there will be more sophisticated equipment available at a lower overall cost;
- gait analysis will be used widely in treatment of neuromuscular disorders, in sports medicine, and in musculoskeletal research;
- information will be gathered by treatment centers in a common format, with linkage between large data bases, forming an information-sharing network that will facilitate statistical pattern recognition and speed the process of determining optimal treatment protocols; and
- computer simulation of treatment will be possible through mathematical modeling and artificial intelligence techniques.

When all of this has occurred, the orthopaedic management of cerebral palsy will have evolved from a rather poor art to a more exact science.

## REFERENCES

Bleck, E.E. (1987). *Orthopaedic management in cerebral palsy.* Oxford: Mac-Keith Press.

Dempster, W.T. (1955). *Space requirements of the seated operator.* (WADCTR 55-159). Wright-Patterson Air Force Base, Ohio.

Peacock, W.J., Arens, L.J., & Berman, B. (1987). Cerebral palsy spasticity: Selective dorsal rhizotomy. *Pediatric Neuroscience, 13,* 61-66.

Rang, M., Silver, R., & de la Garza, J. (1986). Cerebral palsy. In W.W. Lovell & R.B. Winter (Eds.), *Pediatric orthopaedics* (2nd ed.) (pp. 345-395). Philadelphia: J.B. Lippincot.

Winters, T.F., Hicks, R., & Gage, J.R. (1987). Gait patterns in spastic hemiplegia in children and young adults. *Journal of Bone & Joint Surgery, 69-A,* 437-441.

Ziv, I., Blackburn, N., Rang, M., & Koreska, J. (1984). Muscle growth in normal and spastic mice. *Developmental Medicine & Child Neurology, 26,* 94-99.

# Index

Townsend, M.A., 7
TRACK (computer program), 12, 15
Tracking software, 51-54
Trujillo, D.M., 90
Tsai, R.Y., 29-30
Turner-Smith, A.R., 137

**U**

Uicker, J., 181
Unix (operating system), 108
Upper extremity data bases, 221, 227, 239
Upper limb analysis applications, 306

**V**

van Dijk, R., 137
Van Gheluwe, B., 330, 344
van Ingen Schenau, G.J., 27
van Langelaan, E.J., 137
Vaughan, C.L., 296
Vaughan, C.L.A., 345
Veress, S.A., 137
Vernon, A., 238
Video-based systems
  evolution of, 41-43, 57
  future of, 54
  hardware for, 43-50, 316
  lens distortion with, 35, 36
  software for, 26-27, 43, 50-54, 64-65
  specific applications of, 296, 315-316, 317, 318-319, 357
  video images vs. computer graphics, 111-112
Video cameras, 43-48, 61
Video monitors, 48-49
Video processors, 49-50
Video recorder-playback units, 48
Virtual work, 162, 169-172
Volume rendering and tracking software, 54
Voss, H. von, 238

**W**

Wahba, G., 90-92
Warwick, D., 211
Waterskiing applications, 331-334
WATSMART optoelectronic system, 15
Weber, E., 208
Weber, W., 208

Weijs, W.A., 219, 246
White, A.A., 136
White, S.C., 219, 220, 231
White, S.P., 137
Whittle, Michael W., 295-309
Wickiewicz, T.L., 233, 238
Wiener filtering, 88-89
Wilder, D.G., 134, 136
Willner, S., 137
Wilson, B.D., 345
Window software, 108-109
Winter, D., 344
Winters, Jack M., 215, 232, 257-292
Winther, D.A., 336
Woittiez, R.D., 235
Wolff, E., 238
Woltring, Herman J., 39, 79-99
Wood, G.A., 39
Wood, J.E., 227, 239
Workstations for computer graphics, 102, 106-110
Wu, C-H., 272
Wu, G., 15

**X**

X-ray photogrammetry
  advantages/disadvantages of, 138
  applications of, 136-137, 219
  basic radiographic principles for, 125-128
  calibration for, 130-131, 133-134
  components of system for, 127, 129-131
  definition of terms, 125
  markers for, 131-133
  reconstruction procedures with, 133-135

**Y**

Yamaguchi, G.T., 215, 277
Yeadon, M.R., 344-345
Young, K-Y., 272

**Z**

Zajac, F.E., 232, 233, 234, 235
Zchakaia, M., 244
Zeltzer, D., 118
Zernicke, R.F., 235

# About the Editors

**Paul Allard, PhD,** is a professional engineer, an associate professor at the University of Montreal in Canada and an adjunct professor in the department of mechanical engineering at École Polytechnique de Montréal. He is also a senior scientist at the Sainte-Justine Hospital in Montreal, where he is the co-founder and director of the Human Motion Laboratory. His work on three-dimensional analysis of human movement led him to develop an asymmetrical keel foot prosthesis for lower limb amputees, for which he was honored with a Canada Awards Certificate of Merit in 1992. Dr. Allard co-organized the first North American Congress on Biomechanics, which was held in 1986 in Montreal, and the First International Symposium on 3-D Analysis of Human Movement in 1991 in Montreal.

**Ian A.F. Stokes, PhD,** has over 20 years' experience as a researcher in orthopaedic biomechanics. He is currently a research associate professor in the Department of Orthopaedics and Rehabilitation at the University of Vermont and has been an editorial advisor for the *Journal of Biomechanics* since 1980. Dr. Stokes chaired the Scientific Committee at the First International Symposium on 3-D Analysis of Human Movement, held in Montreal in 1991. He was awarded the Volvo Award for Low Back Pain Research in 1980 and received a Fogarty Senior International Fellowship in 1993.

**Jean-Pierre Blanchi, PhD**, is a professor at the Université Joseph Fourier in Grenoble, France, where he was instrumental in creating engineering courses on health, safety, and ergonomy as well as a new laboratory for engineering of sport ergonomy. His work on 3-D analysis of human movement and quadrupedal posture led him to develop a new quadrupedal ergometer for the studies of gravitational effect, handicap, and climbing. Dr. Blanchi, a member of the French-Speaking Biomechanics Society, was a co-organizer of the First International Symposium on 3-D Analysis of Human Movement.